ADVANCES IN PERMANENT MAGNETISM

ADVANCES IN PERMANENT MAGNETISM

Rollin J. Parker

President, Parker Associates.
Senior Consultant, Walker Magnetics Group.
Former Manager Advanced Design and Development, Hitachi Magnetics Corp.

A Wiley-Interscience Publication

JOHN WILEY & SONS

New York / Chichester / Brisbane / Toronto / Singapore

Library of Congress Cataloging-in-Publication Data:

Parker, Rollin J.
Advances in permanent magnetism/Rollin J. Parker.
p. cm.
"A Wiley-Interscience publication."
Bibliography: p.
1. Magnets, Permanent. I. Title.
QC757.9.P37 1989
537'.3—dc20 89-16461
ISBN 0-471-82293-0 CIP

Printed in the United States of America

10 9 8 7 6 5 4 3 2 1

CONTENTS

PREFACE ix

CHAPTER 1 INTRODUCTION AND HISTORICAL PERSPECTIVE 1

1.1 Introduction, 1
1.2 Early History of Permanent Magnets, 2
1.3 Growth of the Industry, 5
1.4 Property Improvement and the Changing Pattern of Use, 6
1.5 Raw Material Impact, 10
1.6 A Look Ahead, 12

CHAPTER 2 MAGNETISM AND THE PERMANENT MAGNET 15

2.1 Introduction, 15
2.2 Unit Systems, Definitions, and Conversion Factors, 15
2.3 The Hysteresis Loop, 19
2.4 Demagnetization Factors and Magnetic Circuit Concepts, 22
2.5 Permanent Magnets and Electromagnetism, 25
2.6 Energy Relationships, 26
2.7 Boundary Conditions and Figures of Merit, 37

CHAPTER 3 THE PHYSICS OF PERMANENT MAGNETISM AND THE ORIGIN OF PERMANENT MAGNET BEHAVIOR 43

3.1 Overview and Perspective, 43
3.2 The Variations of Magnetic Behavior, 44

3.3 Some Features of Ferromagnetic Materials, 46
3.4 Energy Barriers and Coercive Force, 50
3.5 Single Domain Particle Variables, 54
3.6 Coercivity Mechanisms in Rare-Earth Magnets. 56
3.7 Understanding Magnetization and Demagnetization Processes in Rare-Earth Magnets, 58

CHAPTER 4 CLASSIFICATION OF PERMANENT MAGNET PROPERTY SYSTEMS AND PROCESSING TECHNOLOGY 61

4.1 Introduction, 61
4.2 Inclusion Hardened (Early Steel) Magnets, 62
4.3 Fine Particle Magnets Utilizing Shape Anisotropy, 64
4.4 Fine Particle Magnets Utilizing Crystalline Anisotropy, 74
4.5 Matrix or Bonded Magnets, 97
4.6 Semi-Hard Magnets (Hysteresis Alloys), 97

CHAPTER 5 PERMANENT MAGNET STABILITY 101

5.1 Introduction, 101
5.2 Classification of Magnetization Changes, 102
5.3 Theoretical Considerations, 104
5.4 Temperature Effects, 106
5.5 Magnetic Field Effects, 121
5.6 Temperature Compensation, 123
5.7 Mechanical Energy Input and Stability, 126
5.8 Corrosion and Surface Oxidation, 128
5.9 Nuclear Radiation, 129
5.10 Enhancing Stability, 129
5.11 Stabilization Techniques, 130
5.12 Conclusions and Comparison of Materials, 133

CHAPTER 6 DESIGN RELATIONSHIPS AND UNIT PROPERTY SELECTION 135

6.1 Introduction, 135
6.2 Relating Unit Magnetic Properties to Magnet Volume, Magnet Geometry, and Device Parameters, 137
6.3 Determination of Permeance, 143
6.4 Magnetic and Electrical Circuit Analogy, 165
6.5 Use of High Permeability Materials in Permanent Magnet Circuits, 173
6.6 Economic Considerations in Design and Property Selection, 175

6.7 Permanent Magnets and the Laws of Electromagnetic Scaling, 179

CHAPTER 7 APPLICATIONS OF PERMANENT MAGNETS 183

7.1 Introduction and Classification of Applications, 183
7.2 Applications Based on Coulomb Force Law, 185
7.3 Applications Based on Faraday's Law, 194
7.4 Applications Based on Lorentz Force Law, 204
7.5 Applications Based on Lorentz Forces on Free Electron Charges, 236
7.6 Miscellaneous Applications Based on a Variety of Physical Principles, 244

CHAPTER 8 MEASUREMENTS 248

8.1 Introduction, 249
8.2 Apparatus and Techniques for Generating Magnetic Fields, 249
8.3 Measuring Magnetic Fields and Magnetic Potentials, 251
8.4 Instrumentation for Magnetic Field Measurement, 256
8.5 Open Circuit Measurement Techniques, 266
8.6 Instrumentation Systems for Closed Circuit and Hysteresis Loop Measurements, 269
8.7 Search Coil Arrangements and Characteristics, 277
8.8 Special Measuring Techniques for Permanent Magnet Circuit Analysis and Design Optimization, 279
8.9 Calibration, Standards, Precision, and Accuracy, 280
8.10 Measurement Practice in Production Quality Control, 282

CHAPTER 9 MAGNETIZATION AND DEMAGNETIZATION 284

9.1 Introduction, 284
9.2 Theoretical Considerations, 285
9.3 Requirements for Complete Magnetization, 289
9.4 Equipment and Techniques to Magnetize, 296
9.5 Equipment and Techniques to Demagnetize, 303
9.6 Calibration and Stabilization Techniques, 305

APPENDIX 1 GLOSSARY OF TERMS, DEFINITIONS, SYMBOLS, SPECIFICATIONS, AND STANDARDS 307

APPENDIX 2 MAGNETIC AND PHYSICAL PROPERTY TABLES 316

APPENDIX 3 **DEMAGNETIZATION CURVES FOR DESIGN ANALYSIS** **322**

APPENDIX 4 **CHRONOLOGY OF DISCOVERY OF PERMANENT MAGNETS** **331**

INDEX **335**

PREFACE

Permanent magnets can be found in a vast array of consumer, industrial, and defense applications. The author has witnessed the growth of permanent magnetism for some three decades. Over that time, magnets have changed from an interesting speciality component to a widely accepted magnet circuit element that has major influence on the size, efficiency, response, stability, and cost of very important magnetoelectric devices and systems. Today, many aerospace and defense systems are limited only by permanent magnets and thus the U.S. government is a major source of development funding.

The interest in permanent magnets is truly international, with development and production facilities in every advanced industrial country. In three decades, the level of commercial production has grown from about US$50 million to well over US$1 billion in 1987.

My aim in writing this book is to make available, under one cover, a unified and comprehensive treatment of permanent magnetism, from the origin of permanent magnet behavior to the use of magnet components in energy conversion devices. This book is primarily for the engineer who has a permanent magnet device to design and produce. This text, however, will be of interest to the research and development community and as a reference to all who have interest in permanent magnets.

In spite of the growth and acceptance of permanent magnets, there is considerable difficulty in working with permanent magnets. There remains a mysticism; many designers approach a magnetic circuit problem with considerable apprehension. To an extent, this is due to some very inherent difficulties associated with a sophisticated subject. Often, however, it is due to the lack of a unified understandable treatment of the subject.

This book covers all permanent magnet property systems. However, it places a great deal of emphasis on the new rare-earth magnets. The properties of these new magnets have been largely responsible for a rebirth of interest in magnetoelectric devices. Many devices that previously used pneumatics or hydraulics can be redesigned with high energy magnets providing improved efficiency, size, cost, and reliability. Earlier magnets only allowed the replacement of electromagnetism. However, with rare-earth magnets one is able to solve problems not possible or feasible with current carrying conductor systems. Permanent magnets now have an identity of their own. The development of rare-earth magnets is indeed very timely because of the current emphasis on energy conversion and conservation. No energy is required to maintain a permanent magnet field. Therefore, there is the potential of decreasing the energy consumption of many devices and systems. In motion control systems, we are limited by transducer response and high energy density magnets have allowed the development of transducers with improved response and power handling. Rare-earth magnets are the companion components to microelectronics in the changing work pattern. In the automatic offices and factories of the future, rare-earth magnets will find extensive use.

In Chapter 1, the early history of permanent magnetism serves as a fitting introduction to the subject. It gives the reader a historical perspective as well as an indication of the new tempo of today's developments. Milestones in permanent magnet property achievement are noted and related to industry growth. Additionally some projections of future property possibilities are given.

Chapter 2 introduces the terms, definitions, and concepts central to the description and evaluation of permanent magnets. The relationship between an electric current and a magnetic field is developed. Energy concepts and magnetic circuit parameters are also introduced, as well as figures of merit.

One needs some understanding of the origin of permanent magnetism to have a feel for how various properties are developed. In Chapter 3 energy barriers and various coercive force mechanisms are described. Crystal properties, the single domain concept, and fine particle interactions are described to form a framework of understanding of how properties are achieved and how one searches for improved properties.

Chapter 4 covers all major types of permanent magnets with details on magnetic and physical properties. In addition, processing events and their relationship to property achievement are developed. A compelling reason for using a permanent magnet is often the stability of the field over a long period of time in a wide range of environmental conditions. In Chapter 5, the theoretical framework and the measured magnetization changes to allow the designer to predict the stability achievement of a specific device are developed. In this area there have been numerous misconceptions regarding the nature of the permanent magnet and its ability to maintain a field independent of time. The flux output of a permanent magnet is shown to be predictable and consistent with the laws of physics.

In Chapter 6, the design relationships are developed. The linkage between unit properties and the permanent magnet component, which has a specific volume and geometry, to produce a certain function in a device or system is described. It is in this area of the choice of properties to perform a given end product function that the designer can be overwhelmed. This chapter is a very important bridge in communications between how the magnet industry describes its product and the permanent magnet circuit component that performs a specific device function. Determining magnetic permeance and estimating leakage flux is a challenging issue. Numerous techniques are described, and several examples of circuit solutions are shown.

All major applications are described in Chapter 7. The relationship between unit properties and device parameters is developed. In some cases the evolution of the device or system is traced with emphasis on the sensitivity of magnet properties on device function. From the numerous examples the designer will gain insight on how to decide on configuration and properties to achieve that one unique solution.

Chapter 8, on measurements, describes a wide range of instruments and techniques. Magnetism is quite intangible until it is reduced to numbers with measurements. Chapter 9 describes magnetization and demagnetization equipment and techniques. High energy rare-earth magnets are extremely difficult to magnetize and demagnetize. Consequently it is necessary to optimize the transfer of electrical energy to magnetic field energy.

This book represents the work of many scientists and engineers. I am very grateful to my colleagues in universities and in magnet producing organizations around the world for their contributions in advancing permanent magnetism.

Greenville, Michigan
November 1989

Rollin J. Parker

1

INTRODUCTION AND HISTORICAL PERSPECTIVE

1.1 Introduction
1.2 Early History of Permanent Magnets
1.3 Growth of the Industry
1.4 Property Improvement and the Changing Pattern of Use
1.5 Raw Material Impact
1.6 A Look Ahead

1.1 INTRODUCTION

Modern permanent magnets have major influence on the size, efficiency, stability, and cost of magnetoelectric devices and systems. Today the permanent magnet is a vital component in a wide range of industrial, consumer, and defense products.

This introductory chapter is intended to give a perspective on where we are in the evolution and development of permanent magnetism. It is necessary to review early history and examine the techniques used to develop and use early magnets if we are to have a true appreciation of today's technology and position ourselves to appraise the future. The greatly improved properties have allowed magnets to solve many device problems and today, the total world market for magnets is in excess of US $1,000,000,000. The relationship between property advances and markets is an interesting study. The interest in permanent magnet property development is international and magnets are being developed and produced in every industrialized nation. For many sophisticated devices and systems, we find that performance parameters are limited by available permanent magnet properties. This has prompted investment in permanent magnet research and development by governments and leading industrial companies all over the world. The changing patterns of use and the increasing importance of permanent magnets in key devices will be described.

As the energy density of the permanent magnet has improved, the permanent magnet has found use in many non-traditional devices and energy conversion systems.

In this chapter, the historical perspective is concluded by looking into the future regarding the possibility of still better properties and the use of permanent magnets in our changing world where energy conversion and energy conservation are of great importance.

1.2 EARLY HISTORY OF PERMANENT MAGNETS

The early history of the permanent magnet is rather fascinating. Not only are man's early attempts at using the naturally occurring magnet, "lodestone," of major interest, man's early attempts to make artificial magnets yielded surprisingly strong magnets nearly three centuries ago. The permanent magnet industry, the development of today's major property system and the major application areas are all products of the 20th century. However, very early in recorded history, one finds references to man's attempt to use and understand magnets. The only magnetic material available in ancient times was the loadstone or lodestone (literally "way stone" from its use in guiding mariners on their way; cf. lodestone). The lodestone was a form of magnetitie (Fe_3O_4) which in its natural state is magnetic. This material was given the name magnes because it was found in Magnesia, a district in Thessaly. The attractive powers of the lodestone were mentioned by Greek philosophers of the period 100–200 B.C.

The first artificial magnets were iron needile which were "touched" or magnetized by a lodestone. Man's first practical use of magnetism may have been the compass. Around 1200 A.D. there are references in a French poem to a touched needle of iron supported by a floating straw. Figure 1.1 shows a print from approximately 1637 A.D. of magnetic needles for compasses being made. This may be man's first magnet plant. Other references suggest that good magnet steel was available from China about 500 A.D.

The earliest systematic reporting of magnets was a classical paper by William Gilbert in 1600. Gilbert described how to arm lodestone with soft iron pole tips to increase attractive force on contact, but he was careful to note that arming did not increase the attraction at a distance. Figure 1.2 shows examples of capped or "armed" lodestone. Gilbert gave three ways by which permanent magnetism could be given to iron and steel. The first was by touching with a lodestone. The second method was forging with the steel speciman pointing north and south in the earth's magnetic field.

Figure 1.1 Magnetic needles for compasses are being made by craftsmen in this print of 1637. Good steel was manufactured in China from 500 A.D. onwards.

Gilbert also noted that iron wire became magnetized if drawn in the north-south direction, but not in the east-west direction. The third method Gilbert described was to place a red-hot iron bar in the earth's field and "let it cool." The bar would become magnetized. He further noted that unheated bars, if left for a long period of time in the earth's field would acquire magnetism. Method two required some kind of deformation. Methods one and two could also be enhanced if vibration was present.

The next great advance in magnetism came with the invention of the electromagnet by Sturgeon in 1825. Now, a convenient way was in hand to magnetize and Sturgeon and others became quite interested in experimenting with the magnetic properties of alloys. Some workers reported that nonmagnetic brass could be magnetized by touching and hammering. There was much speculation as to the amount of iron present in the brass and as to whether or not non-ferrous elements contributed to magnetism. Having found silver coins then in circulation to be magnetic, Sturgeon concluded

Figure 1.2 Gilbert's capped or "armed" loadstones.

that various alloys of copper, silver and gold with zinc were magnetic. He believed that the magnetic effects were not due to iron. This work was indeed a milestone in the sense that the properties of alloys had been shown to be due to something other than the individual properties of the constituents. This new thinking was fundamental to the development of modern permanent magnets.

By 1867, German handbooks recorded that magnetic alloys could be made of nonmagnetic materials and non-magnetic alloys of magnetic materials, mainly iron. In 1901, for example, the Heusler alloys, which had outstanding properties compared to previous magnets, were reported. The composition of a typical Heusler alloy was 10–30% manganese and 9–15% aluminum, the balance being copper.

In 1917 cobalt steel alloys were discovered in Japan and were very important. Also from Japan in 1938, Kato and Takei developed a very different kind of magnet made from powdered oxides. This development was the forerunner of modern ferrite. However, similar magnets appear to have been made by Gowin Knight in England in perhaps 1760. From information published in 1779 it appears Knight prepared oxides by placing iron filings in water and by working them back and forth for several hours until a suspension of fine iron oxide was obtained. Knight then mixed the fine particles with linseed oil, molded them into various shapes and then baked them in an oven and magnetized them by the best touch methods of the day.

The early history of permanent magnetism is remarkable in the sense that results were obtained in a period when the disciplines we know and use today were in their infancy.

1.3 GROWTH OF THE INDUSTRY

The permanent magnet industry has experienced substantial worldwide growth over the past century. In 1955, an estimated 12,000 tons of permanent magnets were produced with an estimated value of approximately US $100 million. The major portion of the production was alnico type magnets. In 1985, world production had increased to an estimated 180,000 tons having an estimated value of US $1 billion. The annual compound growth in tonnage is approximately 12%. The industry is quite dynamic in shifting to new property systems as estimated in Table 1.1. Ferrites have become the dominant property system due to their value and freedom from the use of strategic materials.

In three decades there has been a substantial shift in production from the United States and Europe to Japan. Table 1.2 shows estimated production in 1985 by area.

There are essentially four major types of magnets in production today: alnico; ferrite; rare-earth cobalt; rare-earth iron; and miscellaneous other types such as vicalloy, iron cobalt lodex cunife, and bonded.

Table 1.1 Estimated production by type (1985)

Type	Value (US $million)	Percent
Alnico	150	15.0
Ferrite	675	67.5
Rare earth	125	12.5
Other	50	5.0
	1000	100.0

Table 1.2 Estimated production by area (1985)

Area	Value (US $million)	Percent
United States	300	30.0
Europe	185	18.5
Japan	415	41.5
Other countries	100	10.0
	1000	100.0

There are over 100 producers with approximately 20 of them in the United States. These producers offer hundreds of different sets of unit properties, many having only slight differences from each other. These grades and their magnetic and physical properties are shown in Appendix 2.

1.4 PROPERTY IMPROVEMENT AND THE CHANGING PATTERN OF USE

It is interesting to trace the history of property improvement and relate it to man's use of permanent magnetism. In Figure 1.3 the property achievement in terms of laboratory MGOe values against time is shown. Three important milestones in the history of permanent magnets are indicated in this figure. The first occurred in the last century when very weak magnets with stability problems were used in devices in which it was imperative to have permanent magnets in order to function. This was the case for the compass, magneto and meter. Rather excessive volumes of early permanent magnetic material were tolerated in these devices. The second milestone occurred in the 1940s when permanent magnet properties improved to the point of being able to compete with electromagnets, both functionally and economically. In loudspeakers and in small specialty d.c. motors, permanent magnets were extensively used and the commercial industry grew rapidly. In 1970, a milestone of profound importance occurred; the properties improved nearly tenfold. Rare-earth magnets were available and for the first time in history, permanent magnets allowed new solutions that were not possible or feasible with electromagnets. Permanent magnets had arrived as unique components. The volumetric efficiency of these new magnets allowed designers to restructure magnetic circuits and devices and to obtain operating parameters that enhanced the value; in several instances, the new magnets were cost justifiable in spite of very expensive raw materials.

As the laws of scaling [3] were explored, it was clear that in small devices the permanent magnet had major advantages over electromagnetism. A permanent magnet's ability to establish field energy remains constant regardless of scale. As we scale down an electromagnet, we find that the conductor current density increases inversely with scaling factor and we run into insurmountable cooling problems. As permanent magnets are projected into larger devices we find that 30–40 MGOe magnets have volume efficiencies superior to that of electromagnetism and no excitation loss. Permanent magnets have traditionally been limited to low wattage devices but now there is a definite trend to apply permanent magnets in larger equipment.

The greatly improved magnetic moments per unit of mass in high energy magnets has led to a new motion control industry. Rare-earth magnets are truly the companion components to microelectronics in changing the speed of doing things.

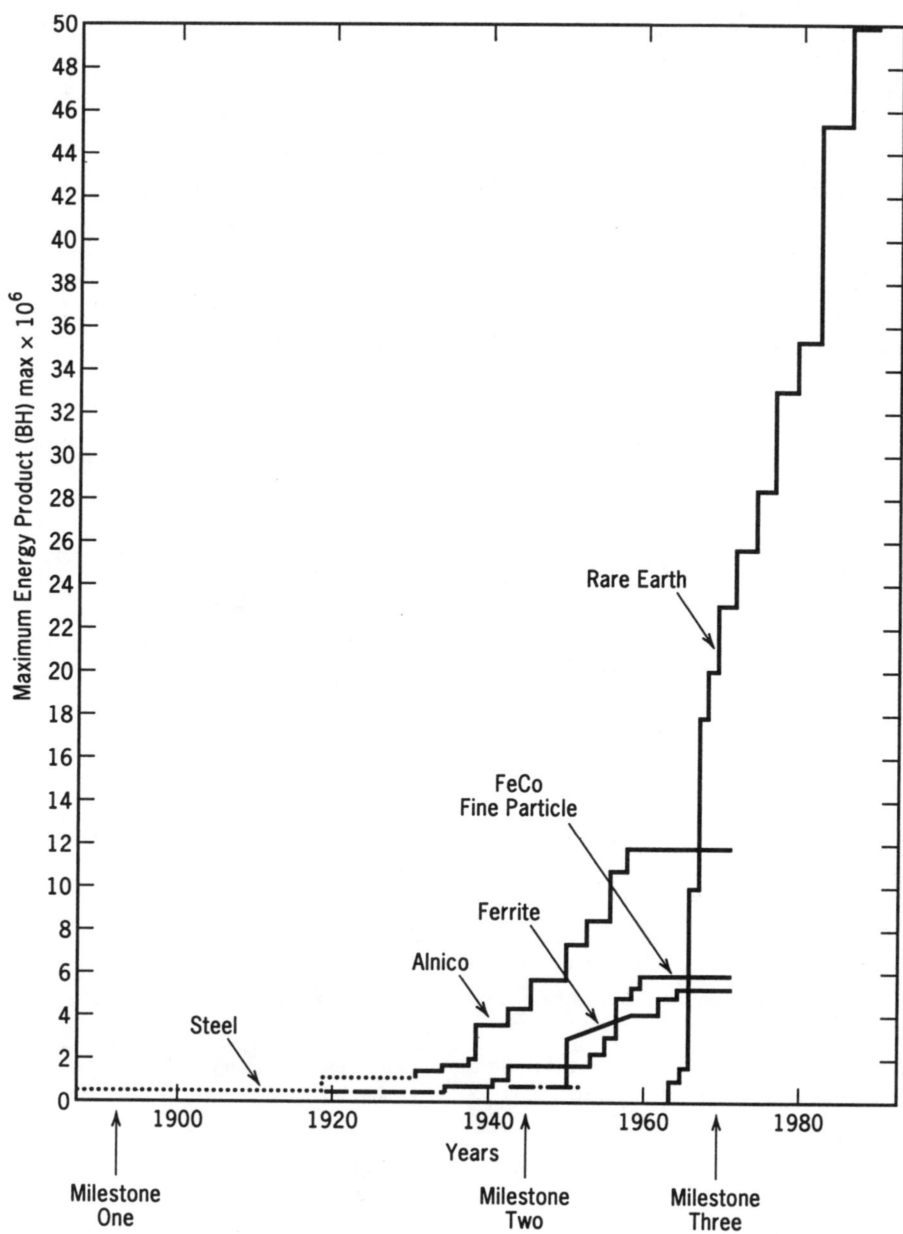

Figure 1.3 Progress in property development.

Some data on patterns of use in various devices are available from producers around the world. Table 1.3 is an estimate of use by application compiled in 1985. Certainly, producing a force or a torque in converting electrical energy to mechanical energy is the dominant area of use for permanent magnets. Motors and loudspeakers account for about 75% of total use.

In Chapter 7, which covers applications, one finds that there are a tremendous number of types of motors, most of which do not use permanent magnets. However, most types could use magnets to improve their functional worth and therefore tremendous growth can be expected due to changes in motor function use and in the electrical input control features. For example, computer peripheral motor requirements may use competing techniques to position a read-write head, such as a stepping motor or a linear voice coil type motor.

We are witnessing the convergence of several technologies impacting on traditional motor markets. Over a long period of time there have been distinctive, nonoverlapping markets for d.c. and a.c. motors. Today, the distinction is often not clear. Indeed, we may well be heading toward a universal type motor in which a moving magnet rotor with a squirrel cage rotates in a distributed stator. This machine could operate as a line start

Table 1.3 The magnet market by device

Use	Percent of production used
Small motors	60
Loudspeakers	15
Communications	8
Electronic tubes	7
Novelties and miscellaneous	6
Mechanical work devices	4

Table 1.4 Summary of use of d.c. permanent magnet motors

	Consumption US $millions		Average % annual growth rate
Equipment	1984	1990	1984–1990
Computer and office equipment	447	1884	27.0
Consumer	759	3547	29.3
Industrial and instruments	367	1253	22.6
Military and aerospace	190	379	12.2
Total	1765	7063	26.0

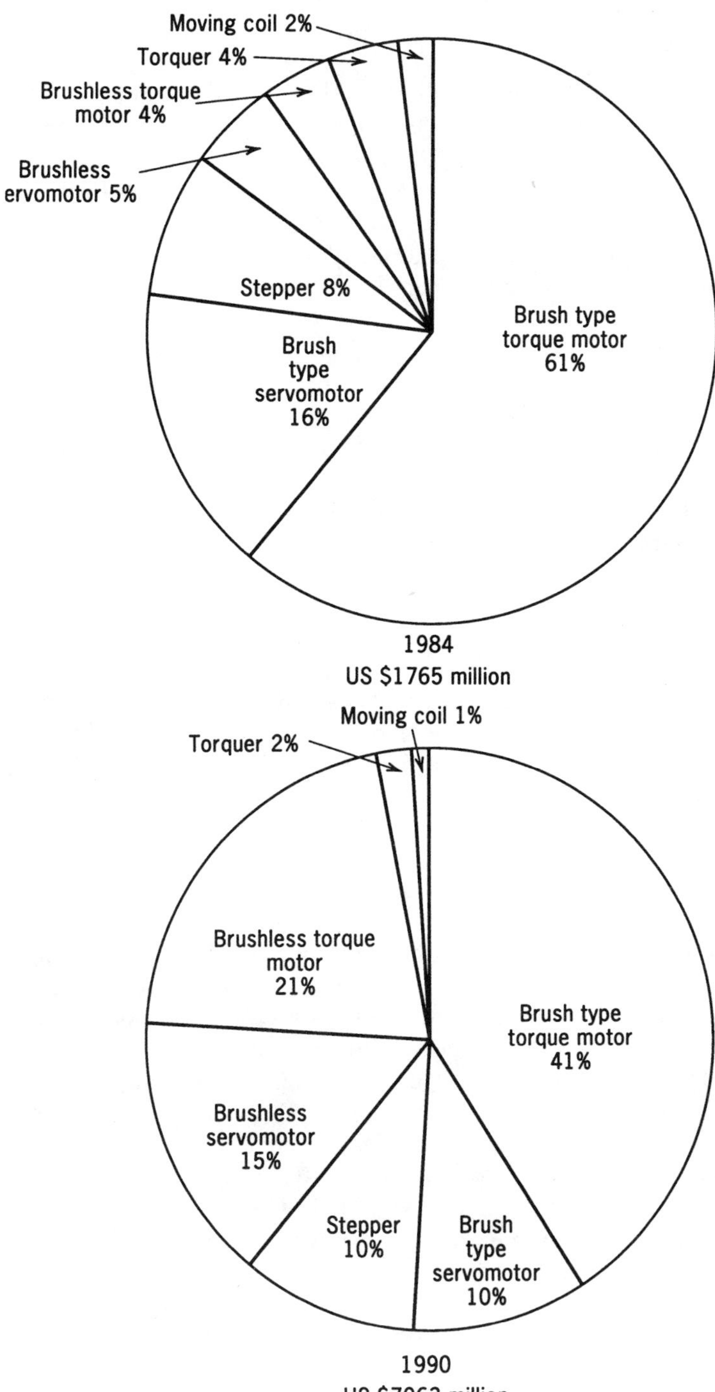

Figure 1.4 Trends in the permanent magnet motor markets.

synchronous machine or, depending on its electrical input, run as a brushless d.c. motor or as a stepper.

As an example of the potential growth and dynamics of the motor markets for permanent magnets, a good analysis of the market was published in 1985 by Neidhart [2]. This analysis estimates the 1985 market for small motors to be in excess of US $5 billion and growing by 5% a year in real terms. Table 1.4 gives a breakdown by motor type of the permanent magnet d.c. motors currently used in 1985 and shows the estimated growth by 1990. The growth for d.c. permanent magnet motors is projected to be 26% overall with some types, such as appliance motors in the consumer category, projected to have an annual growth rate in excess of 100%.

The trend by type of motor is illustrated in the pie charts of Figure 1.4. The trend to d.c. brushless machines is quite apparent. The growth in d.c. permanent magnet machines will, to an extent, be at the expense of the a.c. machine. One can expect that a.c. machine producers can quite easily adapt their facilities to producing the brushless torque motor. The central features possible with the brushless type machine will qualify it for many future applications.

At some point one can expect property/cost relationships in permanent magnets which will allow the induction type a.c. motor to be converted over to an inductor start synchronous machine; a line start machine that has no rotor losses and hence has efficiency and power factor improvements. Then the distinction between a.c. and d.c. machines may be totally lost and permanent magnets may well find themselves in virtually all electrical machines.

1.5 RAW MATERIAL IMPACT

The commercial success of a permanent magnet material is a strong function of the cost, availability and geographic source of its constituent elements. In ferrite magnets, for example, the elements involved are plentiful, low in cost and nonstrategic. In alnico magnets, the cobalt crisis of the 1970s was, to a large extent, responsible for the decline of this type of magnet. In 1972 alnico magnets represented about 40% of the domestic magnet industry. By 1982 alnico magnets were only 7% of the industry.

The commercialization of $SmCo_5$ and $Sm_2(CoFe)_{17}$ were both seriously limited by high percentages of cobalt and the fact that smarium was such a small percentage of rare earth ore. It worked out that total available world smarium production was only enough to support 1–2% of the world's permanent magnet requirement in terms of pounds. However, because of the remarkable energy density, the dollar value of that amount is approximately 15% of the industry.

Rare-earth cobalt introduced us to high energy magnets that could be cost-effective in only a very narrow and specific area of use. They were limited by the raw material and therefore, could not become a major property system. With NdFeB the raw material situation is very favorable. This system is free of cobalt. Industries supplying iron and boron are well established and can easily support the growth of NdFeB magnets. We need only be concerned with sources of neodymium.

Because of electronic shell similarity and position in the periodic table, the rare-earth elements are difficult to separate. It is this characteristic that is responsible for the name rare-earth. In terms of availability, or percentage of the earth's crust, they are more abundant than many elements such as lead and copper that we do not regard as scarce. Rare-earth elements are found in a number of commercially mined ores, the most important being bastnasite and monazite. Although many rare-earth elements are quite abundant, ore concentrates that are economical to mine are not so common. Bastnasite is mined principally in the United States and China, while monazite mining is carried out in Australia, Malaysia, and India. The rare-earth industry reports production in terms of rare-earth oxide rather then the pure industrial elements. When the industry reports known reserves it refers to resources which are economic to mine. Currently, the United States has a lead position in production of rare-earth oxides. The Molycorp Corp. operation at Mountain Pass, CA is the leading producer at this time and is estimated to supply about half of the free world rare-earth oxide market.

In terms of known reserves, China is believed to have 85% of global reserves. In terms of the Nd content of the ore, the Chinese reserves would represent in excess of 6 million tons of Nd_2O_3. China at some point will be a major supplier and is becoming interested in forming joint ventures to attract capital for development of production facilities.

The percentage of the individual elements in each of the two major ore systems is given in Table 1.5, along with estimated production. The relative abundance in the ore can influence the cost of the individual elements. The early product of the rare-earth industry was the family of elements called misch metal. A market for individual elements has only existed for about 20 years. As interest in individual elements developed, the producers needed a balanced market for all of the elements to make separation of any one element feasible. Because of the technical problems of separation, even today, the common practice is to separate all of the elements and to attempt to balance demand with the natural occurring element ratios in the ore.

Although there are now substantial markets for Nd, it is expected that the use of Nd in magnets will represent the largest use. Many rare-earth producers and some magnet producers are actively developing ways to produce Nd metal and NdFe alloys. The methods which appear to offer the most promise are the calciothermic fluoride reduction process and the metallothermic reduction process. In the latter process, metal is produced from Nd_2O_3 by reducing with sodium metal in a calcium chloride bath. The price of Nd to the magnet producer will have major impact on the growth of

Table 1.5 Rare-earth availability (U.S. Bureau of Mines)

Rare-Earth oxide	Bastnasite California		Monazlte Australia	
	% Reo	Prod T	% Reo	Prod T
Lanthanum	32.00	7840	23.90	2390
Cerium	49.00	12,005	46.03	4603
Praseodymium	4.40	1078	5.05	505
Neodymium	13.50	3308	17.38	1738
Samarium	0.50	123	2.53	250
Gadolinium	0.30	74	1.49	149
Dysprosium	0.03	7	.69	69
Other Reo	0.27	65	2.93	293
Total	100.00	24,500	100.00	10,000

NdFeB magnets. Nd metal prices may change greatly because of the number of factors involved; for example, the rate of growth and the demand for the other rare-earth elements, as well as the processing technology which is selected as best from a yield, purity, investment and operational viewpoint.

1.6 A LOOK AHEAD

Clearly there are some compelling reasons to use permanent magnets over electromagnets. The growth of permanent magnets in industry is at least as rapid as the electrical manufacturing industry. However, in certain areas, it appears they are positioned for explosive growth. One such area is the motion control industry. As we automate our offices and factories, very often the key to improving quality and productivity is the electric drive system. Our drive systems are not limited by the speed of digital electronic commands, but by the transducers that produce the work function. High energy permanent magnets determine how rapidly we move a head on a disc or how many operations per minute are realized in a numerically controlled machine tool. Pneumatic and hydraulic devices in aircraft and in industrial plants are being replaced with electromagnets with improved reliability, cost, and function. In transportation and energy conversion, the permanent magnet may well play a critical role in the energy scenarios of the future.

There is a continuing need for better communications in the industry. We must become system thinkers concerning magnetism. In research and development there is a need for collaboration and the role of research institutes should be explored. The United States, still have a lead position in the science of permanent magnetism due, in part, to military funding, but implementation is generally slow.

Today, there is a new tempo regarding development. Traditionally, there was a 15–20 year lapse between the development of new properties of significance. Today, we are seeing new properties every year or two. The number of workers and the number of papers presented at conferences has increased dramatically. This is difficult to manage in the traditional magnet company where, perhaps, the time was so long between major developments that research and development was eliminated because there was nothing happening.

Another trend that seems significant is the building of integrated permanent magnet circuits by magnet producers. With the newer high energy density materials, the sophisticated problems of magnetization, temperature stabilization, and measurements suggests the increasing need for one source of responsibility for the performance of a permanent magnet circuit. For example, in matrix materials it seems feasible to use a progressive tooling system to form both magnet volume and the high permeability flux conducting members of a circuit.

Regarding improvements in the unit properties, we are, as developed in Chapter 2, limited by the saturation magnetization of elements and alloys to perhaps maximum energy products of $(25 \times 10^3)^2/4$ or 150×10^6 $(BH)_{max}$ from a composition viewpoint. At the present time, we have, with NdFeB, achieved one third of this maximum. It seem reasonable to expect that from rare-earth compounds, we will be able to develop materials with B_r levels of perhaps 16 kG. If we can achieve H_{ci} levels of $\frac{1}{2}B_r$ (approximately 8 kOe) then, a very useful permanent magnet of 64 MGOe would result. It would have a breaking demagnetization curve, but this would be very useful for applications where flux density is the figure of merit. Such applications would include speakers and eddy current devices.

For several decades, most of the emphasis in development has been in understanding how the microstructure gives increased levels of H_{ci}. Of equal importance would be a focus on development of very high B_r materials with modest levels of coercivity. Such magnets would be easier to magnetize and they may be acceptable even in motors where the electronic controls limit inrush current at start and shall condition.

When one considers the anisotropic field levels of single crystals of the rare-earth cobalt and iron compounds, we see that we have an opportunity for much higher H_{ci} levels in bulk magnets. As we learn more about the coercive force mechanisms, we can look forward to considerably improved H_{ci} levels in bulk magnets. This will be of interest in applications such as motors where often the magnet is exposed to external fields much higher than its own self-demagnetizing field.

The future for NdFeB magnets appears particularly bright because of its property potential and a very favorable raw material situation, but it is this writer's opinion that ferrites will remain the dominant material for the forseeable future. Ferrite will co-exist with NdFeB which should represent perhaps 40% of the magnet industry in its various forms. Within a decade,

$SmCo_5$, $Sm_2(Co\text{-}Fe)_{17}$ and Alnico magnets will be important only for specialty applications.

In the long-term, high energy permanent magnets may be expected to lead to very interesting new magnetoelectric technology an area of where machine energy density will be well above conventional electromagnetism and more toward what we now have with superconducting techniques. Permanent magnets of 50–100 MGOe are very possible and such magnets, if made from elements that are abundant and of reasonable cost, could allow a new breed of magnetoelectric energy converters; machines involving only permanent magnets and current conducting material.

REFERENCES

[1] E. N. da C. Andrade, Early History of the Permanent Magnet, Endeavour XVII(65) (1958); Reprinted in R. J. Parker and R. J. Studders, Permanent Magnets and Their Applications, (John Wiley, New York, 1962).

[2] H. A. Neidhart, Small Motor Market Trends (IEEE, 1985).

[3] R. J. Parker, Paper No. II-5, 6th International Workshop on Rare Earth-Cobalt Permanent Magnets and their Applications, 1982.

2

MAGNETISM AND THE PERMANENT MAGNET

2.1 Introduction
2.2 Unit Systems, Definitions, and Conversion Factors
2.3 The Hysteresis Loop
2.4 Demagnetization Factors and Magnetic Circuit Concepts
2.5 Permanent Magnets and Electromagnetism
2.6 Energy Relationships
2.7 Boundary Conditions and Figures of Merit

2.1 INTRODUCTION

In the first chapter, the early history described how early workers achieved useful magnet properties. This is a rather remarkable feat when one considers that at that point in time, there were few definitions or measurements so important to communication between contributors.

In this chapter we want to introduce the units, definitions and quantitative measurement procedures to be used throughout this book. Important relationships involving electromagnetism, energy conversion and magnetic circuit concepts are also introduced.

2.2 UNIT SYSTEMS, DEFINITIONS, AND CONVERSION FACTORS

In order to understand, evaluate, and compare permanent magnets, the first requirement is for a quantitative measurement of a magnetic field and the magnetic state of a material. In magnetics, the magnetic pole was chosen by early scientists as a fundamental quantity. However, this is not a physical quantity that can be isolated and measured, as is the case with mass or charge on an electron. We must remember that permanent magnets existed long before electrical currents. A unit pole is a pole, which, when placed a distance of 1 cm from a like pole in a vacuum, would be repelled with a force of 1 dyne. This unit of magnetizing field H is called the oersted in the CGS system.

In a similar manner, the magnetic state in a material may be described by referring to the number of unit poles in a given cross-section. Such a measure is termed the intensity of magnetization and is given the symbol I (or sometimes J). In the bar magnet of Figure 2.1, with m unit north poles at one end and m unit south poles at the other, of length l with cross-section a, the intensity of magnetization I may be seen to be m/a. Faraday considered the space that surrounds such a bar magnet to be filled with endless imaginary "lines of force" threading the magnet poles, with one line arising from each pole pair. The convention is that the lines radiate out from the north pole and return to the south pole. The total lines are termed the total flux and the lines in a unit area, the flux density B. When a magnetic material is placed in a field this induction B is comprised of two components. The lines of force due to the field H and the lines of magnetism due to the presence of the magnetic material termed the intensity of magnetization I. Therefore

$$B = \phi/A = \mu_0 H + 4\pi I \tag{2.1}$$

The factor 4π arises from the fact that at every point, unit distance from a unit pole, a unit field is established. A unit field would exist on the surface of a sphere of unit radius enclosing the pole, the sphere area being equal to 4π. In engineering work with permanent magnets, the term $4\pi I$ is usually expressed as B_i, the intrinsic magnetization. The field stength is usually in

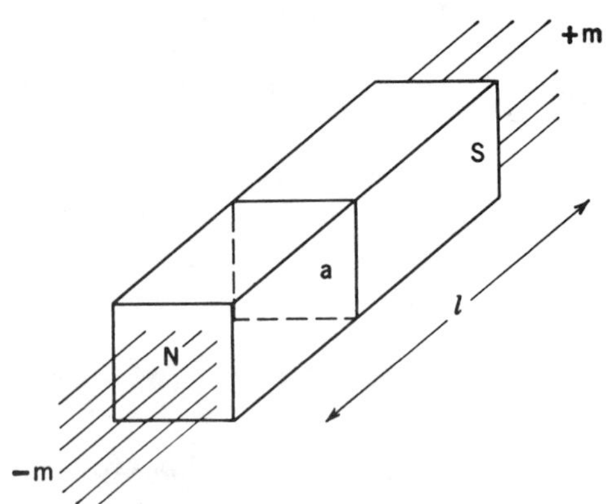

Figure 2.1 The bar magnet poles.

the same or opposite direction as the induction. The general relationship between induction, magnetization due to the material, and field may be expressed as

$$B = B_i \pm \mu_0 H \tag{2.2}$$

The relative contributions of B_i and H greatly influence magnetizing, demagnetizing, stability, and measurements in permanent magnets. The term μ_0 is called the magnetic constant. In the CGS system μ_0 has a value of 1. In the SI system μ_0 has a value of $4\pi \times 10^{-7}$ webers/ampere meter (Wb/A m). B, H, and B_i are all vectors, however, as we use them they are usually parallel or antiparallel, so that we can use the relationships in scalar form. In the SI system, B and B_i have the same units, weber/(meter)^2 (Wb/m^2) or the tesla (T). The unit for H is ampere/meter (A/m). In the CGS system, the common unit is maxwell/cm^2. However, they have different names; B and B_i are in gauss (G) and H is in oersteds (Oe). In free space B and H are not independent but are related by $B = \mu_0 H$. In a magnetic material B and H can vary independently.

At this time the units and terms for magnetism are in transition. For about half a century, the CGS system has been widely used. However, there is now a strong movement toward SI units because of its wider acceptance in science. Already in some countries, the SI system is now required by law. Table 2.1 compare units and symbols in the two systems. Using SI units in magnetism has some basic problems, due in part to the great difference between numerical values of B and H. We find that our instruments relate to the weber and tesla. We cannot plot what we read without conversion on the H axis. Recoil permeability cannot be plotted directly without conversion. Induction and magnetization cannot be plotted on the same scale, as they have the units of tesla and ampere/meters, respectively. In Figure 2.2,

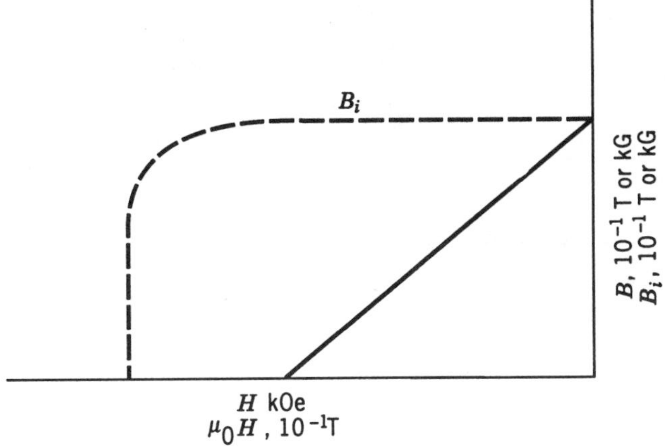

Figure 2.2 A bridge between systems of units. From Martin [1].

Table 2.1 Symbols, definitions, and units

Symbol	Definition	CGS	SI or MKSA
ϕ	Flux or lines of force	Maxwell	Weber, Wb
B	Flux density or induction $B = \phi/A = \mu_0 H + B_i$	Maxwell/cm^2 = gauss, G	Wb/m^2 = tesla, T
H	Magnetic field strength	Oersted, Oe	Ampere-turns/meter, A/m
μ_0	Magnetic constant	Unity, i.e., IG = 1 Oe	$4\pi \times 10^{-7}$ henry/meter or $4\pi \times 10^{-7}$ Wb/Am
$\mu_0 H$	Magnetic field strength or field flux density in space	Maxwell/cm^2 = G	Wb/m^2 = T
B_i	Intrinsic flux density $B_i = B - \mu_0 H$	$B_i = 4\pi J$, G	$B_i = J$, T
B_{is}	Saturation flux density	$B_{is} = 4\pi J_s$, G	$B_{is} = J_s$, T
J	Intensity of magnetization or magnetic moment per unit volume	$J = B_i/4\pi$, G G/4π = e.m.u. = dyne/cm^{2}Oe	J, T
H_d	Demagnetizing field, usually self-demagnetizing field	Oe	A/m
H_a	Applied or external magnetic field strength	Oe	A/m
H_c	Coercivity or coercive force, magnetic field strength to reduce B to zero after saturation	Oe	A/m
H_{ci}	Intrinsic coercive force, magnetic field strength to reduce B_i to zero after saturation	Oe	A/m
$(BH)_{max}$	Maximum $B \times H$ product in the demagnetizing quadrant	Gauss-oersted, usually million gauss-oersteds, MGOe	Kilojoule/meter^3, kJ/m^3
$\mu_{,0}$ $(BH)_{max}$	Maximum $B \times \mu_0 H$ product	Gauss-oersteds	T^2
N	Demagnetizing factor $N = -\mu_0 H_d / B_{id}$		

demagnetization curves are plotted with $\mu_0 H$ instead of H on the X-axis. In the CGS system, μ_0 is unity and hence we have the conventional CGS demagnetization curves. By using the SI unit decitesla (dT) for B, B_i, and $\mu_0 H$ there is now a convenient one to one numerical relationship between units in the two systems for B, B_i, and $\mu_0 H$ [1]. Only H and $(BH)_{max}$ have a numerical difference. However, $(BH)_{max}$ can be reduced to a one to one numerical relationship if $\mu_0 H$ in decitesla instead of H in oersteds is used for

Table 2.2 Conversion factors

Quantity	CGS to SI	SI to CGS
ϕ	1 maxwell = 10^{-8} Wb	1 Wb = 1 V-s = 10^{8} maxwells
B	1 G = 10^{-4} T	1 T = 1 Wb/m^2 = 10^4 G
	1 kG = 1 dT	1 dT = 1 kG
H	1 Oe = 79.58 A/m	1 A/m = 12.57×10^{-3} Oe
	1 kOe = 79.58 kA/m	1 kA/m = 12.57 Oe
$\mu_0 H$	1 Oe = 10^{-4} T	1 T = 10^4 Oe
	1 kOe = 1 dT	1 dT = 1 kOe
(BH) product	1 GOe = 79.58×10^{-4} J/m^3	1 J/m^3 = 125.7 GOe
	1 MGOe = 7.96 kJ/m^3	1 kJ/m^3 = 12.57×10^{-2} MGOe
$\mu_0(BH)$ product	1 GOe = 10^{-8} T^{-2}	1 T^2 = 10^8 GOe
	1 MGOe = 1 dT^2	1 dT^2 = 1 MGOe

the demagnetizing field. A magnet of 30 MGOe in CGS units would have 30 dT^2 for energy product in SI units. This technique of plotting so that we have a numerical relationship is quite helpful for many people during what appears to be a very long period of transition between the systems. Today it appears that most engineering work is with CGS units. The majority of papers are still presented in CGS units. Perhaps the simplicity of using CGS units so far has outweighed the fundamental worth of the SI system. For these reasons this book will use CGS units. However, one may refer to Table 2.2 for the conversion factors to use in working a problem in SI units.

2.3 THE HYSTERESIS LOOP

Consider a specimen of permanent magnetic material as shown in Figure 2.3. The magnet is arranged in a magnetizing yoke so that the field H can be controlled and reversed. If the specimen in unmagnetized, we will start at point O and apply a positive H. If B is plotted as a function of H, we will obtain the curve OP, called the initial magnetizing curve. If the field is reduced to zero, B will not fall to zero but retain a value B_r. It is necessary to apply a negative field to reduce B to zero. This value of H is defined as coercive force H_c. If the specimen is cycled several times by applying positive and negative values of H, a symmetrical loop is obtained: $PB_rH_cP'B_r'H_c'P$. This curve is the induction or normal hysteresis curve or loop and is the basis of evaluation in magnetic materials. If we arrange the sensors for determining H and B correctly, we can just as easily obtain the intrinsic hysteresis loop $QB_rH_{ci}Q'B_r'H_{ci}'Q$. Note that as H is increased we obtain a larger and larger loop area up to the point where H is high enough to saturate the material. At saturation, B_r and H_c are constant values called the remanence (B_r) and coercive force H_c. At saturation, the intrinsic

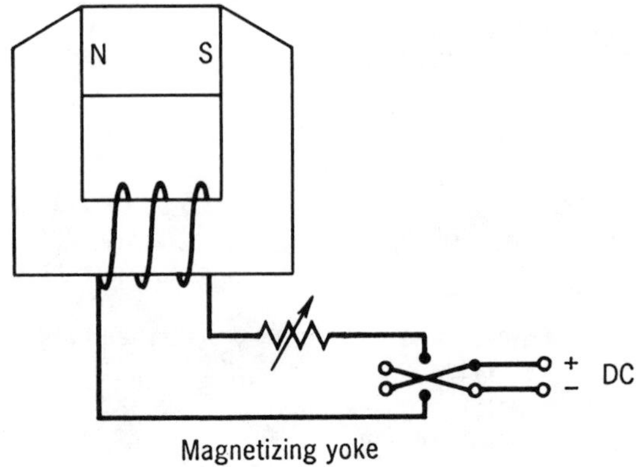

Magnetizing yoke

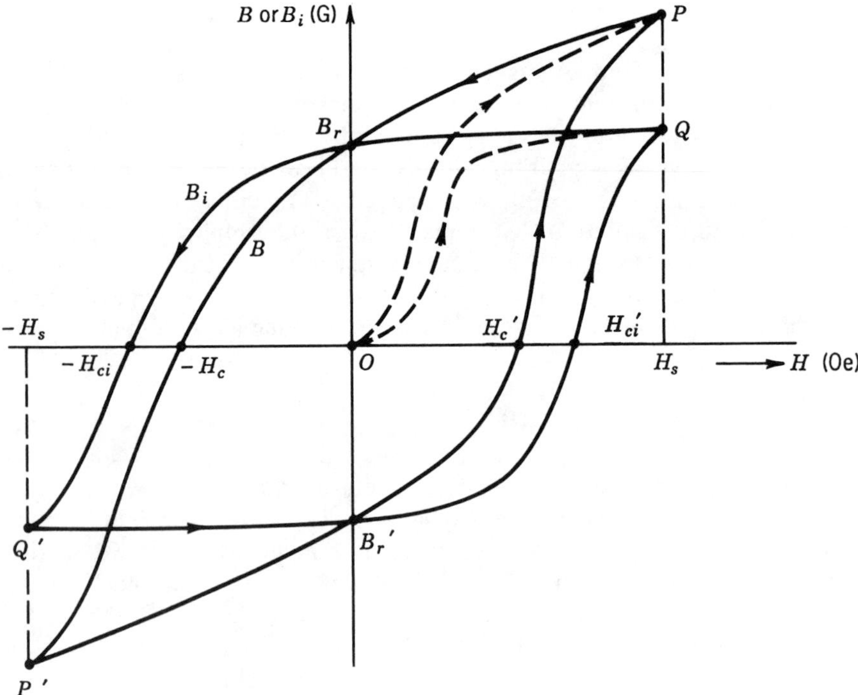

Figure 2.3 Hysteresis loops.

magnetization (level of Q) becomes constant. At every point these two curves differ by the value of H, or $B = B_i \pm H$.

In most devices and systems, the permanent magnet operates in the second quadrant of the hysteresis loop between B_r and H_c. However, we need first quadrant data for magnetization problems and at times a magnet is driven into the third quadrant as it interacts with external fields. Additionally, magnets are sometimes used in hysteresis devices where the total loop area is of significance in the design process.

The ratio of induction B to magnetizing force H is termed permeability μ. The permeability can vary widely at various points on the magnetizing and demagnetizing portions of the hysteresis loop. In particular, we are interested in μ_i the initial permeability, μ_d the differential or maximum permeability and μ_r the reversible permeability. The initial permeability is the initial slope of the magnetizing curve in the first quadrant as shown in Figure 2.4. Differential permeability is the ratio of B to H for a small change in H at any point. Along the magnetizing curve there is a point A where B/H is a maximum, as the curve exhibits maximum inflection. Sometimes μ_d is used to estimate second quadrant properties. As we have seen in considering the hysteresis loops in Figure 2.3, when a demagnetizing field is applied to a saturated magnet, the induction B will decrease along the major loop in the second quadrant. If, when the induction reaches point C in Figure 2.4, the demagnetization field is reduced, the induction will generally not retrace the major loop, but follow a new path CD_1. Alternately, varying the demagnetizing field at some intermediate strength will cause the induction to trace a small interior loop such as DD_1. An infinite number of interior loops could be formed, depending on the magnitude of the demagnetizing field. These interior loops are often referred to as minor loops to distinguish them from the outer or major hysteresis loop. The area of the minor loop is small and a line drawn through the tips of the loop may be used to represent the minor loop. Such a line has the average slope of the loop and is termed recoil or reversible permeability μ_r. The recoil permeability is approximately equal to the slope of the major loop at $H=0$ (at remanence B_r).

Having gained some insight as to how the permanent magnet functions, it probably would be well to mention at this point that magnetic materials are

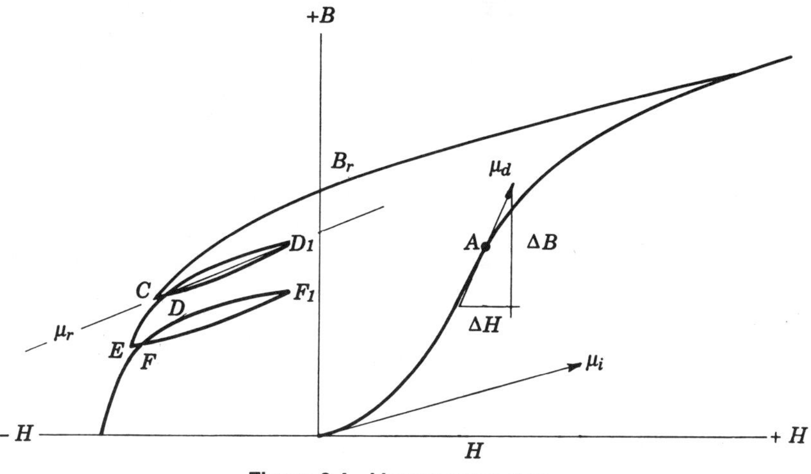

Figure 2.4 Magnet permeability.

divided into two general categories; hard magnetic materials and soft magnetic materials. There is no rigid line between the categories, but in general it seems that below 200 Oe for coercive force, we tend to think of it as a soft magnetic material. Such materials find use in transformers or for conducting flux as in a return path circuit element. Above 200 Oe we tend to think we are working with a permanent magnetic material. These low H_c materials find use in hysteresis motors and torque drives. Appendix 1 shows a chart of hard and soft materials published by the International Electro-Technical Commission (IEC). This is the international group responsible for standards and definitions regarding electricity and magnetism.

2.4 DEMAGNETIZATION FACTORS AND MAGNETIC CIRCUIT CONCEPTS

A very important concept in permanent magnetic materials is the concept of self-demagnetization. In electromagnets we establish a field H by passing a current through a winding. If a ferromagnetic material is placed in the coil, magnetic lines are born due to the contribution of the material. H and B vectors are parallel and additive. A permanent magnet is inherently different. We apply a field H, and B_i the intrinsic magnetism is born. If the permanent magnet is removed from its magnetizing yoke free poles are established and a field potential $-H_d$ exists between the poles. In this case, the potential results from some of the intrinsic magnetization returning internally across the magnet. The useful potential $-H_d$ is a product of the intrinsic B_i and it is 180° opposed to B_i (see Figure 2.5). If an external field H_a is applied to the magnet, as was necessary to obtain the hysteresis loop of Figure 2.3, then we have to consider the combined influence of the applied and internal field

$$H = H_a + H_d \tag{2.3}$$

H_d is proportional to the magnetization or

$$H_d = \frac{NB_i}{\mu_0} \tag{2.4}$$

where N is the demagnetization factor which is dependent on geometry of specimen. We may then write

$$H = H_a - \frac{NB_i}{\mu_0} \tag{2.5}$$

In determining H, the magnets true internal field, we have to consider the field due to the free poles and provide for elimination of the free poles effect

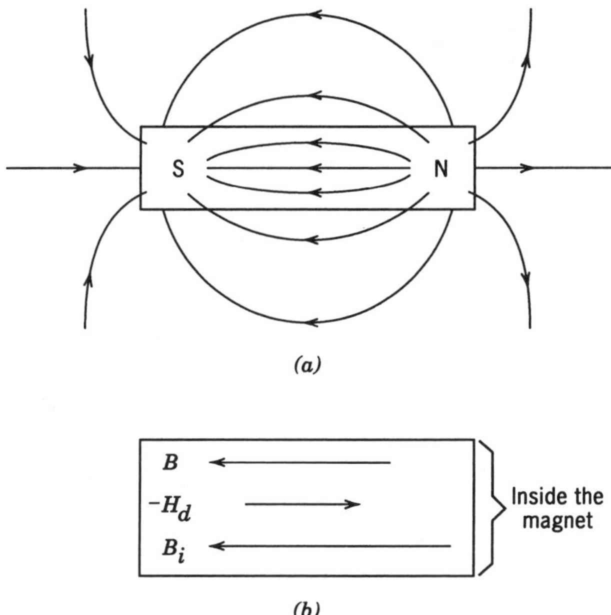

Figure 2.5 Permanent magnet field relationships.

or correction for H_d in any measurement technique. For example, in a closed magnetic circuit, H_d would be zero. In the special case of the ellipsoid, where the magnetization is uniform from point to point, N can be calculated. Joseph [2] has derived both ballistic and magnetometric demagnetization factors for uniformly magnetized cylinders. For material with high coercive force, or square loop demagnetization curves, the magnetization of simple geometries approaches the conditions of uniform magnetization at all points in the specimen. This leads to poles only at the terminals or end surfaces. Joseph's data for cylindrical, rod shapes are shown in Figure 2.6. The factor N is plotted versus length to diameter ratio. N_b is the ballistic demagnetization factor to be used when the measurement involves the central or neutral cross-section of the cylinder. N_m is the magnetometric demagnetization factor to be used when the magnetic moment of the entire cylinder is involved.

If we go to equation (2.4) and substitute $(B_d, -\mu_0 H_d)$ for the magnetization B_i, we have an expression relating the induction B_d and the self demagnetizing field H_d to the magnet geometry and the factor N

$$H_d = \frac{B_d}{1-(1/N)} \tag{2.6}$$

By rearranging, the very useful load lone slope can be determined

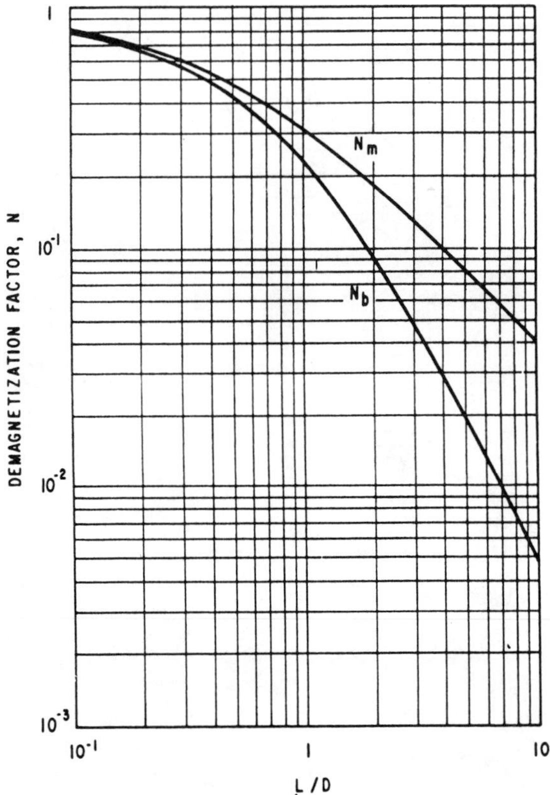

Figure 2.6 Ballistic (N_b) and magnetometric (N_m) demagnetizing factors for cylindrical rods with invariant B_i. From Joseph [2].

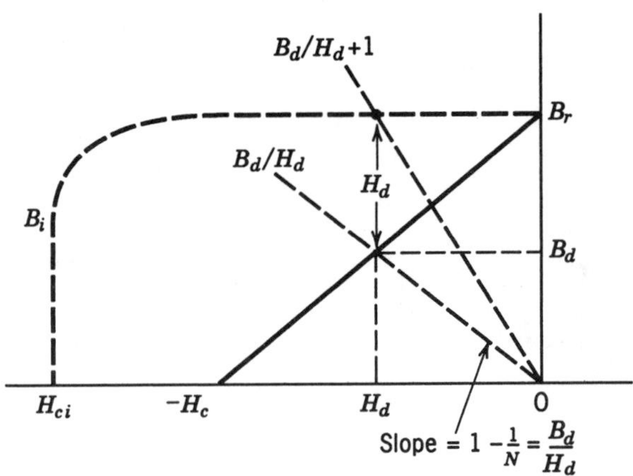

Figure 2.7 Unit properties and the load line.

$$\frac{B_d}{\mu_0 H_d} = 1 - \frac{1}{N} \tag{2.7}$$

In magnet design work, this ratio is used extensively. It is also referred to as unit permeance or the coefficient of self-demagnetizaiton. Figure 2.7 shows the graphical relationships for a high coercive force permanent magnet and the load line slope.

2.5 PERMANENT MAGNETS AND ELECTROMAGNETISM

The relationships between a magnetic field and electric currents are a vital part of permanent magnet technology. When magnetic flux linking an electrical coil or circuit is changed, a voltage is generated equal to $n(d\phi/dt) \times 10^{-8}$, where n is the number of turns and $d\phi/dt$ is the rate of flux change with respect to time. If the electric circuit is complete, a current (i) will flow and power will be developed. Since power is the product of current and voltage, the power may be expressed as $ni(d\phi/dt) \times 10^{-8}$. In a given time, if the flux is changed by a definite amount (ϕ), the work done will be $W = ni\phi \times 10^{-8}$ J.

Let us now consider a long straight wire (Figure 2.8a) and the field intensity around it. The magnetic lines of force surrounding a long straight wire will be concentric circles in a plane perpendicular to the wire. If a unit pole is carried around one of the lines of force, each of the 4π lines from the unit pole will link the electric circuit and the work done will be $4\pi i$. The total distance moved is $2\pi r$ where r is the radius of the concentric path followed. The field intensity is then

$$H = \frac{0.4\pi i}{2\pi r} = \frac{0.2i}{r} \text{ dynes per unit pole} \tag{2.8}$$

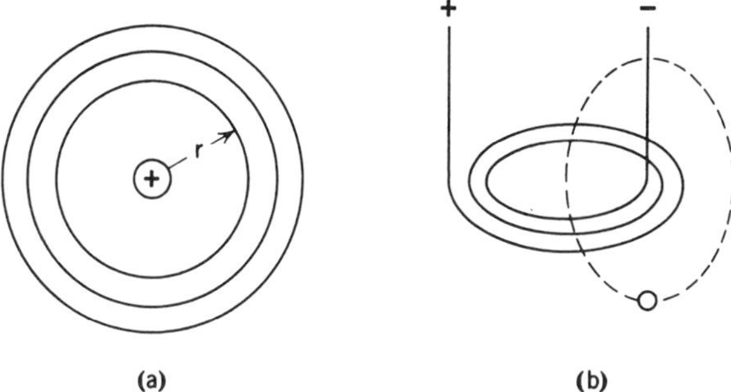

Figure 2.8 (a) Cross-section of a long straight wire with current flowing creating magnetic lines of force. (b) A coil of *n* turns carrying a current *i*; dotted line is closed path of unit pole.

The ability of an electric current to establish magnetic flux is termed magnetomotive force, F. Consider a coil of n turns carrying a current i (Figure 2.8b). If a unit pole is moved from any point through the coil and back to the starting point the work done will be $4\pi ni$ erg. This is the common expression for a magnetomotive force developed by an electric current. The unit of F is the gilbert and is the magneto-motive force which will produce a field intensity of 1 dyne per unit pole in a path 1 cm long.

The work done in carrying a unit pole along any path between two points in a magnetic field is defined as the line integral between the points. The work done in carrying a unit pole around a conductor carrying a current is 4π times the current and independent of path taken or

$$\int H\,dl = 4\pi i \tag{2.9}$$

2.6 ENERGY RELATIONSHIPS

When a volume of magnetic material is magnetized to a certain induction level, energy is expended. When the magnetizing force is removed, a portion of the energy is returned to the source of magnetizing energy. Let us first consider the energy relationships when the source magnetizing energy is electrical in nature [3]. Figure 2.9a shows a ring magnet and magnetizing winding, and the hysteresis loop of the magnetic material. If a current flows in the winding on the ring, magnetic flux will increase from zero to a certain maximum depending on the ampere turns and the reluctance of the core. As flux increases in the core, a voltage will be induced depending in magnitude on the rate at which the flux is increasing. If the flux changes an amount, $d\phi$

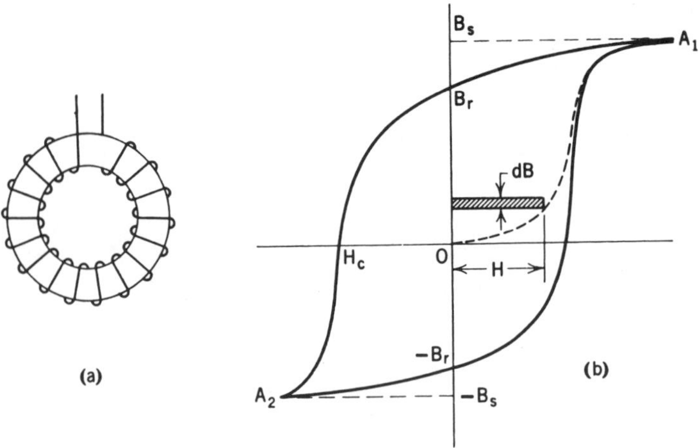

Figure 2.9 (a) Uniformly wound ring magnet. (b) Hysteresis loop of the ring magnet material.

in a time interval dt and turns are expressed by n, then the voltage induced will be $n(d\phi/dt) \times 10^{-8}$ V, and if the current at a particular instant is i A the instantaneous power or rate of doing work is $ni(d\phi/dt) \times 10^{-8}$ V. In keeping with the conservation of energy and Lenz's law, the voltage is in such a direction as to oppose the increasing current, consequently energy is absorbed from the power supply. If dW represents work done in the time dt, to magnetize the core by the amount $d\phi$, then we may write

$$\frac{dW}{dt} = ni\,\frac{d\phi}{dt} \times 10^{-8} \tag{2.10}$$

or

$$dW = ni\,d\phi \times 10^{-8} \tag{2.11}$$

If $\phi = Ba$ and $d\phi = a\,dB$, where B is flux density and a is core area, also $ni = Hl/4\pi$ where H is field intensity and l is core length in centimeters (from the line integral theorem). Substituting these for ni and ϕ in the preceding expression for dW we have

$$dW = \frac{(a\,dB)H}{0.4\pi \times 10^8} = \frac{VH\,dB}{0.4\pi \times 10^8} \tag{2.12}$$

since $V = al$. The work done in magnetizing the core of volume V to a density B is given by

$$W = \frac{V}{0.4\pi \times 10^8} \int H\,dB\,, \quad \text{J} \tag{2.13}$$

In Figure 2.9b the area OA_1B_s represents $\int H\,dB$. With decreasing field intensity, the voltage generated is in such a direction as to oppose the decreasing current and consequently energy is returned to the power source. In decreasing, the flux density follows path A_1B_r, the returned energy is represented by area $A_1B_sB_r$, and the net energy dissipated is represented by area OA_1B_r. This energy manifests itself in the form of heat. This is work done in overcoming the forces which hold the magnetic domains in a randomly oriented condition. As the magnetizing current is reversed, the voltage and current relationship is in the sense that additional work is done and energy is dissipated in reducing the flux to zero and magnetizing to the $-B_s$ level. Area $OB_rA_2{-}B_r$ represents the net energy dissipation, and similarly in the third and fourth quadrants the energy loss is area $-B_rA_1B_r$. The total energy loss in one cycle is represented by the area between curves or the total loop area. The significant changes as far as the permanent magnet is concerned, of course, occur in the first two quadrants, and we see that work is required to magnetize the permanent magnet material. With the magnet operating at B_r additional energy input is required in order to demagnetize and reduce the magnetism to zero. The second quadrant area OB_rH_c represents the approximate energy required.

In order to continue the energy relationships let us consider the same ring of Figure 2.9 carried through its hysteresis cycle with a mechanical rather than electrical source of energy. Figure 2.10 illustrates the ring being driven by mechanical torque through two air gaps of opposite polarity. The direction of magnetization is now axial; however, we shall consider the same magnetic material as in the electrical case and assume that it is isotropic in nature (i.e. exhibits the same magnetic properties in all directions). In order to develop the relationships between mechanical energy and magnetic field energy we will consider the fundamental law governing force between two isolated magnetic poles.

$$F = mm^1/r^2 \tag{2.14}$$

where m is the strength of one pole; m^1 is the strength of another pole and r is the distance between them. The force that m^1 would exert on a unit pole ($m = 1$) is by definition the magnetic field intensity H.

$$H = \frac{m^1}{r^2}, \qquad \vec{F} = m\vec{H} \qquad \text{(vectorial)}$$

We will now consider the potential energy of a magnetic dipole in a magnetic field. A dipole is a pair of magnetic poles of equal strength spaced a given distance from each other. The potential energy of a dipole in a field (H) is the work required to move each pole separately from a region of zero field intensity to a region of constant and uniform field intensity (H) or

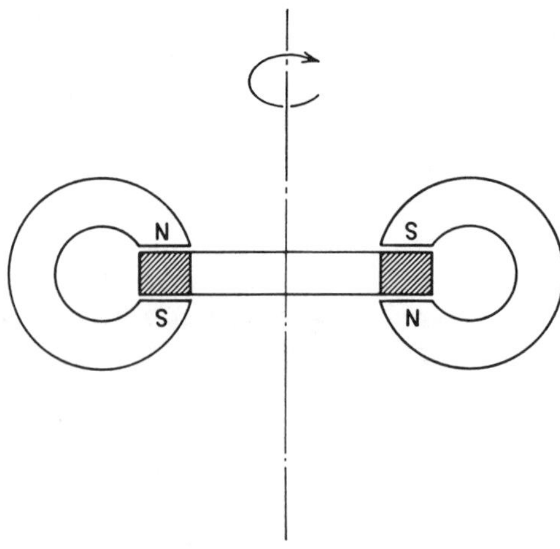

Figure 2.10 Ring magnet mechanically driven through two air gaps.

$$E \text{ (potential energy)} = (ml)H \cos\theta$$

where m is pole strength, l is the distance between poles, θ, the angle between axis of dipole and the direction of the field and ml is the magnetic moment of the dipole.

We may consider our ring magnet as comprised of a great number of magnetic dipoles randomly oriented, which under the influence of a field tend to align themselves in the direction of the field. In a unit volume of the ring magnet the intensity of magnetization is $I = m/a$ from the definition of the magnetic moment $\vec{M} = mlI = V\vec{I}$, and the potential energy of a volume of the ring may be written $E = \vec{M}\vec{H} = V\vec{I}\vec{H}$. If I and H are uniform and in the same direction, $E = VIH$.

Let us now follow the energy changes of volume (V) during one revolution of the ring magnet. Since

$$E = VIH$$

$$dE = VI\,dH + VH\,dI \tag{2.15}$$

The term $VI\,dH$ is the work done by the external mechanical force in moving the volume (V) from a field H to a field of $H + dH$, assuming the volume (V) to act as a real dipole. The term $VH\,dI$ is the change in potential energy caused by a reduction in I in the field of H. Actually some of the elementary dipoles in volume (V) reach a state of instability and change their orientation. The dipoles are held by certain internal force systems. As a dipole changes its position, a certain amount of kinetic energy is imparted to its neighbors and in this manner the total kinetic energy of all the atoms in volume (V) is increased or heat is produced. If we integrate the expression dE around a complete hysteresis loop the initial values of I and H are encountered and the final and initial potential energies will be the same. If we consider the whole ring then

$$\oint dE = 0 = +V\oint I\,dH + V\oint H\,dI$$

or

$$-V\oint I\,dH = +V\oint H\,dI \tag{2.16}$$

Work done to drive ring around one hysteresis loop	=	Heat produced in the ring in one hysteresis cycle

Or, since the flux density inside the ring is

$$B = H + 4\pi I$$

we have

$$I = \frac{1}{4\pi}(B - H)$$

Substituting for I in (2.16)

$$-V \oint \frac{(B-H)\,dH}{4\pi} = +V \oint \frac{H\,d(B-H)}{4\pi} \tag{2.17}$$

or

$$-\frac{V}{4\pi}\left[\oint B\,dH - \oint H\,dH\right] = \frac{V}{4\pi}\left[\oint H\,dB - \oint H\,dH\right] \tag{2.18}$$

Since H returns to its original value after a complete cycle, $\oint H\,dH = 0$ in which case the equation reduce to

$$-\frac{V}{4\pi}\oint B\,dH = \frac{V}{4\pi}\oint H\,dB$$

and per cubic centimeter of ring material

$$-\frac{1}{4\pi}\oint B\,dH = \frac{1}{4\pi}\oint H\,dB \quad \text{in erg/cm}^3 \text{ per cycle} \tag{2.19}$$

In order to clarify the relationships consider, the various points on the hysteresis loop of Figure 2.11 (a–h) and the areas associated with (i) the potential energy, (ii) the work done by the external force, and (iii) the heat produced in the ring.

Between any two points such as (a) and (b), the above expression relating energy loss per cycle may be expressed as follows

$$-\frac{1}{4\pi}\int_a^b B\,dH = -\frac{1}{4\pi}[BH]_a^b + \frac{1}{4\pi}\int_a^b H\,dB \tag{2.20}$$

Work done by external torque = Potential energy change + Heat produced

In Figure 2.11 we find that although the potential energy at each point as given by the BH terms is varying, when considered over the whole cycle these terms cancel out and the potential energy returns to its original value at (a).

When all the $-\int B\,dH$ terms are added, we find the summation of these terms is represented by positive area of the hysteresis loop and indicates the work done by the external torque. Similarly, the summation of the $\int H\,dB$ terms is the same positive loop area, representing the heat produced due to the changing kinetic energy.

Having followed the cyclic energy changes in a magnet material with both electrical and mechanical sources of energy, we will now focus on energy

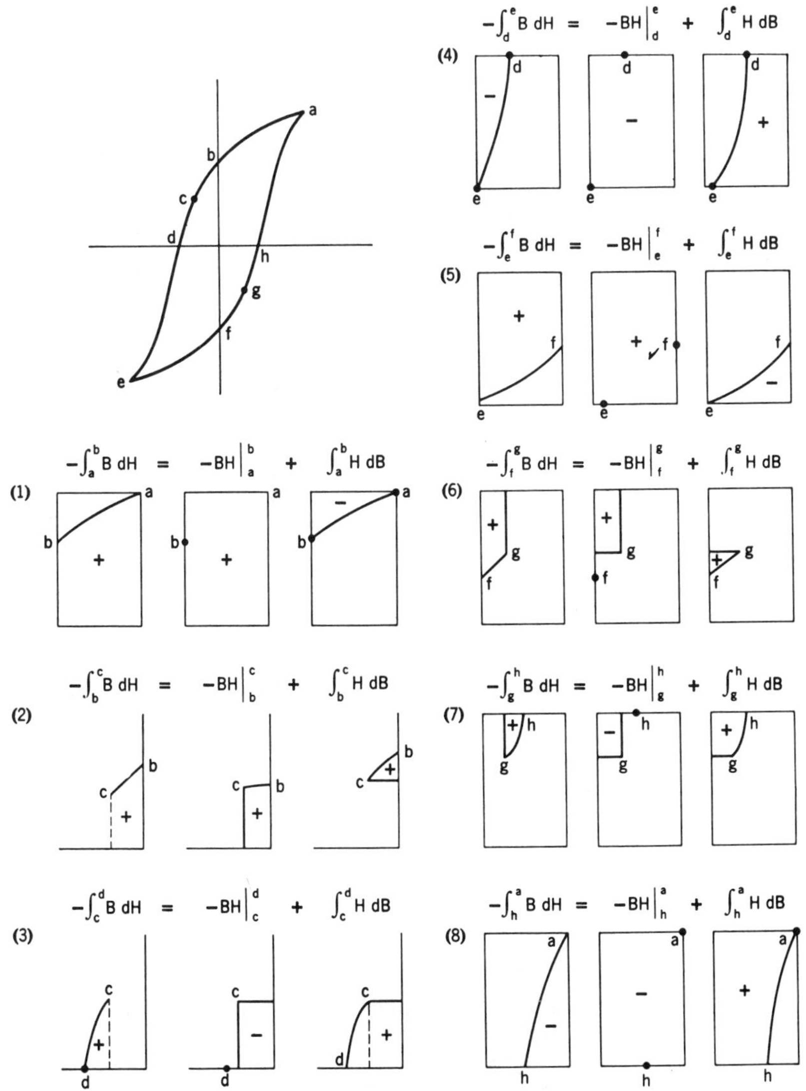

Figure 2.11 Energy relationships in the magnetization process.

changes as the magnet is normally used in the first and second quadrants. In Figure 2.12a, the energy stored by a field in volume (V) of a magnetic material is OAC or $E = V \int H\,dB$. If $B = \mu_0 H$ then $E = V \int B\,dB = VB^2/2$. In an air gap

$$E_g = \frac{V_g B_g^2}{2} = \frac{V_g H_g B_g}{2} \tag{2.21}$$

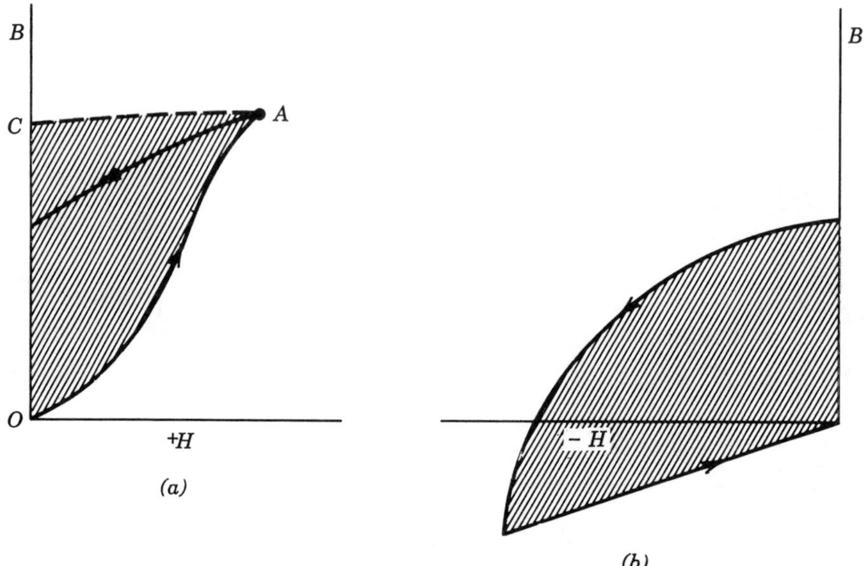

Figure 2.12 Energy to magnetize and demagnetize.

With a magnet in a closed path the energy to magnetize is

$$E = \int H \, dB \tag{2.22}$$

To demagnetize this approximate amount of energy must be removed with a reverse field (Figure 2.12b). We now consider a magnet and air gap as shown in Figure 2.13. In order to find the total F_m to magnetize with gap present, we will have to apply $F_m = H_s L_m + H_g L_g$ where H_s is the field to magnetize the material to saturation. We need to assume that there is no leakage and that magnet area A_m equals gap area Ag. We can then write

$$\frac{F_m}{L_m} = H_s + B_s \frac{L_g}{L_m} \tag{2.23}$$

where H_s is the potential for the magnet and $B_s L_g / L_m$ is the potential for the air gap.

The load line can be moved across the figure as the applied H is varied as shown in Figure 2.14. For the magnet to operate at the remanence point, the equivalent of short circuit or zero gap condition, some positive field is required. If, as shown in Figure 2.14, we reduce the applied field to zero, energy will be supplied by the magnet for the air gap. There will be a field balance as the magnet's operating point represents the intersection of the air gap or load line with the demagnetization curve. Area OAC represents the energy stored per unit volume of magnet. Area OCK is the energy in the air

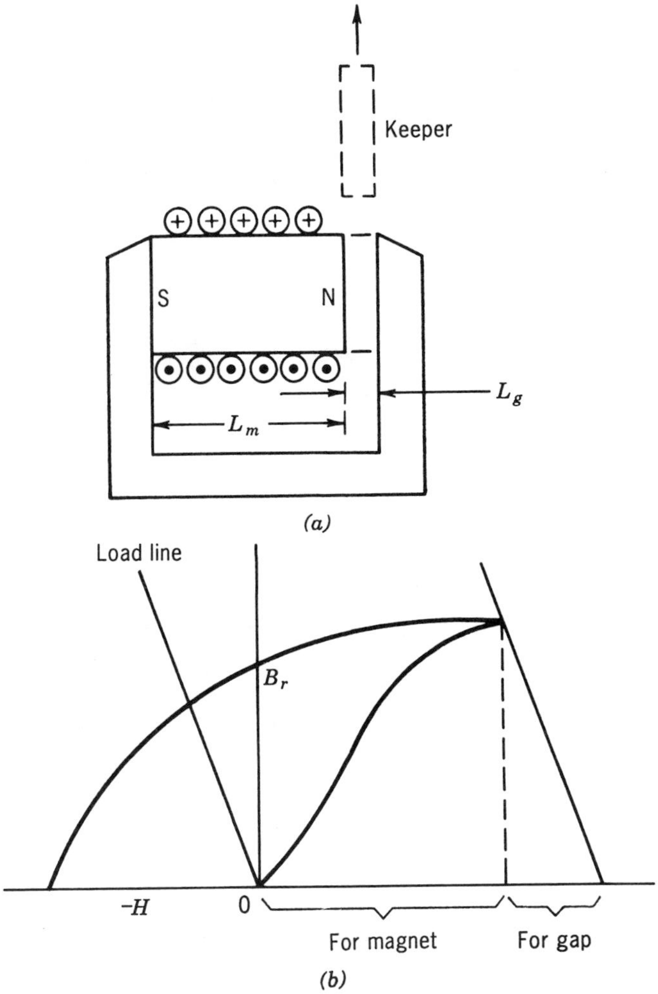

(a)

(b)

Figure 2.13 Magnetization with air gap.

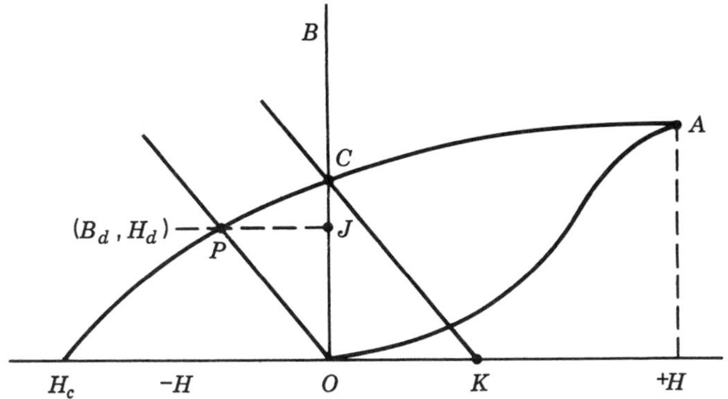

Figure 2.14 Magnetizing relationships.

gap per unit volume. Area *CJP* represents an energy change within the magnet material. Referring again to Figure 2.13a, the energy requirement to magnetize and establish air gap energy could have come from both the electrical power supply and mechanical energy input. By pulling the keeper from the gap we could have reduced the total energy input from the power supply.

We will now consider the important energy relationships when the magnet is used in a mechanical work function. The energy stored in a medium permeated by a magnetic field is expressed as

$$E = \frac{1}{4\pi} \int H\,dB \quad \text{erg/cm}^3 \text{ in air } B = H$$

therefore

$$E = \frac{H^2}{8\pi} \quad \text{erg/cm}^3$$

Consider two equiportional surfaces shown in Figure 2.15. The area of the surfaces is A and the spacing between them is L; then the energy in the space between the plates

$$E = \frac{LA}{4\pi} \int H\,dB \quad \text{erg} \tag{2.25}$$

If one surface is moved with respect to the other, the stored energy will be changed and mechanical work will be done or mechanical energy consumed depending on the direction of motion.

Suppose the length of the gap between the plates is increased by an infinitesimal distance, dl, then the energy will increase by an amount

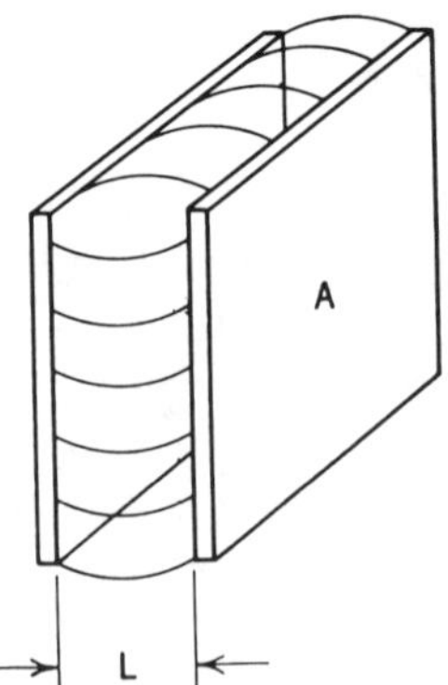

Figure 2.15 Energy stored between equipotential plates.

$$dE = \frac{A\,dl}{4\pi}\int H\,dB$$

since $B = H$ numerically

$$dE = \frac{A\,dl}{4\pi}\int B\,dB = \frac{B^2 A\,dl}{8\pi}$$

$$\frac{dE}{dl} = \frac{B^2 A}{8\pi} \tag{2.26}$$

dE represents the mechanical work done in moving one plate a distance dl and dE/dl = the average value of force, $F = (B^2 A/8\pi)$, the units being dynes and lines per square centimeter. The formula points out the fact that two surfaces joined by magnetic induction attract each other by a force depending on the flux density squared and the area. Consequently, the force can be increased by concentrating the flux into small areas. This technique is used extensively in holding magnets.

In order to relate mechanical work to permanent magnet properties, consider Figure 2.16a–c. When we have a holding magnet and a keeper or armature, the magnet is fully magnetized and operates at point A on the demagnetization curve. As the armature is brought near the poles a force is experienced, work is done, and the magnetic field energy is decreased. The total work done in attracting the armature to the magnet is represented by the area under the pull curve (Figure 2.16b), and the energy converted in terms of magnetic units is represented by area OAG for each unit volume of the permanent magnet.

Equating the magnetic field energy change to the mechanical work done

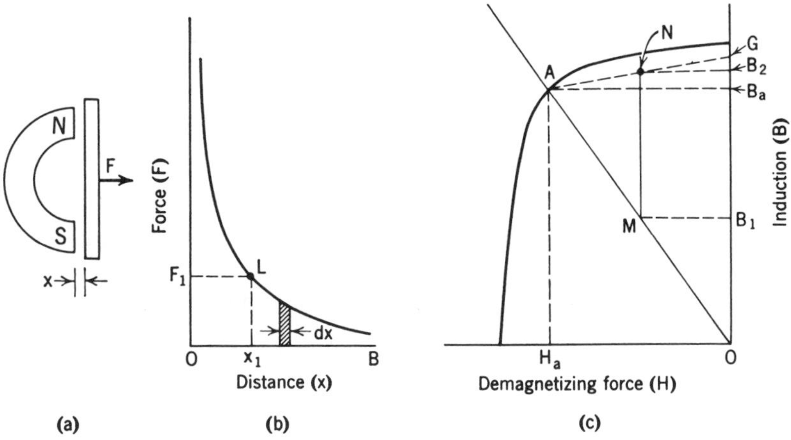

Figure 2.16 (a) Magnet and moving armature. (b) Force–distance characteristic curve. (c) Energy changes in terms of unit magnetic properties.

$$\frac{\text{(Volume magnet) Area } OAG}{8\pi} = \int_0^B F\,dx \tag{2.27}$$

there is a point on the pull curve F_1x_1 where maximum work is done. In terms of magnetic units this corresponds to point N half way between A and G. In a triangular area, the inscribed rectangle is a maximum when AN is equal to AG. The equivalent field energy is represented by MNB_1B_2. The external field energy in the gaps between magnet and armature is a maximum at this particular point as represented by the unit permeance line through point N. If dw is the mechanical work done during a small displacement of the armature dx, then

$$\frac{dw}{dx} = \frac{1}{2}\,\phi_m\,\frac{dF_m}{dx}$$

where

$$\phi_m = BA_m\,, \qquad F_m = HL_m$$

Assuming constant flux in the circuit

$$\phi_m = F_mP_g\,, \qquad \frac{dF_m}{dx} = \frac{\phi_m}{P_g^2}\,\frac{dP_g}{dx}$$

$$\text{Force} = \frac{dw}{dx} = \frac{1}{2}\,\phi_m\,\frac{\phi_m}{P_g^2}\,\frac{dP_g}{dx} = \frac{1}{2}\,F_m\,\frac{dP_g}{dx}$$

$$P_g = \mu_0\,\frac{S}{x}$$

where S is the combined gap area, μ_0 is the permeability of space

$$\frac{dP_g}{dx} = \mu_0\,\frac{S}{x^2}$$

or

$$F = \frac{F_m^2}{2}\,\mu_0\,\frac{S}{x^2} \tag{2.28}$$

which is a quantative expression of the force distance relationship. This general relationship is widely used in problem solving.

In summary the significant energy transformations in permanent magnet behavior are as follows.

1. Remembering the basic physical concept of a volume of permanent magnet material, it is necessary to do work on the force system which in the demagnetized condition gives a minimum energy situation. Energy is dissipated in changing the balance of forces which hold the elementary magnet

groups in a randomly oriented condition. The energy input may be either electrical or mechanical in nature.

2. Additional energy input must be supplied either by mechanical work input or electrical energy input during magnetization to create the demagnetizing influence—the creation of free poles and external field energy. Physically this is the one effort required to change the orientation of the domains near the gap, the influence of which gives rise to the permanent magnet potential and external field energy.

3. Energy is involved only in changing magnetic field energy, not in maintaining it. Field energy established via the permanent magnet is independent of time in the sense that no energy is required to maintain it. It is not subject to change unless additional work is done on the magnet by demagnetizing energy input. (Heat and adverse fields are typical forms of demagnetizing energy to be considered under stability).

4. A permanent magnet in a stabilized condition is a reversible medium of energy transformation in that mechanical work may be done at the expense of the stored potential field energy and the field energy may be re-established by mechanical energy input. The total potential energy of the permanent magnet is composed of both internal and external field energy.

2.7 BOUNDARY CONDITIONS AND FIGURES OF MERIT

There are some useful observations that can be made regarding the geometry of permanent magnet demagnetization curves. For those concerned with property development and evaluation, these observations should be of particular interest. The first observation made by Hoselitz [3] is that the intrinsic magnetization cannot be increased by a demagnetizing field. The (B_i, H) curve can never rise above a horizontal line parallel to the H axis and through the point B_r (see Figure 2.17). Working from the relationship valid in the second quadrant, namely $B = B_i - H$, a limiting induction demagnetization curve can be plotted. The resulting curve is a straight line connecting the B_r and H_c points with H_c numerically equal to B_r. That the coercive force H_c can never exceed B_r is apparent, for this would result in the intrinsic induction at the point $H = H_c$, being higher than that at the point $H = 0$ where it equals B_r, and is an impossible condition (see Figure 2.17a,b).

From this limiting BH curve we can deduce a certain theoretical upper limit on the performance of permanent magnet materials. The maximum rectangle which can be inscribed under this straight line demagnetization curve is $B_rH_c/4$, but with $H_c = B_r$ becoming

$$(BH)_{\max}(\text{limit}) = \frac{B_r^2}{4} \qquad (2.29)$$

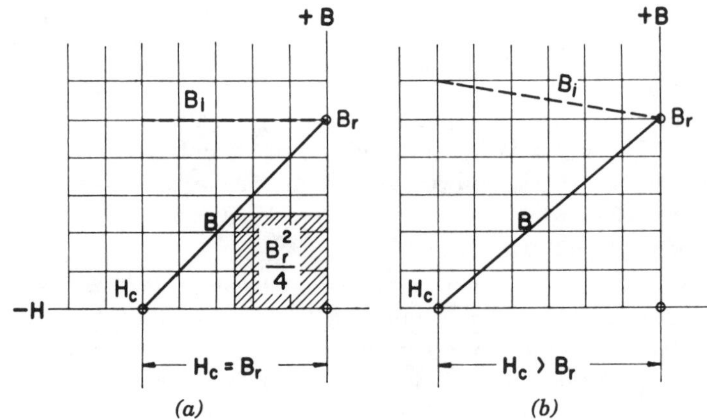

Figure 2.17 (a) Limiting (ideal) demagnetization curve as a consequence of nature. With an intrinsic induction curve B_i parallel to the field axis, H_c will equal B_r and the normal demagnetization curve may be seen to be a straight line, with the maximum available energy equal to $B_r^2/4$. (b) An impossible set of demagnetization curves with H_c larger than B_r resulting in the intrinsic induction increasing with a demagnetizing field.

This quantity is an upper limit to the $(BH)_{\max}$ energy product of any magnetic material and, as described by Hoselitz, this fact is unalterable by any heat treatment or other manipulations. Although the induction coercive force H_c is limited by the value of B_r, the intrinsic coercive force H_{ci} is under no such restraint and can be very large in some materials. We may assume that the remanence B_r can approach the intrinsic saturation B_{is} as the magnetic anisotropy of a specimen reaches a maximum. Our formula then becomes

$$(BH)_{\max}(\text{limit}) = \frac{B_{is}^2}{4} \tag{2.30}$$

Therefore, we need to know only the value of intrinsic saturation of a material to establish its absolute maximum capabilities as a permanent magnet.

It is also of interest to note that in a given property system the full potential $(BH)_{\max}$ is achievable for H_{ci} levels of half of the magnetization. The resulting induction or normal demagnetization curve, would of course be nonlinear and would exhibit a knee.

Depending on how the magnet is applied, there are several figures of merit that are relevant. Without question, the best known of these is the maximum energy product $(BH)_{\max}$ or the largest rectangle which can be drawn under the normal demagnetization curve. This energy product curve can be constructed by plotting (B_d, H_d) products against either B or H. It is now common practice to construct constant energy contours on demagnetization curves so that by inspection one can estimate $(BH)_{max}$ for a given material or compare a group of materials. Both ways of displaying $(BH)_{max}$ are shown in Figure 2.18. For the static gap permanent magnet application developed earlier, the volume of magnetic material required is inversely proportional to $(BH)_{max}$. In Figure 2.19a, this figure of merit is shown in comparison to other ways of evaluation of permanent magnets. In Figure 2.19b, the useful recoil energy concept is shown. In this situation the magnet is used in a dynamic load application. For example, converting magnetic field energy to mechanical energy. As the air gap changes the magnet's operating load line will change from P_2 to B_p, the triangular area OP_2B_p will represent the total BH units converted to mechanical energy. At point P_3 midway along the recoil line, we will have the condition of maximum useful energy (shaded rectangular area) $(BH)_u$. It is the largest rectangle that can be drawn within area OP_2B_p. This point will represent the largest force distance product under a mechanical work curve. The useful recoil energy is always less than $(BH)_{max}$ but in high coercive force materials it can approach $(BH)_{max}$ as a limit. For each demagnetization curve there will be an optimum recoil line and optimum load line slopes OP_2 and OP_3. As a matter of convenience it is possible to construct recoil energy contours for the various materials so that the desired operating points can be located graphically. In Chapter 6 on design, the construction of useful recoil energy contours is described.

Returning to Figure 2.19c, the concept of an intrinsic energy product is shown. $(B_i, H)_{max}$ is the largest rectangle that can be drawn under the intrinsic demagnetization curve. This figure of merit is of importance when the magnet is subjected to large demagnetizing influence in its operating environment. An example of its use is in the d.c. motor application, where

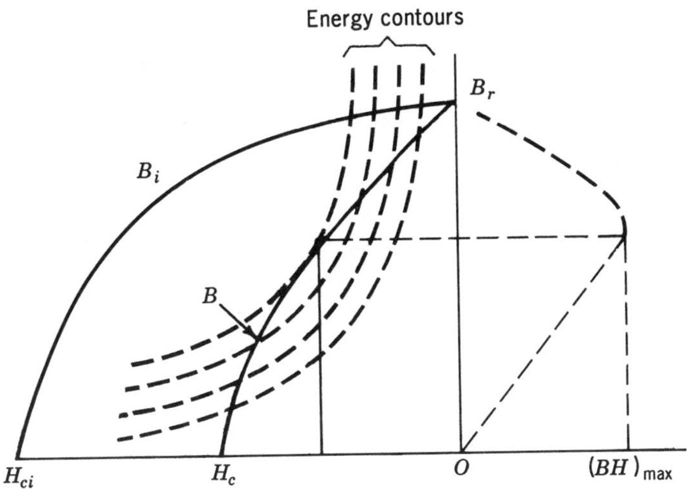

Figure 2.18 Energy product displays.

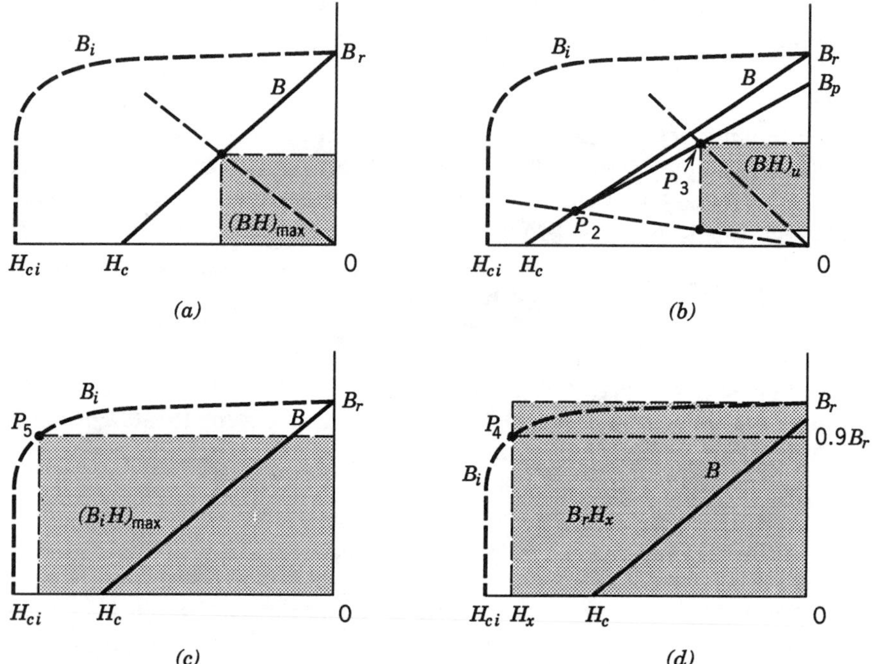

Figure 2.19 Permanent magnet figures of merit.

the magnet is exposed to very high adverse fields during start and stall conditions. The length of magnet is inversely proportional to H_{ci} and the area inversely proportional to B_r, hence magnet volume is approximately inversely proportional to $(B_iH)_{max}$. For earlier magnets like Alnico 5 $(B_iH)_{max}$ and $(BH)_{max}$ would be quite similar; however, for ferrites, $SmCo_5$ and $NdFe_B$ magnets $(B_i, H)_{max}$ will be much larger than $(BH)_{max}$.

Another figure of merit proposed first by Ireland [5] is shown in Figure 2.19d. H_x is the value of H which will decrease the magnetization 10%. Thus B_rH_x is now the useful product where the magnet sees high adverse fields. In Chapter 7, in the discussion on motors, the various figures of merit are considered further.

It may be useful to consider two basic types of permanent magnet behavior based on the level of H_{ci} with respect to the magnetization. In Figure 2.20, type I and type II demagnetization curves are indicated. With type I magnets we have H_{ci} considerably less than B_r, while type II magnets have H_{ci} levels equal to or greater than B_r. Listed below are some of the different patterns of behavior.

1. In type II magnets, the point of irreversibility, R, has moved to the third quadrant. This means the magnet can be exposed to fields greater than its own maximum self demagnetizing field without suffering irreversible loss.

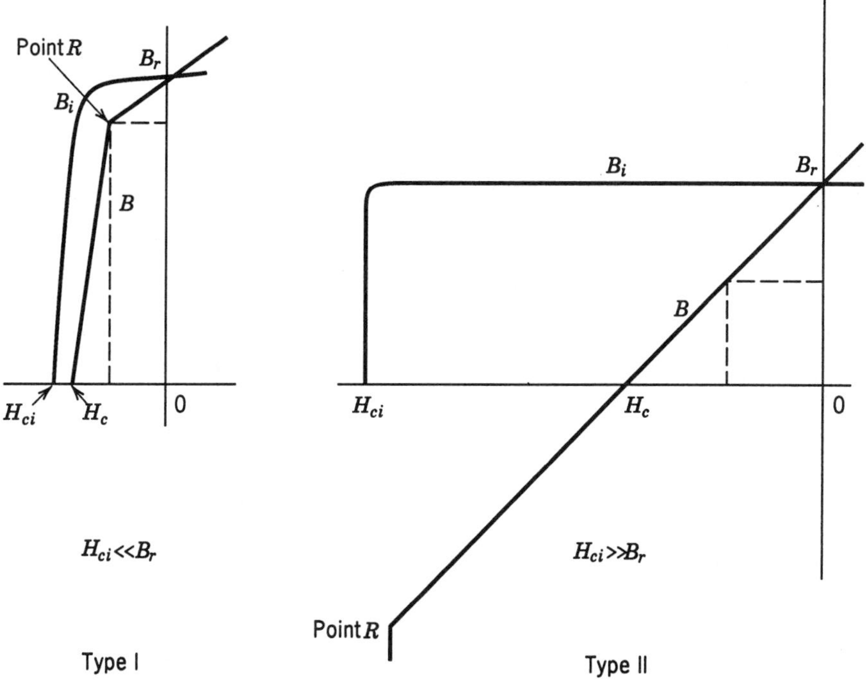

Figure 2.20 Comparison of characteristics.

2. Type II magnets have permeability close to unity in the CGS system. The uniform magnetization in the magnet reduces leakage. All of the magnetic flux tends to come to the terminals of the magnet. In Figure 2.21, the relationships between permeability and limb leakage are displayed. At the top of the figure we have the situation with a type I magnet. There is an appreciable difference between μ_1, the permeability of air space, and μ_2, the internal permeability of the magnet. Under these conditions μ_1/μ_2 will be small and for any value of θ_2, θ_1 will be close to zero or lines of flux will leave the magnet limb perpendicular. For magnets having μ_r values of 3 or above it appears that the flux distribution is similar to an electrostatic pole distribution. The poles of the magnet are not at the terminals but appear to be spaced at about 0.7 of the length of the bar. For the case of the type II magnet μ_1/μ_2 is close to unity and hence θ_1 will be a large angle, that is, flux lines tend to come to the terminal and the pole spacing is the full length of the bar. In a long bar of alnico 5, only 45% of the lines in the neutral section of the magnet get to the end of the bar. A ferrite bar has about 95% of the flux at the neutral section reaching the end of the magnet.

3. A type II magnet stores potential energy largely within its volume which leads to useful energy $(BH)_u$ levels approaching $(BH)_{\max}$. In type I

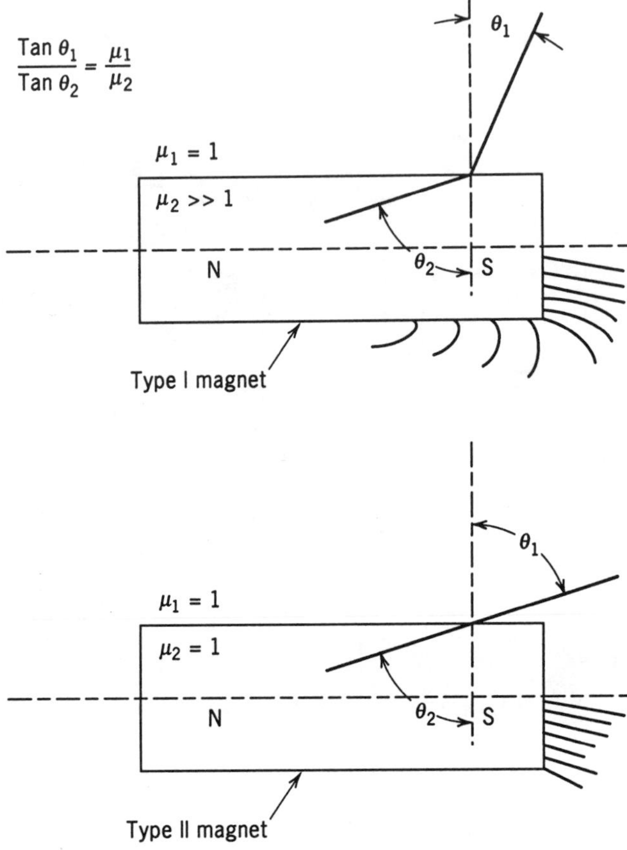

Figure 2.21 Permeability and leakage.

magnets, a high percentage of the total energy is stored in the leakage fields and this portion cannot be converted efficiently into another form of energy.

REFERENCES

[1] D. L. Martin, Permanent magnet characterization measurements, General Electric Report No. 81CRD086, 1981.

[2] R. I. Joseph, Ballistic demagnetization factors in uniformly magnetized cyclinders, J. Appl. Phys. 37 (1966) 4639.

[3] R. J. Parker and R. J. Studders, Permanent Magnets and Their Application (John Wiley, New York, 1962) p. 38.

[4] K. Hoselitz, Philos. Mag. (1944) 35,91.

[5] J. R. Ireland, Ceramic Permanent Magnet Motors, (McGraw-Hill, New York, 1968).

3

THE PHYSICS OF PERMANENT MAGNETISM AND THE ORIGIN OF PERMANENT MAGNET BEHAVIOR

3.1 Overview and Perspective
3.2 The Variations of Magnetic Behavior
3.3 Some Features of Ferromagnetic Materials
3.4 Energy Barriers and Coercive Force
  3.4.1 Single Domain Particle Concept
  3.4.2 Shape Anisotropy
  3.4.3 Crystal Anisotropy
  3.4.4 Other Forms of Anisotropy
3.5 Single Domain Particle Variables
  3.5.1 Size of Particle Versus H_{ci} and B_r
  3.5.2 Particle Interactions
  3.5.3 Particle Alignment and Properties
3.6 Coercivity Mechanisms in Rare-earth Magnets
3.7 Understanding Magnetization and Demagnetization Processes in Rare-earth Magnets

3.1 OVERVIEW AND PERSPECTIVE

Although the processes and flow charts of Chapter 4 appear to be quite different for each material, there is the common task in each of forming fine particles with high magnetization and having an energy barrier associated with the fine particles, to make the rotation of the magnetization difficult. In this chapter we are concerned with these fine particles, their size, orientation, and interaction with their neighbors. We are also concerned with some rather unique and fundamental concepts which have evolved as our understanding of permanent magnet behavior has progressed. Experimental techniques and tools such as the electron microscope and X-ray diffraction have allowed us to link the microstructure features with properties to some extent.

Since the 1950s there are examples of theory predicting how processing can gain important new property levels. However, there are many observations which are only partially understood. There are many models and attempts to explain coercivity. In real materials the models are undoubtedly too simple. Very complex structure is involved. By and large, magnetism remains an experimentalist domain. We have accumulated a rather impressive framework of knowledge which guides our search for new properties. The trial and error methods of the past are at least supplemented by an enlightened empiricism. Hence, the efficiency of the search has been greatly improved and the tempo of development has changed.

3.2 THE VARIATIONS OF MAGNETIC BEHAVIOR

In this chapter we are concerned with the physics of permanent magnetism and the origin of permanent magnet behavior. However, first we should consider how the permanent magnet relates to matter in general, in terms of response to a magnetic field. All matter consists of atoms and consequently, moving charges. Therefore, all matter reacts in some way to an external field. The categories of different forms of magnetic behavior as illustrated in Figure 3.1 are as follows.

1. Diamagnetism

If the net magnetic moment of each atom in a material is zero because of mutually cancelling electronic movements within the atom, then the net flux density within the material, due to an applied external field, is slightly less than it would be in space for the same field. Such a material is termed diamagnetic. Examples are Cu, Bi, Pb, and Ga. Bismuth is the most pronounced diamagnetic element known.

2. Paramagnetism

If individual atoms have a net moment caused by unbalanced spin of orbital electrons, then there is a tendency for alignment of spins when an external field is applied. If there is no ordering arrangement of spins between atomic

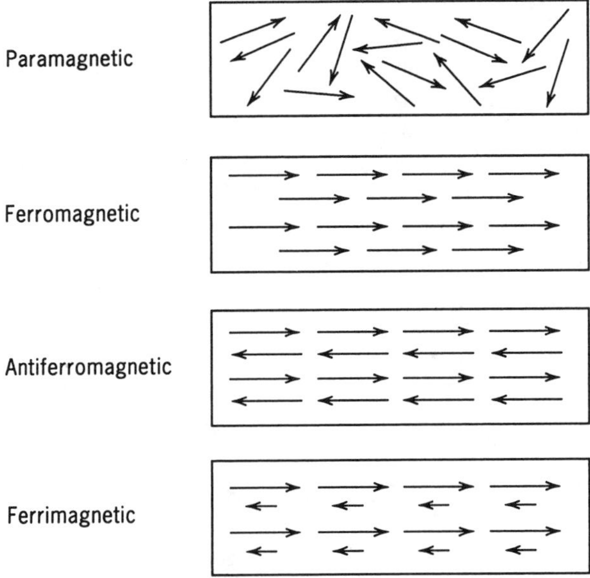

Figure 3.1 Different forms of magnetic behavior.

near neighbors, then this alignment is unbalanced by thermal agitation and the overall average alignment is very near random for most field levels. Such a response is termed paramagnetic. Many salts and metals exhibit paramagnetism as do nickel and cobalt at high temperatures.

3. Antiferromagnetism

In some salts and metals atoms have a net spin moment but exhibit a quantum interaction causing antiparallel ordering of the moments of neighboring atoms. Such a substance shows a slightly positive susceptibility. This susceptibility increases with temperature up to a critical temperature, the Néel point, at which the ordering breaks down, and normal paramagnetic response is observed. Examples are Cr and Mn.

4. Ferrimagnetism

This is a special case of antiferromagnetism. In a crystalline substance, antiferromagnetism may be considered to result from two sublattice atomic moments being parallel in each sublattice, but anti parallel between the two sublattices. If, in such an arrangement, the spin moments in each sublattice are unequal (due to different ions) then there will be an appreciable net moment when a field is applied. Such substances can show high permeability, but the change of saturation magnetization with temperature can be complex and quite different from that of a ferromagnetic material. Magnetite and the whole range of ferrites are ferrimagnetic.

5. Ferromagnetism

Iron, nickel, cobalt, and many other compounds show very large and fairly complex changes in magnetization with moderate magnetizing field levels. This is the main characteristic of a ferromagnetic substance. In ferromagnetic materials, individual atoms have a net magnetic moment due to unbalanced electron spins, as for paramagnetism and antiferromagnetism, but in addition the interatomic spacing lies within a critical range, which gives rise to exchange forces (or the Weiss molecular field) causing parallel alignment between neighboring atoms. In saturation level fields all moments are aligned. As the temperature is increased a point is reached where thermal agitation overcomes the exchange forces. This critical temperature is the Curie point above which the material becomes paramagnetic.

3.3 SOME FEATURES OF FERROMAGNETIC MATERIALS

About 60 years ago, French physicist Pierre Weiss postulated that a ferromagnetic body must be composed of some regions or domains, each of which is magnetized to saturation level, but the direction of the magnetization from domain to domain need not be parallel. Thus a magnet, when demagnetized, was only demagnetized from the viewpoint of an observer outside the material. Man-made fields only serve as a control in changing the balance of potential energy within a magnet. This theory still provides the basis of our highly sophisticated body of knowledge, which explains quite satisfactorily the observed properties of ferromagnetic materials and provides an intelligent guide for the search for improved materials.

The inherent atomic magnetic moment associated with such elements as iron, cobalt, nickel, and many compounds is believed to originate from a net unbalance of electron spins in certain electron shells. For example, in iron in the third shell there are more electrons spinning in one direction than in the opposite direction. Having an inherent atomic magnetic moment is a necessary but not a sufficient condition for ferromagnetism to be exhibited. Additionally, there must be cooperative interatomic exchange forces that maintain neighboring atoms parallel. Little is known of the exact nature or magnitude of these forces but observations suggest they are electrostatic. It has been pointed out that in ferromagnetic materials the ratio of interatomic distance to the diameter of the shell in which the unbalance exists is unusually large compared to this ratio in materials that do not exhibit ferromagnetism.

In Figure 3.2 an exploded view of a ferromagnetic volume is shown. The relative dimensions of the atom, domain, crystal, and a measurable volume are noted in the figure.

The atomic exchange force also produces magnetostrictive effects and is associated with the crystalline structure of magnetic materials in a way that

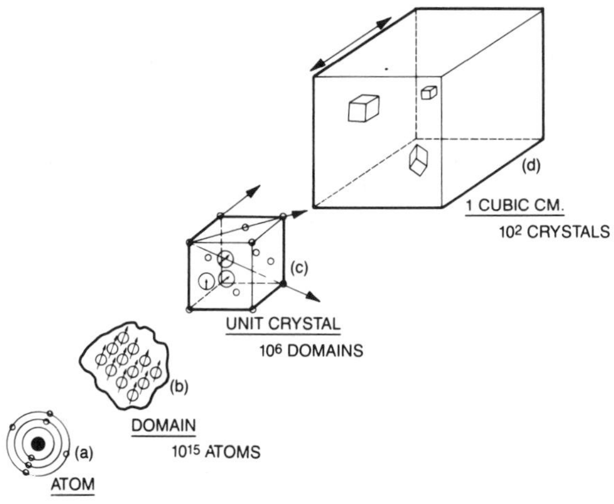

Figure 3.2 Exploded assembly of ferromagnetic volume.

exhibits anisotropy or directional dependence with respect to the crystal axis.

In Figure 3.3 the directional dependence is shown for iron. The easy axis of magnetization is the cube (100) edge. We can view the magnetic domain as a region in which the atomic moments cooperate to allow a common magnetic moment which may be rotated by externally applied fields. Domain size, not a fundamental constant of physics, varies widely depending on composition, purity, and state of strain of the material as well as some very important energy relationships. Figure 3.4 shows a boundary region

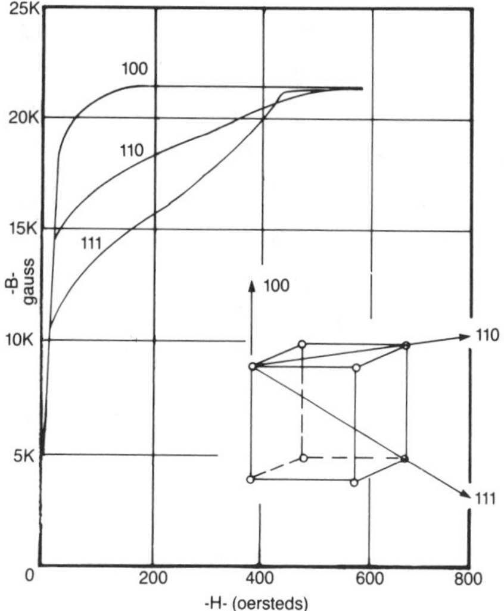

Figure 3.3 Directional dependence (iron).

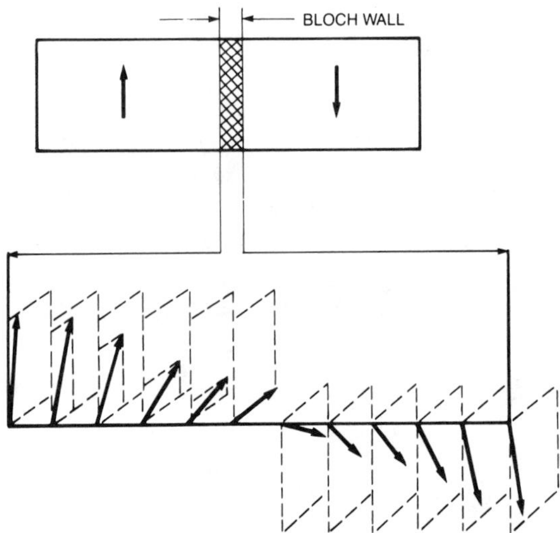

Figure 3.4 Bloch wall between ferromagnetic domains.

between two domains. This boundary region and its significance was first proposed by Bloch. The Bloch wall is a transition region containing many atomic planes. The 180° change in magnetization must occur over a considerable distance to minimize the potential energy in the wall. However, the width of the wall will be restricted because of the restraining influence of crystal anisotropy (directional dependence of magnetism with respect to crystalline axis). Figure 3.5 illustrates an additional energy relationship which influences the size of the domain and involves the magnetostatic or field energy surrounding a magnetized volume. A magnetized volume tends to subdivide itself. It will be energetically possible for subdivision to occur as shown in Figure 3.5 until the decrease in magnetostatic energy is less than the potential energy associated with the Bloch wall foundation. At this point we might say that the magnetization vector arrangement associated with domain volumes in a ferromagnetic material results from a complex energy

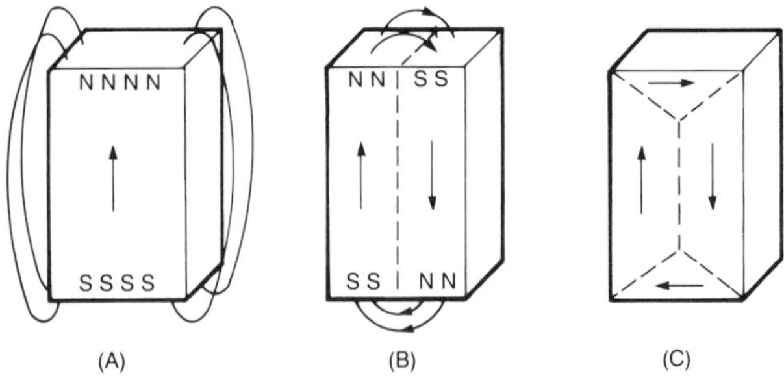

Figure 3.5 Domain subdivision.

balance so arranged that the total potential energy of the system is a minimum. Externally applied fields to magnetize or demagnetize only disturb the balance of the potential energies involved and our familiar S-shaped magnetization curves are records of the change in balance with respect to the external influence. Figure 3.6 shows the action as a bar of ferromagnetic material is magnetized. The demagnetized condition (A) results from an internal arrangement with cancelling directions of magnetization vectors. In region (B) with low values of external field, the action is primarily one of domain boundary stretching, usually around imperfections. This is a reversible process (reversible magnetization process is one in which the magnetization vectors re-orient to their original position after the field H is removed). As the field is increased, region (C) domain boundaries break away and move through the material. The more favorably oriented regions grow at the expense of their less favorably oriented neighbors. As a result, a large increase in magnetic induction occurs. This is an irreversible process in which the magnetization vectors tend to keep their new position after a field H is removed.

In region (D) at still higher values of magnetizing force, the magnetization vectors are rotated against the forces of strain and crystalline anisotropy into alignment with the direction of the applied field, and saturation occurs. Removing the magnetizing force causes some relaxation; the domains rotate back to the easy direction of magnetization (a reversible process). This relaxation can be minimized by making the direction of easy magnetization coincident with the desired direction of magnetization.

Subjecting the magnet to a demagnetizing force returns the domain boundaries to a condition similar to their original positions in (A) and hence the magnet is demagnetized.

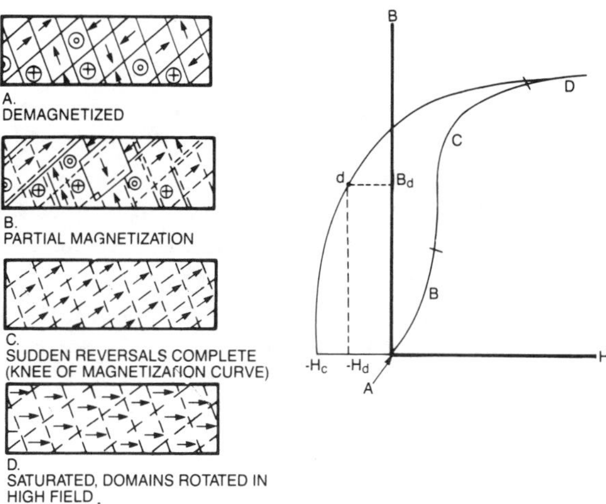

Figure 3.6 Pictorial explanation of magnetization curve in a ferromagnetic bar.

3.4 ENERGY BARRIERS AND COERCIVE FORCE

In our search for improved permanent magnets we are concerned with forming regions exhibiting high saturation magnetization and with some kind of an energy barrier that makes changing the magnetization difficult. The picture presented in Figure 3.6 was satisfactory for explaining coercive force in early permanent magnets of perhaps 50–250 Oe. The coercivity resulted from impeding domain wall motion by introducing structural inhomogeneities into the material. Domain wall motion was impeded due to the energy required to pass a boundary through a nonmagnetic inclusion. In the early steel magnets, the inclusions were in the form of precipitated carbides. Todays best permanent magnets have coercive forces of several thousand oersteds and we must look for coercive force mechanisms quite different from wall motion impedance in terms of energy content.

3.4.1 Single Domain Particle Concept

A significant milestone in understanding permanent magnet properties occurred with the suggestion of Frenkel and Dorfman [1] that, if small particles were prepared with dimensions less than the width of a domain boundary, such particles would contain no boundaries. This concept is central in the fine particle magnet theory and provides a satisfactory explanation of modern high coercive force magnets in use today.

In bulk magnetic materials the wall boundary energy is negligible compared with the external field energy, but this situation reverses at a sufficiently small particle diameter. This is because the magnetostatic, or external field energy is a volume effect that is proportional to the cube of the particle radius, while the domain wall boundary energy is proportional to the square of the radius. Below a critical particle size there is less energy associated with the particle's external field than with a domain boundary; this is the critical radius below which single domain particles exist and it is energetically impossible for a domain wall to form in a particle. For iron the critical particle diameter is approximately 100 Å and for barium ferrite approximately 10,000 Å or 1 μm.

Single domain particles make possible the preparation of materials without domain boundaries, material whose magnetization can only be changed with the simultaneous rotation of all the atomic moments in each particle. This can be a much more difficult process than domain wall motion. The obstacles to the rotation of the magnetization vectors are related to the anisotropy energy. Anisotropy is a term used to describe a material property that is directionally dependent. Magnetic anisotropy is the preference of magnetic moments to lie along a specific direction. For example, an applied field created a unidirectional anisotropy. Strain, shape, magnetocrystalline, surface, and exchange are other forms of anisotropy.

For any source of anisotropy the H_{ci} may be calculated by following the changes in the equilibrium positions of the magnetization. For example, consider a uniformly magnetized particle with an anisotropic axis K at angle θ to the applied field as shown in Figure 3.7. The torque T on B_i is $T = B_i \times H$. Equilibrium occurs when T is equal and opposite to the torque exerted by the anisotropy and is found by minimizing the total energy. As an example, for uniaxial anisotropy and energy E per cm^3 is

$$E = K \sin^2 \alpha - HB_i \cos(\theta - \alpha) \tag{3.1}$$

For the case of oriented particles where H is directed at $\theta = 180°$, that is, applying a reverse field after saturation, for equilibrium

$$\partial E / \partial \alpha = 0 \tag{3.2}$$

we find only $\sin \alpha = 0$ ($\alpha = 0$ or $180°$) are stable positions for B_i as indicated in Figure 3.7. The field at which B_i goes from 0 to 180° can be obtained from (3.2) or can be easily seen by rewriting (3.1) as

$$E = HB_i + 2 \sin^2 \frac{\alpha}{2} \left(2K \cos^2 \frac{\alpha}{2} - HB_i \right) \tag{3.3}$$

The $\alpha = 0$ position is no longer stable when the factor within the bracket goes negative. This occurs at a critical field

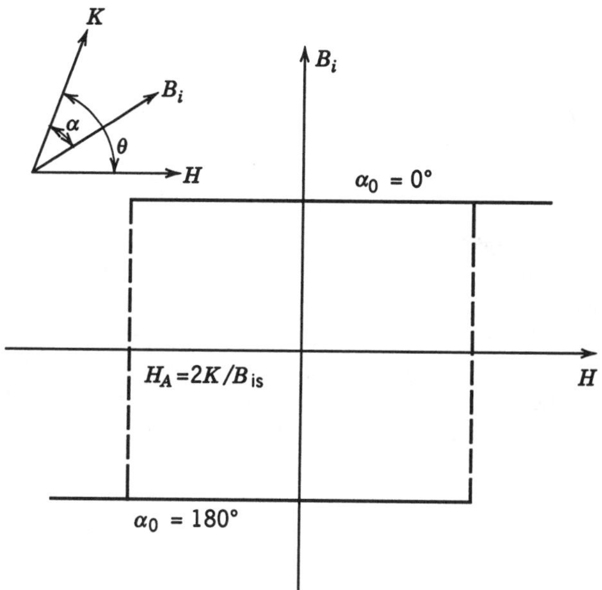

Figure 3.7 Calculated hysteresis loop for single domain particle with anisotropy K and $\theta = 0$. At equilibrium $\alpha = \alpha_0 = 0$, 180°.

$$H_{ci} = H_A = \frac{2K}{B_i} \tag{3.4}$$

The value of K depends on its physical origin. These various anisotropies are briefly described.

3.4.2. Shape Anisotropy [2]

Shape anisotropy arises from dipole-dipole effects in various particle shapes. Exact calculations are possible for ellipsoids. For example, for a prolate spheroidal particle it can be shown that for coherent magnetization reversals, where the moments rotate together while remaining parallel with each other

$$K = \frac{1}{2}(N_b - N_a)B_{is}^2 \tag{3.5}$$

where N_a and N_b are the demagnetization factors along major and minor axis, respectively. Substituting into (3.4)

$$H_{ci} = (N_b - N_a)B_{is} \tag{3.6}$$

In real magnet systems the moments will not necessarily rotate coherently, since there may be a lower energy rotation process. The basic interaction of ferromagnetism, the spin-spin or exchange interaction favors parallel alignment of moments. The comparatively weak dipole-dipole coupling or magnetostatic interactions, giving rise to shape anisotropy effects, favors non uniform magnetization rotation because of their long-range effects. With no anisotropy, the competition between the exchange and dipole effects leads to what is called magnetic "curling" during reversal. With anisotropy present, and in the case of large particles, this curling reduces to actual domain wall motion. For sufficiently small particles where exchange forces between atoms are dominant, domain walls are energetically not favored, and thus the particles may be in a state of uniform magnetization, both in zero field and during reversal. In an important intermediate size range between these extremes, it has been found that the lowest energy process is neither coherent reversal nor wall motion, but rather the "curling" process. This is shown schematically [3] in Figure 3.8. The H_{ci} of such elongated cylinders, reversing by "curling" is

$$H_{ci} = 2\pi KA/b^2B_{is} - N_aB_{is} \tag{3.7}$$

where A is the exchange constant, b the semiminor axis, and K is a constant varying from 1.08 for an infinitely long cylinder to 1.38 for a sphere. For an infinite cylinder of iron

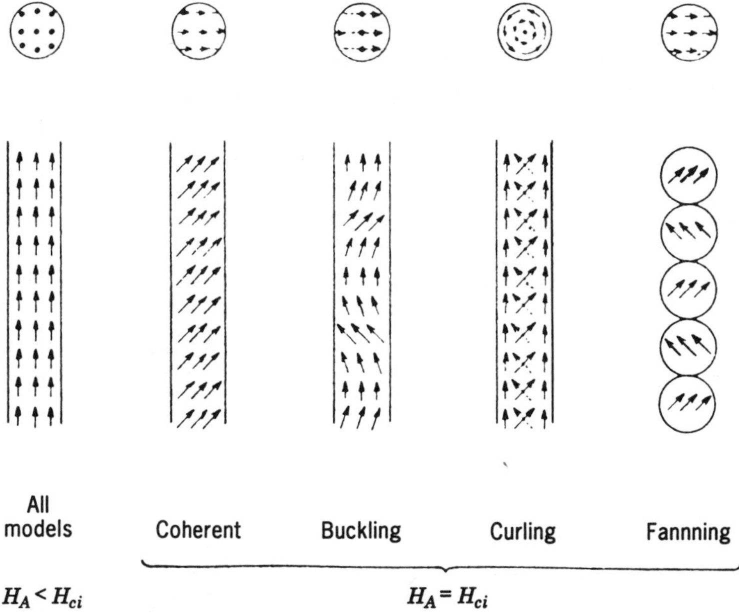

Figure 3.8 Schematic representation of magnetization reversal configurations. The cross-sections show only the transverse flux components.

$$H_{ci} = 3.116 \times 10^{-8}/d^2 \tag{3.8}$$

where d is the diameter in centimeters.

In real magnet systems the shaped particles often deviate considerably from the ellipsoidal or rod-like shape. For example, the precipitate structure in alnico looks "peanut-like" in shape, with many cross-ties. Exact calculations of such cases are often, but not always [4] intractable. It has been shown that a fanning mode [3, 5] of magnetization reversal becomes more favorable if the irregularities are such that the "chain of spheres" model as shown in Figure 3.8 is reasonable. Experimentally it has been demonstrated that the size dependence of properties of almost perfect elongated particles [6] agrees quantitatively with (3.8) for diameters less than 1000 Å while irregular shaped particles follow the fanning relationship, namely

$$H_{ci} = \frac{1}{2} \pi B_{is} \tag{3.9}$$

For diameters less than 350 Å it appears that a necessary condition for optimum magnetic hardness is that the elongated particles be free of surface irregularities, infinitely long, small enough to be single domain and possess the highest possible B_{is}. The highest B_{is} is found in a 60:40 FeCo alloy.

3.4.3 Crystal Anisotropy

Magnetocrystalline anisotropy, or simply crystal anisotropy is an intrinsic effect. It arises because the electron orbits with their small contributions to the magnetic moment, are coupled to the crystal lattice, causing the moments to prefer to align themselves along certain crystallographic axes. The orbital motion then couples to the spin moments. This effect is quite sensitive to crystal symmetry and is often largest in crystals of lowest symmetry.

The coercive forces of single domain particles are simply given by (3.4) where K is now an appropriate crystal anisotropy constant which can be determined experimentally, for example, through torque measurements. Today the search for improved properties is to a great extent the search for compounds with high crystal anisotropy.

3.4.4 Other Forms Of Anisotropy

Other forms of anisotropy which have been identified, but which are not of as much significance as shape and crystal anisotropy are stress, exchange [7], and surface [2] anisotropy. Uniaxial strain in a single domain particle develops an axis along which the moments prefer to align. Exchange anisotropy is an interfacial effect betwen two magnetic materials. It is most striking when developed between an antiferromagnet and a ferromagnet.

3.5 SINGLE DOMAIN PARTICLE VARIABLES

The effect of particle size, particle interaction, packing density and particle alignment on intrinsic coercive force H_{ci} and remanence B_r are considered at this point.

3.5.1 Size of Particle Versus H_{ci} and B_r

The size dependence of single domain particles in very small size range is fairly well understood [8, 9]. In larger size ranges, experimental results typically follow a $1/d$ relationship, where d is particle diameter.

Regarding particle size versus B_r, measurements indicate that B_r remains at its maximum up to larger sizes than H_{ci}.

3.5.2 Particle Interactions

To obtain high saturation magnetization, a high volume fraction of magnetic precipitate or synthetic magnetic particles must be held in a nonmagnetic matrix. The influence of packing on properties is very important in all kinds of processing. The description of magnetic particles in a nonmagnetic or lesser magnetic matrix is intended to cover all processing approaches whether developed by precipitation or formed separately by mechanical means and then mixing with a nonmagnetic matrix as in bonded magnets.

Two extremes of behavior have been observed. In all cases, the magnetization increases linearly with increased packing, as shown in Figure 3.9a or b. However in case (a), H_{ci} is independent of p. This is expected for particles with only crystal, strain or exchange anisotropy or for "curling" reversal in infinite cylinders [10], since no free poles are created at any time. In case (b), H_{ci} decreases linearly with increasing p or

$$H_{ci} = [H_{ci}]_{p=0}(1-p) \tag{3.10}$$

This was first derived for an isotropic assembly of elongated particles whose moments are parallel and magnetization is uniform. Later it was shown to hold also for a random assembly of parallel infinite cylinders reversing coherently. Both types of behavior have been observed experimentally at lower values of p. At higher values of p it is difficult to achieve uniform packing, and to prevent deformation of the particles; thus H_{ci} generally decreases somewhat faster than predicted by (3.10).

3.5.3 Particle Alignment and Properties

Maximum properties are achieved in a magnet when B_r is a maximum. This occurs when complete orientation of the easy anisotropy axes are produced. This state corresponds to $B_r/B_{is} = 1$. For a random assembly this ratio may

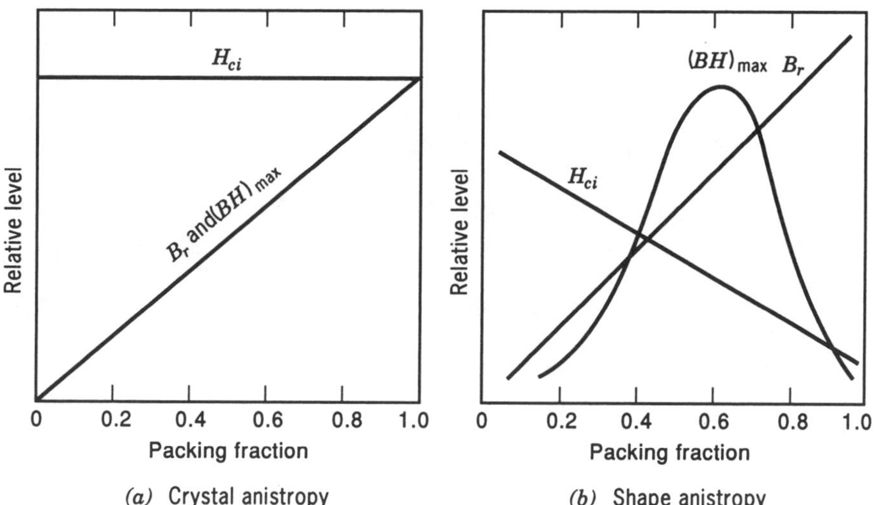

Figure 3.9 The theoretical effect of packing on fine particles with (a) crystal anisotropy and (b) shape anisotropy.

be reduced to 0.5. For random assemblies of particles with different kinds of anisotropies, the values of H_{ci} and B_r will vary considerably. In a permanent magnet $(BH)_m$ may increase up to a factor of 4 on alignment depending on the type of particles and the reversal mechanism. For the energy product limit to be achieved, one also needs coercivity H_{ci} levels of at least $B_r/2$. For type I magnets where $H_{ci} \ll B_r$, the maximum energy product cannot exceed the $H_{ci}B_r$ product. There are several ways to achieve crystal orientation along a preferred axis. In transformer laminations and in Cunife, Vicalloy and CrFeCo permanent magnets, mechanical deformation is used to orient. Most permanent magnet materials are not ductile enough to be workable. In several materials, crystals can be oriented by having the material solidify from the melt with heat extracted along one selected axis. Such is the approach used in Alnico magnets to enhance properties. In several materials the approach is to produce a single crystal by grinding. The single crystals are oriented in a magnetic field and then densified by pressing and sintering. In the case of bonded ferrite magnets, the shape of the single crystal particle allows partial orientation by mechanical means.

3.6 COERCIVITY MECHANISMS IN RARE-EARTH MAGNETS

In rare-earth magnets, experience has shown the optimum grain size to be considerably greater than for a single domain model. There is every evidence that domain walls exist in grains of rare-earth cobalt and rare-earth iron material. In these materials the anisotropy field H_A is very large [11], however, the best coercivities achieved are of the order of 15–25% of the anisotropy field levels. In attempting to explain the coercive force mechanism, two models are involved. The first is the nuculation model. Figure 3.10 shows diagrammatically the course of the initial or virgin magnetization curve and interior hysteresis curves for this case. In essentially single phase permanent magnets, such as $SmCo_5$, NdFeB and ferrite, the grain boundaries are effective in prohibiting domain wall displacement. The coercive force of such magnets is determined by the average value of all of the individual grains. Coercivity is determined by the level of field required to form nuclei for domain reversal or perhaps to release walls anchored in surface regions. Local structural inhomogeneities, misalignment, porosity, etc. can lead to magnetic reversals in much smaller fields than the anisotropy field in this type of material.

The relationship between coercive force and the process variables of sintering and heat treatment has been studied for $SmCo_5$ [12]. Basically the goal is to preserve grain boundaries so that they serve as barriers and prevent wall propagation over the entire magnet volume. By optimizing the process variables a phase structure at the grain boundaries can be established which impedes the formation of inverse domains.

The second coercive force mechanism model to describe observed prop-

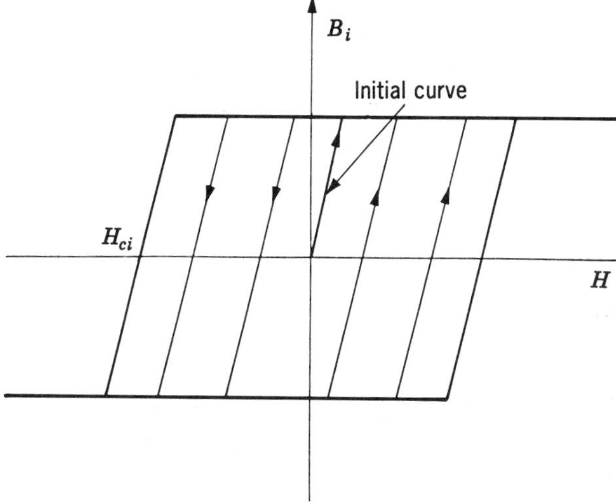

Figure 3.10 Schematic representation of initial curve and inner hysteresis loops in case of a nucleation mechanism effecting the coercive force.

erties is the volume pinning model. Figure 3.11 shows the virgin curve and inner hysteresis curves for the volume pinning mechanism. This type of behavior is due to the existence of a network of pinning sites for domain walls which cover the entire crystal volume. H_{ci} is that field strength at which the walls are released so that a reversal can take place because of wall displacement. This is the basic mechanism for understanding coercive force

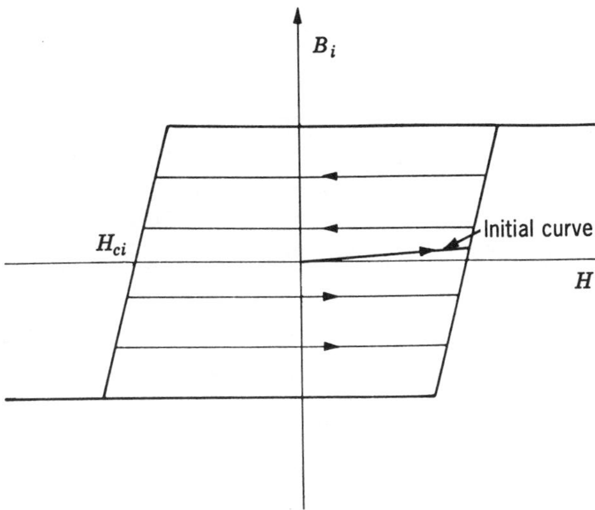

Figure 3.11 Schematic representation of initial curve and inner hysteresis loops in case of a pinning mechanism effecting the coercive force.

levels in both Cu bearing 1:5 alloys and 2:17 alloys having the composition $RE(Co, M)_z$ with $6 \leq Z < 8$ and M = Cu and/or a combination of transition metals [13–15]. In these precipitation hardened alloys, the optimum properties are developed by the tempering treatment. The alloy decomposes into 2:17 and 1:5 phases. The 2:17 structure is surrounded by the 1:5 boundary phase. There are a number of examples which describe the detailed pinning network responsible for H_{ci} in these systems [16].

3.7 UNDERSTANDING MAGNETIZATION AND DEMAGNETIZATION PROCESSES IN RARE-EARTH MAGNETS

The author in unpublished work has used a grain modeling approach to explain complex events in magnetizing and demagnetizing $SmCo_5$ as detailed in Chapter 9. More recently Haberer [17] and Adler and Fernengel [18] have expanded the use of this type of modeling to explain magnetizability and internal hysteresis loops in nucleation type rare-earth magnets. In essentially single phase permanent magnets such as $SmCo_5$, NdFeB and ferrite, the grain boundaries effectively block domain wall displacement [19]. To a large extent the grains behave quite independently.

The central concept to be used in explaining the magnetization and demagnetization processes is that the grains have two distinct magnetic states and that these states exhibit quite different behavior when exposed to fields. We will call grains that have domain walls state A and will show these grains pictorially as ⇅, grains which are single domain are state B and are shown pictorially as ↑ or ↓, depending on their direction. In the case of nucleation type magnets, the virgin magnetization curve Figure 3.12 is a record of the type A grains response to the magnetizing field. In the thermally demagnetized condition, all of the grains are type A having domain walls. These walls are easily driven from the grains and at saturation, all of the grains become single domain or state B. This type of magnet can be fully magnetized in field levels less than H_{ci}. However, even at high fields a few grains will still have walls. This is due to regions having high magnetostatic energy or leakage fields. Such regions result from porosity, nonmagnetic inclusions and misoriented grains. The local magnetostatic fields oppose the applied field. For example, in $SmCo_5$, 15 kOe will magnetize to about 98% of saturation. However, full saturation may require 100 kOe.

As the applied field is reversed and the demagnetization curve is drawn, the fraction of A grains compared to B grains does not change, the fraction having been established by the field to which all of the grains were exposed during magnetization. As the demagnetizing field is applied, approximately 50% of the state B grains will reverse and at H_{ci} the magnet will be demagnetized. Those grains that reverse will be those of lowest coercive force. Each point in the demagnetization process expresses the fraction of B type grains that have inverted, the fraction having been determined by the

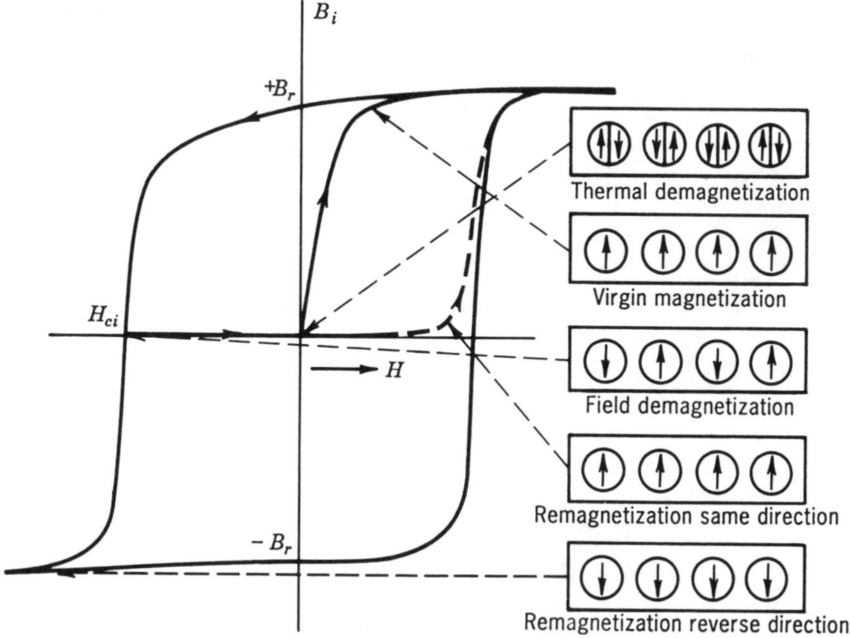

Figure 3.12 Magnetization and demagnetization events in nucleation type magnet.

maximum reverse field to which all grains were exposed. If we attempt to remagnetize the magnet with the original polarity we find that the shape of the magnetization curve has changed significantly from the virgin curve. It will now take a higher field (about the H_{ci} level) to remagnetize. The process is one of domain rotation which is a higher energy process than that of domain wall motion. If we were to reverse the polarity of the magnet we would need still higher fields since to saturate in the other direction, all of the magnetization vectors must be reversed, even those with the highest anisotropy. We must remember that H_{ci} is a statistical average of all of the grains and the grains have a wide distribution of coercivity. In Chapter 9 a detailed example of $SmCo_5$ magnetization and demagnetization is discussed in terms of single and multidomain grains.

For nucleation type magnets one can make a qualitative comparison concerning slopes of the minor loops and the fraction of grains that are of state A and state B. For maximum slope the specimen would be comprised of all single domain grains.

Regarding recoil hysteresis loop shape, again the variation in slope can be interrupted in terms of the relative population of grains having single or multidomain states.

In magnets where the coercivity is determined by pinning such as $Sm(Co, Fe, Cu, Hf)_7$ ($H_{ci} = 6$ kOe), the grains are largely of single domain state. There are no walls to drive out, the magnetization and demagnetization curves are symmetrical and the field to rotate the magnetization is essentailly H_{ci}. The magnetization is pinned and must be rotated against the anisotropy.

A later development in rare-earth magnets formed by precipitation was the development of magnets having a composition $Sm(Co, Fe, Cu, Zr)_{7.5}$ ($H_{ci} = 15$ kOe). By step aging, a two phase structure was produced; a 2:17 structure surrounded by a 1:5 boundary [20]. Now the magnetization field levels are far in excess of the coercivity. We also find that this field is only required in the virgin state and that we can pretreat this kind of magnet by magnetizing and field demagnetizing and then the second magnetization will require a field of about 15 kOe less than the virgin magnetization.

REFERENCES

[1] J. Frenkel and J. Dorfman, Nature 126 (1935) 274.

[2] G. Rado and H. Suhl (eds.), Magnetism, Vol. 3 (Academic Press, New York, 1963) Chapters 6, 7.

[3] F. Luborsky, J. Appl. Phys. 32 (1961) 1715.

[4] C. Kittel, Rev. Mod. Phys. 21 (1949) 541.

[5] I. S. Jacobs and C. P. Bean, Phys. Rev. 100 (1955) 1060.

[6] F. Luborsky and C. R. Morelock, J. Appl. Phys. 35 (1964) 2055.

[7] W. Meiklejohn, J. Appl. Phys. 33 (1962) 1328S.

[8] E. Kneller and F. Luborsky, J. Appl. Phys. 34 (1963) 656.

[9] C. Bean and J. Livingston, J. Appl. Phys. 30 (1959) 120S.

[10] A. Aharoni, J. Appl. Phys. 30 (1959) 70S.

[11] J. Livingston, Paper No. VI-I, 8th International Workshop on Rare Earth Magnets and Their Applications, Dayton, Ohio, 6–8 May, 1985.

[12] H.A. Leupold, F. Rothwarf, J. J. Winter, A. Tauber, J. T. Breslin and A. Schwarz, in: Proc. 2nd International Symposium on Magnetic Anisotropy and Coercivity in RT Metal Alloys, K. J. Strnat (ed.) (University of Dayton, Dayton, Ohio, 1978) p. 87.

[13] J. J. Becker, Rep. Nav. Res. Off. SRD-77-081, 1977.

[14] J. D. Livingston and D. L. Martin, J. Appl. Phys. 48 (1977) 1350.

[15] J. D. Livingston, IEEE Trans. Magn. 14 (1978) 668.

[16] H. Nagel, A. J. Perry and A. Menth, J. Appl. Phys. 47 (1976) 2662.

[17] P. Haberer, 5th French Workshop on Permanent Magnets Paper F1, Performance and Fabrication of Rare Earth Magnets.

[18] E. Adler and W. Fernengel, Paper No. SP2.6, International Symposium on Magnetic Anisotropy and Coercivity, Bad Soden, F.R.G., September 3, 1987.

[19] E. Adler and P. Hamann, Proc. 8th International Workshop on Rare Earth Magnets (University of Dayton, Dayton, Ohio, 1985) p. 747.

[20] H. Kronmuller, K. Durst, W. Ervens and W. Fernengel, IEEE Trans. Magn. 20 (1984) 1569.

Note: The author acknowledges that parts of the fine particle magnet theory in Sections 3.4 and 3.5 are from a General Electric Information Report No. 66C 252, October 1966, by F. Luborsky and R. J. Parker.

4

CLASSIFICATION OF PERMANENT MAGNET PROPERTY SYSTEMS AND PROCESSING TECHNOLOGY

4.1 Introduction
4.2 Inclusion Hardened (Early Steel) Magnets
4.3 Fine Particle Magnets Utilizing Shape Anisotropy
  4.3.1 Alnico Magnets, Aluminium-Nickel-Iron-Cobalt Alloys
  4.3.2 Elongated Single Domain Magnets
  4.3.3 Iron Chrome Cobalt Magnets
4.4 Fine Particle Magnets Utilizing Crystalline Anisotropy
  4.4.1 Ferrite Magnets
  4.4.2 Manganese-Aluminum-Carbon System
  4.4.3 Platinum Cobalt
  4.4.4 Rare-Earth Cobalt Magnets
  4.4.5 NdFeB (Sintered)
  4.4.6 NdFeB (Melt Spun)
4.5 Matrix or Bonded Magnets
4.6 Semi-hard Magnets (Hysteresis Alloys)

4.1 INTRODUCTION

For convenience, the many types of magnets we use today have been divided into five categories: (i) inclusion hardened early steel magnets; (ii) magnets primarily utilizing shape anisotropy to exhibit coercivity; (iii) magnets primarily utilizing crystalline anisotropy to exhibit coercivity; (iv) matrix magnets; and (v) semi-hard magnets.

In terms of our basic understanding of the origin of their behavior, there are only two categories, (a) the inclusion hardened early steel magnets and (b) fine particle magnets that include both synthetic structure and metallurgically formed structures developed by precipitation or order-disorder transformations.

In this chapter, magnets that are currently commercially important are described in terms of their composition, preparation, properties and the physical origin of their properties. Additionally, key processing and manufacturing issues are described. For reference, several older materials that may not be important commercially are also described because they bring out interesting examples of how properties are achieved. They add to the total body of information.

In general we find a sizeable number of different types in production at a given time. The design requirements are too diverse for one material to exclude all others. It is also true that once a permanent magnet material is made in production, it remains in production. This is undoubtedly due to the high cost of tooling magnetic devices and to the general sophistication of magnetic circuits. New improved properties are strong incentives to redesign and retool, but there is a reluctance to scrap what one knows has worked. For example, even some cobalt and tungsten steel magnets are still being used today.

4.2 INCLUSION HARDENED (EARLY STEEL) MAGNETS

Quench hardening steels were the first materials to be extensively used as permanent magnets. The development started with plain carbon steel and ended with the highly alloyed cobalt steels of Honda [1]. They contain up to 1.25% carbon as a necessary constituent and generally have one or more of the following elements: manganese; chromium; tungsten; cobalt.

The origin of the coercive force in these quench hardened steels is due to the difficulty of domain boundary movement resulting from the combined effects of nonmagnetic inclusions, internal strains, lattice defects and submicroscopic inhomogenities in the material. The addition of cobalt raises the saturation magnetization, while the other elements mentioned lead to a decrease.

Conventional melting practice is used to prepare these magnet steels, which may be cast into final form or hot worked into bar and sheet stock and annealed. Magnet shapes may be hot formed at 900°C from annealed stock. The heat treatment for the magnets consists of heating to 800–900°C and rapid oil or water quenching. The precise temperatures depend upon composition and are carefully controlled to obtain consistent magnetic properties.

Table 4.1 Properties of quench hardened magnets

	3.5 Chrome	36.0 Cobalt
Chemical composition % (nominal):		
Chromium	3.50	3.50
Carbon	1.25	0.75
Manganese	0.45	0.50
Cobalt	–	36.00
Tungsten	–	4.00
Iron	Remainder	Remainder
Mechanical properties (heat-treated):		
Rockwell hardness	C58–65	C56–64
Electrical properties:		
Resistivity (μohm/cm per cm^2) temp. 25°C	32	76
Magnetic properties (heat-treated test bars, temp. 25.5°C):		
Peak H (Oe)	300	1000
Peak induction B (G)	14,500	15,500
Residual induction B_r (G)	9800	9600
Coercive force H_c (Oe)	50	240
Maximum energy product $(BH)_{max}$	224,000	936,000
Loss (W-s/cyc)/lb	1.03	4.45
Recoil permeability μ_r	35	12
Additional properties (heat-treated):		
Weight (lb/in^3)	0.280	0.296
Density g/cm^3	7.73	8.2

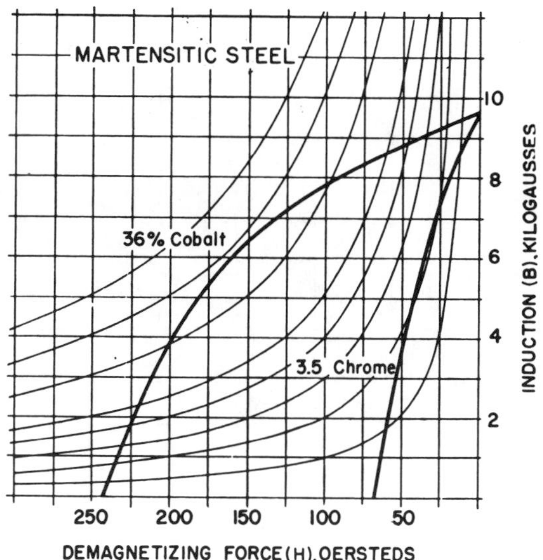

Figure 4.1 Demagnetization curves of two martensitic steels.

A serious limitation of these steels with respect to permanent magnet use is the poor metallurgical stability at normal use temperatures. By use of temperature cycling, the stability of these magnets was improved to the point of being tolerable for some uses. These magnets are rarely used today except in some hystersis type applications (see Section 4.6) Table 4.1 lists the typical properties of two, once popular, quench hardening steel magnet materials, 3.5% chrome and 36.0% cobalt. The demagnetization curves are shown in Figure 4.1.

4.3 FINE PARTICLE MAGNETS UTILIZING SHAPE ANISOTROPY

4.3.1 Alnico Magnets, Aluminum-Nickel-Iron-Cobalt Alloys

A major advance in permanent magnet use and technology began in 1932 with the discovery, by Mishima [2], of the excellent magnetic properties of an aluminum-nickel-iron alloy. Many investigations added to Mishima's work and many variations of composition, proceeding details, and properties resulted. The single domain behavior of these alloys is a result of the size and shape of the magnetic phase developed in a weakly magnetic or nonmagnetic phase. The rapid development and success of these early precipitation-hardening alloys was due both to the superior properties and to the excellent metallurgical stability. Typical properties of early isotropic Alnico 1, 2, 3, and 4 are shown in Figure 4.2.

The physical properties of Alnico alloys are rather poor. High coercivity is closely accompanied by extreme hardness and brittleness. Forming is by casting or sintering as close as possible to the desired size and shape. For close tolerances it is necessary to cut or wet grind the magnets.

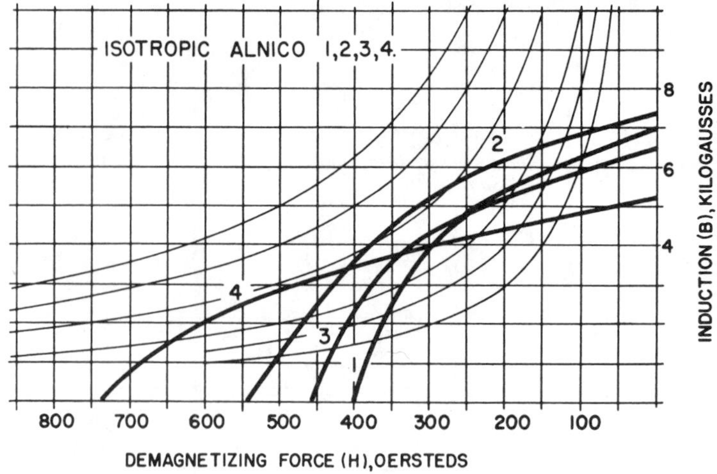

Figure 4.2 Demagnetization curves of isotropic grades of Alnico.

Following the discovery by Mishima, and perhaps of greater importance, was the announcement by Jonas [3] of a process to secure anisotropic magnetic properties in an Alnico alloy with a high cobalt content. On cooling the magnet from an elevated temperature through the region of its curie point, under the influence of a magnetic field, it was found that a preferred axis of magnetization, parallel to the field axis, was developed in the magnet. When later magnetized along this preferred axis, about a fivefold increase in available energy product was obtained. This field orientation, the patent of which is owned by Philips, led to great growth in permanent magnets. Alnico 5 properties dominated the magnet industry from the mid 1940s to about 1970 when ferrite became the most widely used material. Figure 4.3 shows typical cast Alnico magnet configurations.

Alnico magnets are normally prepared in an induction melting furnace and the metal is poured into baked sand molds. Small heats of 150–1000 lb are generally used. Speed of melting and pouring are essential to prevent excessive oxidation losses and metal segregation. Figure 4.4 shows a block diagram of the process for making Alnico 5. The heat treatment is in three steps, a high temperature solution treatment followed by a controlled cooling in a magnetic field and finally, a low temperature aging.

Figure 4.3 Typical cast Alnico magnet configurations.

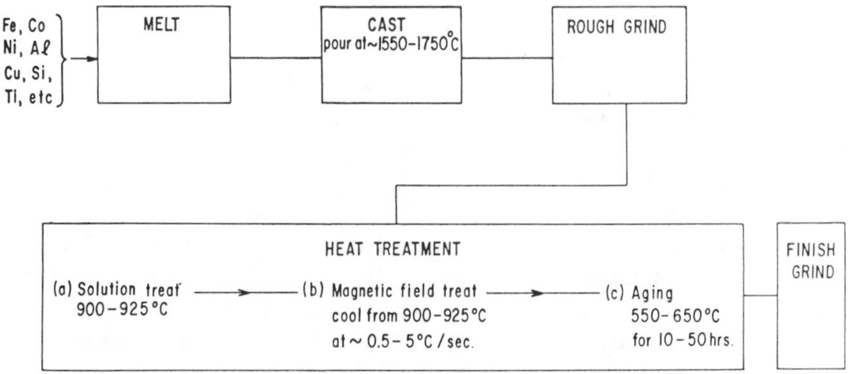

Figure 4.4 Block diagram of process for Alnico 5.

The ability of the magnetic field to influence the geometry, and hence the anisotropy, of the decomposing phases lies in the mechanism of decomposition. The critical phenomenon in this process is the decomposition of the high temperture α-phase into the FeCo-rich α-phase. This occurs by a spinodal decomposition mechanism [4]. In a spinodal decomposition, a discrete precipitate does not form; rather, the two phases develop by gradual fluctuations in composition. The resulting structures are periodic and crystallographically oriented. Figure 4.5 illustrates how the process produces elongated regions of FeCo spaced in a lesser magnetic Fe-Ni-Al phase. Because the magnetic structure is influenced by crystallographic orientation during its formation, achieving the best properties requires careful development of the proper ⟨100⟩ crystal texture. The sensitivity to crystal orientation is shown in Figure 4.6 for Alnico 5.

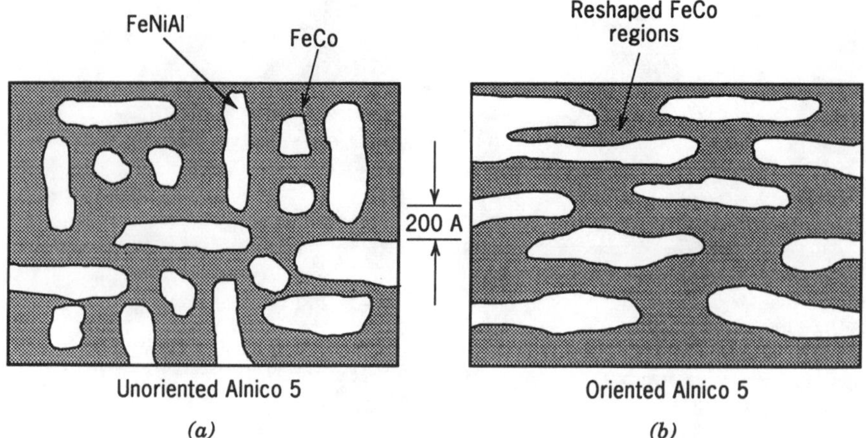

Figure 4.5 Field orientation of Alnico 5.

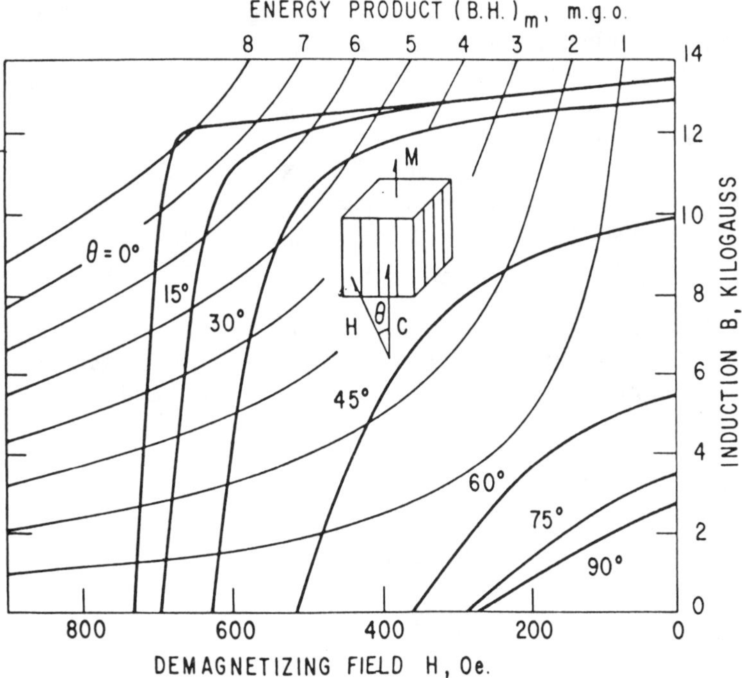

Figure 4.6 Effect of crystal orientation on the magnetic properties of Alnico 5.

In Figure 4.7 the structures of typical samples from commercial processing are shown. These structures may be achieved by a variety of techniques, all of which severely limit the size and shapes available for design.

One must use simple cross-sections over quite limited surface to volume ratios. Alnico 5 DG is achieved by pouring the metal against a shill plate. Better properties are obtained by using hot molds of exothermic material and suitable chilling (Alnico 5–7).

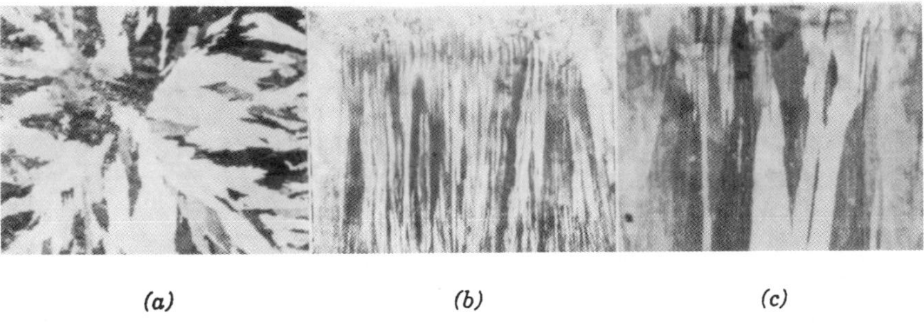

Figure 4.7 Crystal structure of Alnico 5 samples: (a) normal cast section from long rod; (b) DG speaker slug cast on a chill plate; (c) completely columnar structure from section of a long rod continuously cast.

Alternately, random oriented castings may be recrystallized in a separate step [5]. A continuous casting procedure reportedly also gave excellent crystal orientation [6]. Property comparisons for Alnico 5 magnets are shown in Figure 4.8.

There are many compositional variations of Alnico magnets. Compositional changes to alter the shape and intercepts of the demagnetization curve or to improve the physical properties are extensive as shown in Appendix 2. A high titanium alloy, Alnico 8, and its directional grain counterpart, Alnico 9, are in extensive use. The unit magnetic properties are shown in Figure 4.9.

The difficulty in forming small magnets by casting, coupled with the poor physical properties led Howe [7] to develop Alnico alloys made by powder metallurgy techniques. A mixture of the constituent metal powders and a lubricant are compacted in a forming die. Approximately 8% shrinkage occurs during sintering so this must be considered in the die design. The

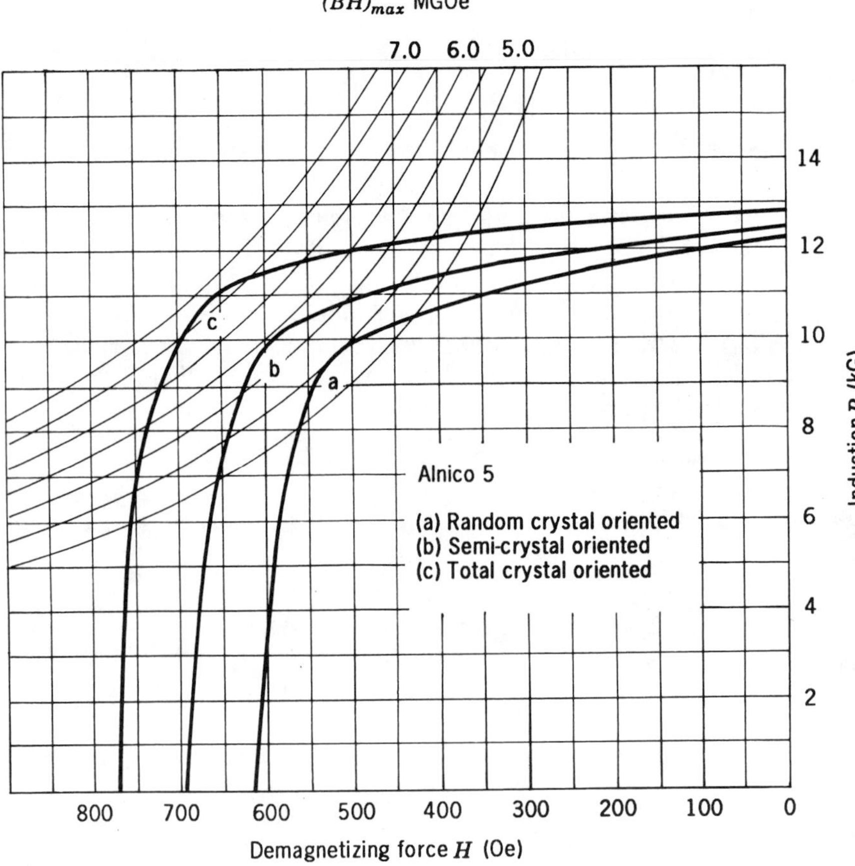

Figure 4.8 Comparison of Alnico 5 properties.

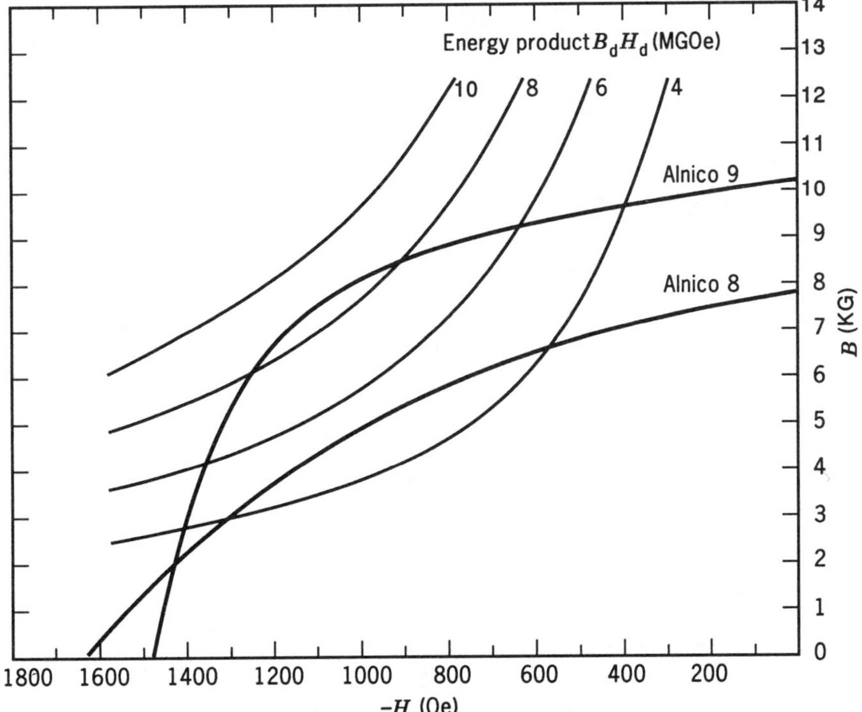

Figure 4.9 Alnico 8 and Alnico 9 properties.

compacts are sintered in a pure, dry hydrogen atmosphere at temperatures very near the melting point of the alloy. A heat treatment is used to optimize the magnetic properties. Available sintered properties are shown in Figure 4.10.

Sintered Alnico magnets exhibit a uniformly fine grain, which greatly enhances the physical properties of these magnets. Sintered magnets are often required for high speed rotors in electrical machines. In the production of such rotors, it is common practice to allow the powder to shrink around a machinable steel insert which facilitates mounting the rotor magnet on a shaft. Complex small parts are readily produced to close tolerances. Typical sintered magnet shapes are shown in Figure 4.11. A disadvantage of sintered magnets is the high cost of tooling which is a serious problem if the number of parts required is small.

4.3.2 Elongated Single Domain Magnets

Elongated single domain magnets rely on shape anisotropy for coercivity. These magnets developed by General Electric are sold under the trade name Lodex. The sequence of processing steps is shown in Figure 4.12 [8, 9]. Iron

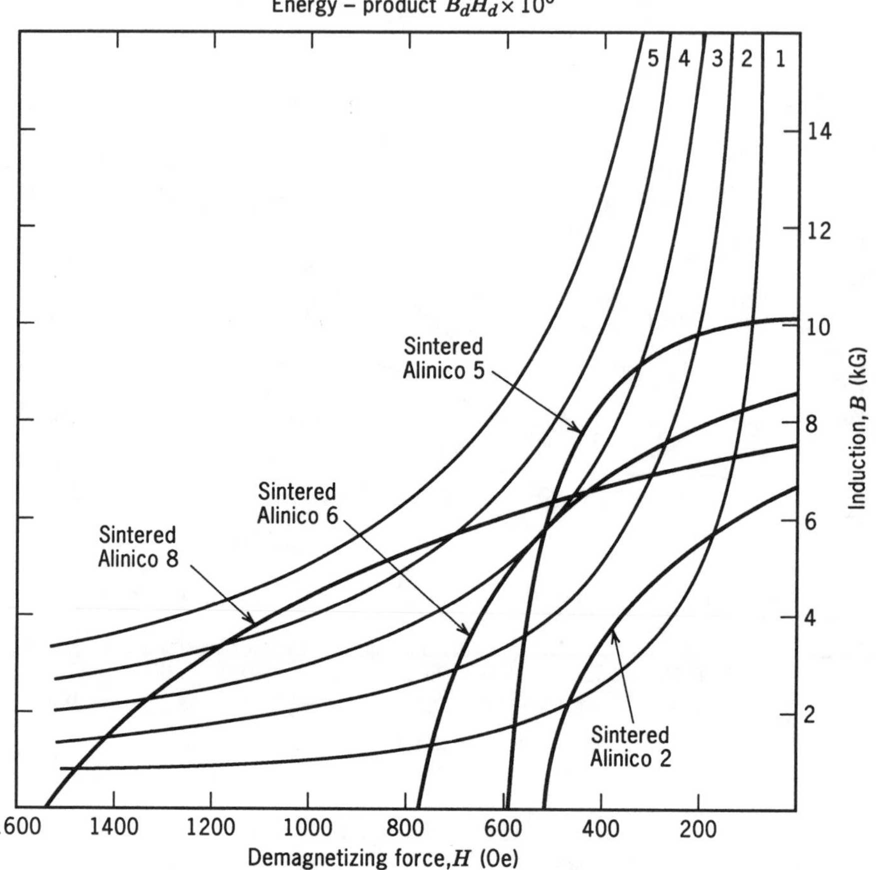

Figure 4.10 Sintered Alnico 2, 5, 6, and 8.

and cobalt are electrodosited in mercury to form elongated structures and then subjected to the following processes: aging to remove the dendritic branches; additition of antimony to form a protective layer; addition of lead-antimony alloy to provide the metal matrix; pressing the slurry into a large block with field alignment; vacuum distillation to remove mercury; grinding to about 10 μm powder and finally pressing at room temperature in a die in a field to the final shape. A series of unit property variations are obtained by varying the packing fraction of iron cobalt particles in the nonmagnetic alloy. Figure 4.13 shows the variations from this process for oriented magnets. Lodex magnets may also be extruded, in which case the orientation develops along the axis of extrusion. Lodex magnets find extensive use in small electromagnetic devices where close physical and magnetic tolerances are vital. Since the pressing is at room temperature, no shrinkage is encountered in this process. Typical configurations of Lodex production magnets is shown in Figure 4.14.

Figure 4.11 Sintered Alnico 2, 5, 6, and 8.

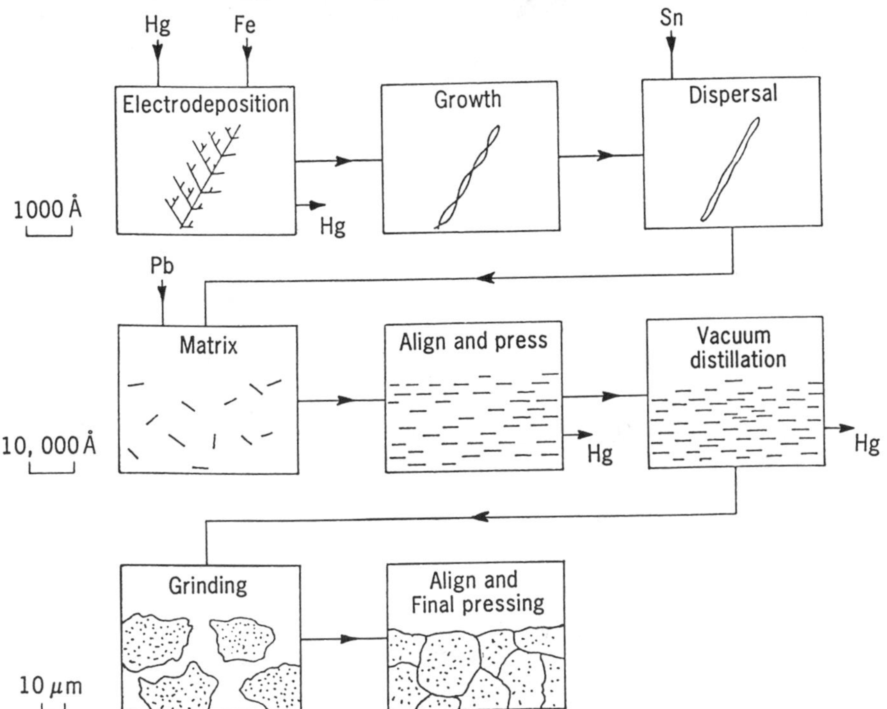

Figure 4.12 Lodex process steps.

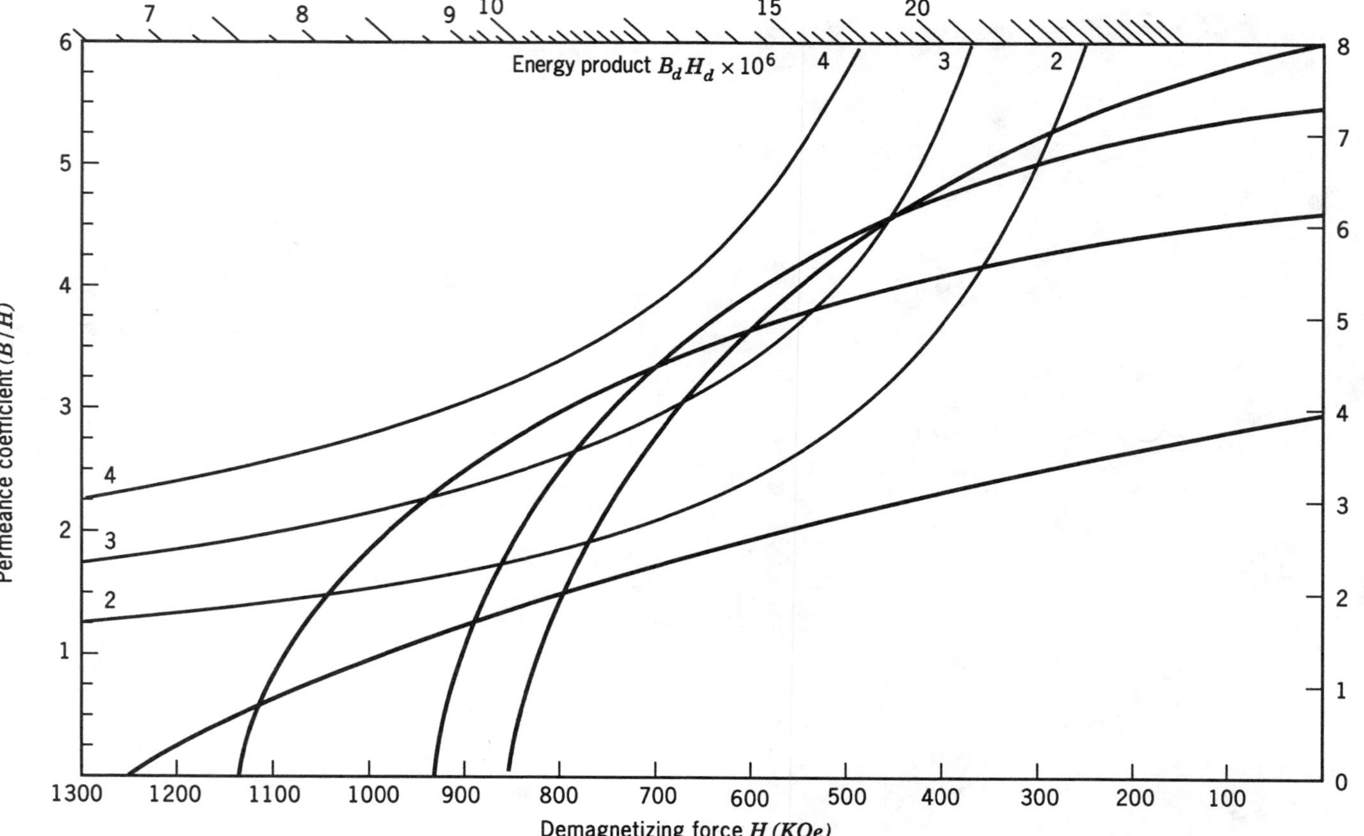

Figure 4.13 Anisotropic Lodex. From Hitachi Magnetics Corp.

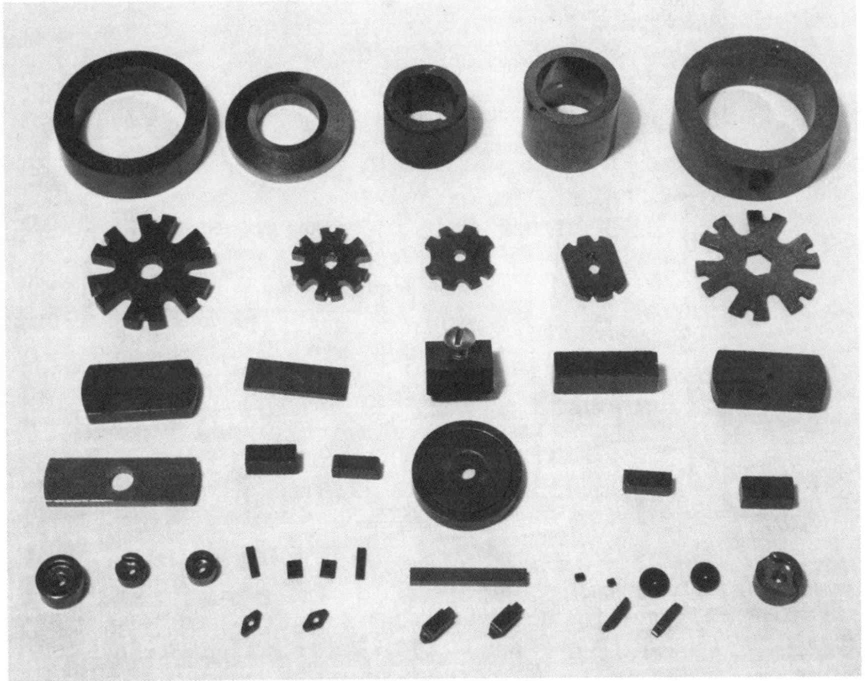

Figure 4.14 Typical Lodex production magnets.

4.3.3 Iron Chrome Cobalt Magnets

In 1971, Kaneko and his associates [10] at Tohoku University announced the development of FeCrCo magnets. Properties vary much like those of Alnico 5 and are obtainable with less cobalt and in ductile form. These magnets can be made by milling and casting, sintering or, as a wrought product in wire and sheet form. Fabrication by coining or stamping is also possible.

Typical heat treatment for FeCrCo is shown in Figure 4.15. A solution treatment is followed by a magnetic field treatment and finally, an aging treatment. At solution temperature, a body centered cubic phase (α) is formed. This phase is retained by quenching to room temperature. By reheating to 630° the α-phase transforms into α_1- and α_2-phases. This part of the heat treatment is in a magnetic field which allows formation of elongated particles in a direction parallel to the applied field. The aging treatment is used to increase coercive force by decreasing the magnetic properties of the α_2-phase.

The process flow sheet for this alloy system is shown in Figure 4.16. Orientation by deformation has been described by Jim et al. [11]. The comparison of the two process approaches to alignment as a function of alloy percent Co, is shown in Figure 4.17.

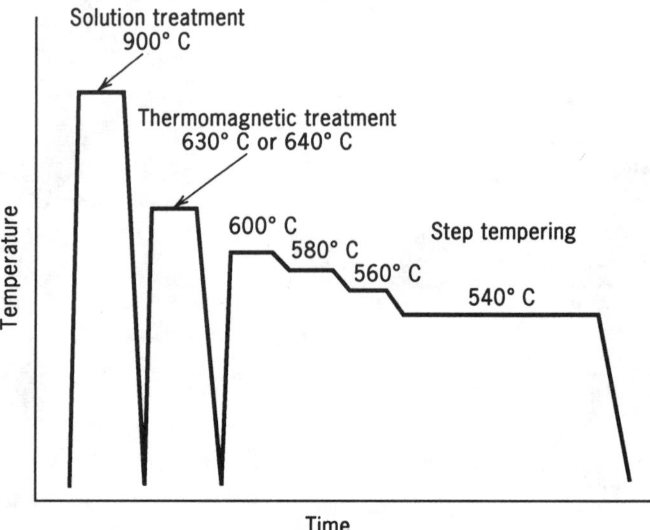

Figure 4.15 Heat treatment schedule for an FeCrCo alloy.

The FeCrCo system can be used to replace many older type magnets such as Alnico 5, Cunico, Cunife, and Vicalloy as well as many steel magnets or semi-hard materials for hysteresis applications (see Appendix 2 for reference). The ductivity and orientation by deformation adds greatly to the versatility of this magnet system. There is some evidence of ability to improve the coercive force of this system with metal additions to the alloy such as Al, Ti, Si, Nb, Ta, Zr, Mo, Cu, Ni, and V.

4.4 FINE PARTICLE MAGNETS UTILIZING CRYSTALLINE ANISOTROPY

A new class of permanent magnetic materials was announced in 1952 by Went et al. [12] which was based on the crystal anisotropy of barium oxide. This class of magnets is generally known as ferrite, but are sometimes referred to as oxide or ceramic magnets. Today, ferrite magnets are by far the most widely used magnets. The success of ferrite is due to several reasons. The raw materials are inexpensive and nonstrategic. The high coercive force combined with reasonable induction has allowed permanent magnets to move into many types of small motors.

Ferrites are produced by powder metallurgy. Their chemical formulation may be expressed as $MO \cdot 6(Fe_2O_3)$ where M is Ba, Sr, or Pb. Strontium ferrite has higher coercive force than barium ferrite and is the larger production ferrite. Lead and barium ferrite both have some production disadvantages from an environmental point of view. Figure 4.18 indicates the size and shape of component parts produced. Table 4.2 shows the range of properties available.

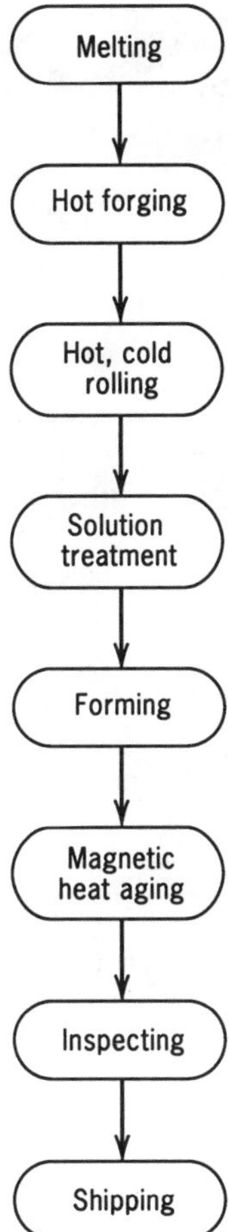

Figure 4.16 Process flow sheet, FeCrCo magnets.

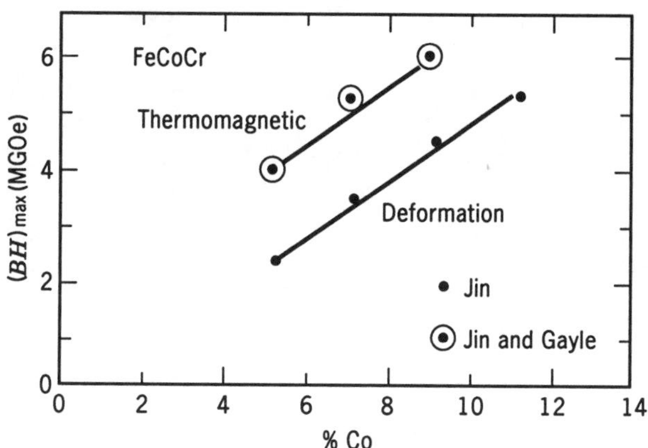

Figure 4.17 Comparison of properties of magnets made by thermomagnetic treatment and by deformation alignment. From Jim et al. [11].

Figure 4.18 Typical ferrite magnet components.

Table 4.2 Commercially available ferrite properties

				Magnetic properties (nominal)			
MMPA Brief Designation	Original MMPA class	IEC Code Reference	Chemical Composition[a]	Max. Energy Product $(BH)_{max}$ (MGOe)	Residual Induction B_r (G)	Coercive Force H_c (Oe)	Intrinsic Coercive Force H_{ci} (Oe)
1.0/3.3	Ceramic 1	S1-0-1	$M0 \cdot 6Fe_2O_3$	1.05	2300	1860	3250
3.4/2.5	Ceramic 5	S1-1-6	$M0 \cdot 6Fe_2O_3$	3.40	3800	2400	2500
2.7/4.0	Ceramic 7	S1-1-2	$M0 \cdot 6Fe_2O_3$	2.75	3400	3250	4000
3.5/3.1	Ceramic 8	S1-1-5	$M0 \cdot 6Fe_2O_3$	3.50	3850	2950	3050
3.4/3.9	–	–	$M0 \cdot 6Fe_2O_3$	3.40	3800	3400	3900
4.0/2.9	–	–	$M0 \cdot 6Fe_2O_3$	4.00	4100	2800	2900

[a] M represents barium, strontium or a combination of the two.

Source: from MMPA Std.

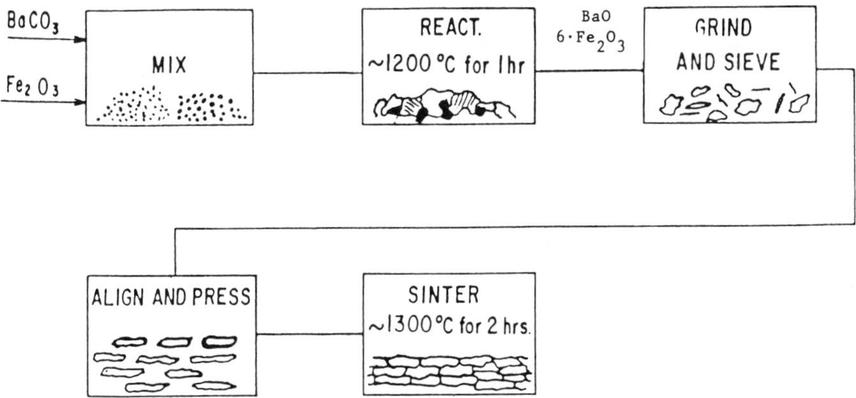

Figure 4.19 Critical events in making ferrite magnets.

Ferrite magnets are available in isotropic and anisotropic grades. Figure 4.19 shows the steps used in preparing barium ferrite. The anisotropic grades are prepared by pre-firing the raw materials and milling the compound to single crystals of the order of single domain size (about 1 μm). The milled powder is then wet or dry pressed under the influence of a magnetic field. The pressed compacts are then sintered in air. The shrinkage is about 15% and ferrite magnets must be ground to maintain close tolerance parts. The ferrite particles develop as platelets with the preferred C-axis of magnetization perpendicular to the plane of the plate. Pressing promotes mechanical anisotropy and somewhat better properties are achieved in the pressing direction than in a transverse direction. In making the oriented magnet, the field should be applied in the direction of pressing. A field of 5 kOe is required to produce oriented magnets. Wet pressing gives the particles increased mobility and improved B_r and $(BH)_{max}$. Dry pressing, however, is faster since the water does not have to be removed during the pressing. Figure 4.20 shows a general flow chart of the ferrite production process.

By combinations of pre-firing time and temperature and sintering temperatures, a wide range of properties can be obtained. Also, grain growth inhibitors may be added to influence B_r/H_{ci} relationships. Appendix 2 lists the property variations. Figure 4.21 gives a summary of progress in property development with respect to time. In each time frame, the points represent commercially available properties with trade-offs between B_r and H_{ci}.

The saturation magnetization of a ferrite single crystal is approximately 5 kG. H_{ci} measurements on single particles have been reported as high as 11.3 kOe [13]. One can only expect small improvements in B_r since we are already close to the limits imposed by the saturation magnetization. Improvements in processing techniques are more likely to be made. These improvements could well have impact on the costs and plant investment.

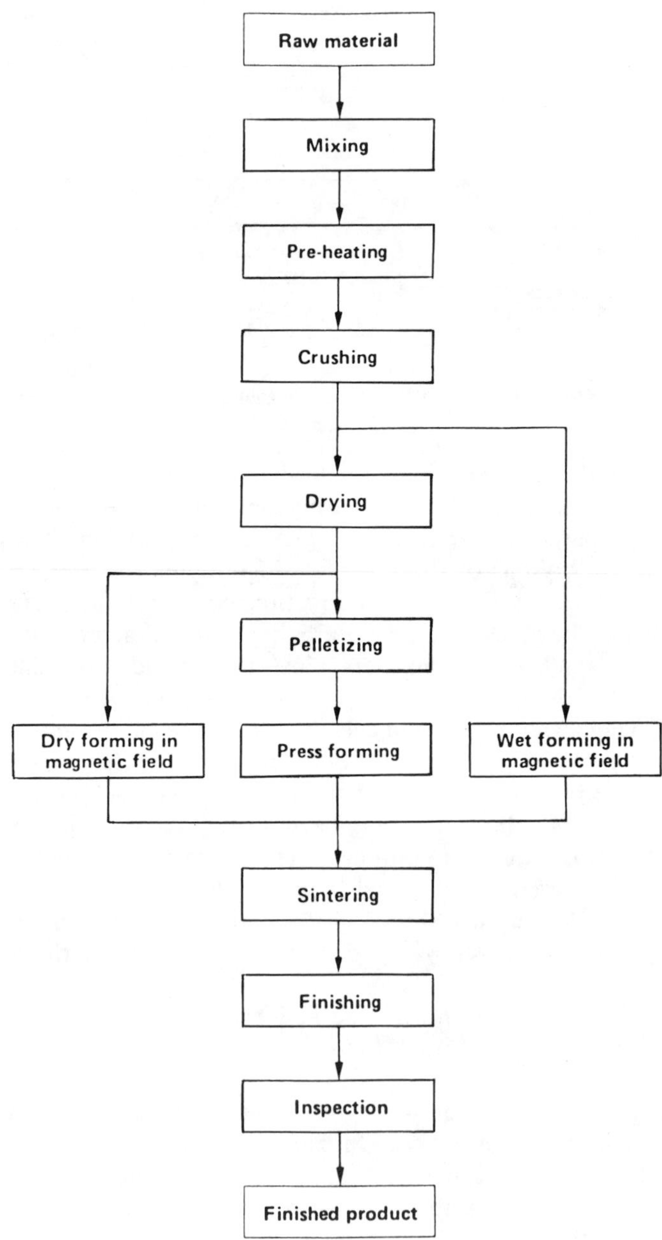

Figure 4.20 Ferrite magnet production process.

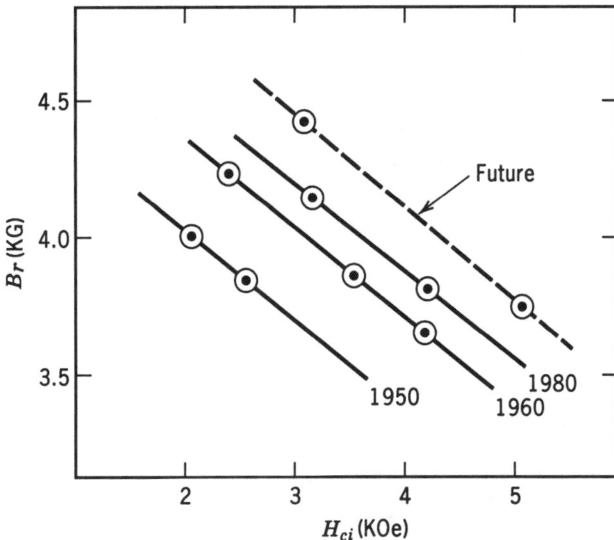

Figure 4.21 Ferrite property development trends.

One interesting variation in processing is shown in Figure 4.22. Fuji Electrochemical Co. has developed a process for making a radially oriented cylinder shape magnet. The unique orientation is possible due to the shape of the ferrite particles. Orientation is achieved without the use of a magnetic field. $(BH)_{max}$ values up to 4.0 MGOe have been reported from this process.

4.4.2 Manganese-Aluminum-Carbon System

Koch et al. [14] developed the first anisotropic alloy in this system. Properties as high as $(BH)_{max} = 3.5$ MGOe were obtained as a result of cold working by swaging. The extreme brittleness and difficulties in swaging limited the usefulness of these magnets. In 1962, Kubo et al. [15] found that the addition of carbon to the MnAl alloy improved the stability of the meta-stable ferromagnetic phases and also improved the coercive force. Kubo later clarified the nature of the crystallographic transformation process which resulted in a new process for making MnAlC magnets. From a detailed study of applying pressure in various directions, it was learned that crystal structure could be formed presumably by an alignment effect of atoms due to external stress. The direction of the C axis of the ferromagnetic phase was changed and a hot plastic deformation process led to the formation of highly anisotropic magnets [16]. Figure 4.23 illustrates the extrusion process used to improve MnAlC magnets. The improvement in alignment by extrusion is shown in Figure 4.24.

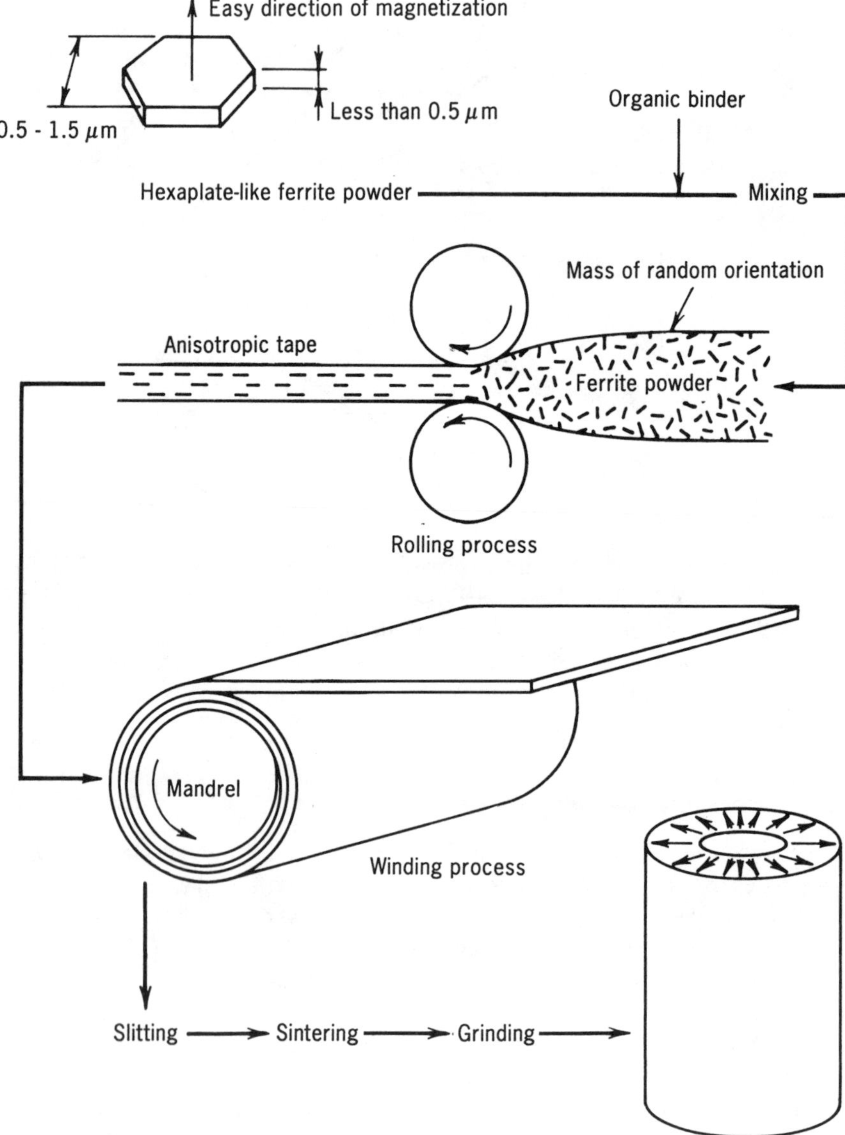

Figure 4.22 Process for making radially oriented cylinders.

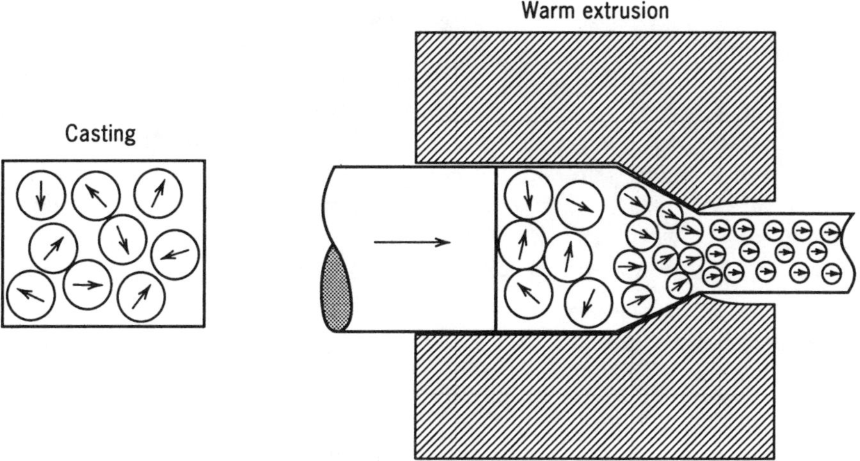

Figure 4.23 MnAlC alignment by extrusion.

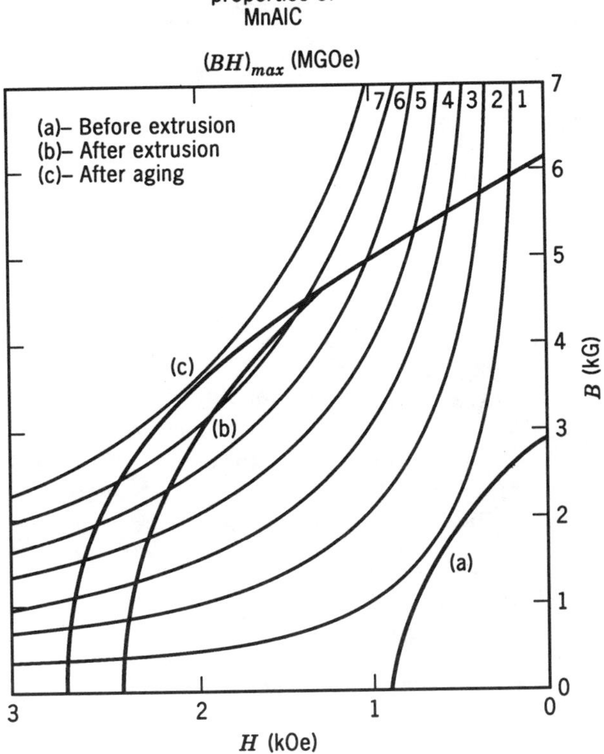

Figure 4.24 MnAlC property improvement due to extrusion. From Matsushita Electric Co.

Matsushita Electric Co. has produced properties of $B_r = 5500$ G, H_{ci} 3000 Oe and $(BH)_m = 7.5$ MGOe. The nominal composition is 70 Mn, 28.5 Al, 0.8 Ni, and 0.5 C. Unfortunately it has a low Curie temperature and hence a high reversible temperature coefficient (0.12%/°C). Additionally, its coercive force drops rapidly with increasing temperature. It would seem that MnAlC magnets are limited to applications in which the maximum temperature is not over 100°C. An advantage is that the material is machinable. However, the high investment in deformation equipment for orientation is a serious drawback.

4.4.3 Platinum Cobalt

This alloy has very unique properties, but its high cost limits its use. The best properties are obtained in the 50:50 atomic ratio. A 10 MGOe material is obtained with the following thermal events. Heat to 1000°C to yield a disordered structure, cool at a controlled rate to room temperature, and finally, age at 600°C for 5 h. The high H_{ci} may be attributed to the high crystal anisotropy of tetragonal single domain regions. PtCo is isotrophic, It has a high ratio of B_r/B_{is} (0.86) which suggests that the final partially ordered material has cubic magnetocrystalline anisotropy. PtCo is a ductile and machinable material. See Appendix 2 for typical magnetic and physical properties.

4.4.4 Rare-Earth Cobalt Magnets

Early processing approaches to make rare-earth cobalt magnets were quite varied. Important characterization and measurements of saturation magnetization and anisotropy fields of powder, as well as projection of the permanent magnet potential of many rare-earth cobalt compounds was done by Strnat and Hoffer [17].

The making of real magnets was a long and difficult development. A densified magnet having 20 MGOe properties was reported by Buschow et al. [18] in 1969. The densification process involved a combination of isostatic and uniaxial pressure of about 20 kbar. An industrial process for densification by pressing and sintering of $SmCo_5$ was reported by Das [19] in 1969. Martin and Benz [20] in 1970 described the use of liquid phase sintering to produce $SmCo_5$ magnets with excellent magnetic properties and high physical density. The processing involved vacuum melting of Sm and Co. The flow chart of the process is shown in Figure 4.25. The earliest $SmCo_5$ magnets were aligned in a high magnetic field and then hydropressed and sintered. Later developments led to die pressing in a magnetic field of about 10 kOe. Die pressing has proven to be a satisfactory compromise between property achievement and processing cost.

In 1974, Cech [21] developed a reduction diffusion process that allowed Sm-oxide powder and cobalt powder to be reduced with Ca to form the

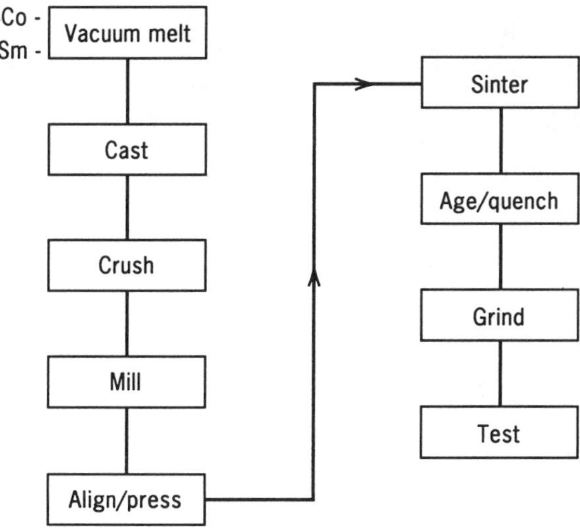

Figure 4.25 $SmCo_5$ production process.

compound directly without the need to start with pure Sm metal. This process (Figure 4.26) and a similar process developed in Germany [22] has had a profound economic impact on the development of the rare-earth cobalt magnet industry. Since Sm oxide costs considerably less than pure Sm metal. The $SmCo_5$ process has, over the years, been modified to reduce the Sm content. Ce misch-metal, being of less cost than Sm, has been used extensively. Up to half of the Sm was replaced and properties of about 14 MGOe were achieved. Pr has also been substituted for Sm to improve properties. Property levels of 26 MGOe have been achieved. At the present time $SmCo_5$ is the most widely used rare-earth magnet.

A totally different processing approach to rare-earth cobalt magnets was announced almost simultaneously in 1968 by Nesbitt et al. [23] and Tawara and Senno [24]. They achieved very high coercivity in bulk SmCo and CeCo alloys in which a two phase microstructure was created by substituting Cu for part of the Co. This was a precipitation type cast structure with properties as high as 8.0 MGOe reported. A later development direction was to use powder metallurgy methods, including sintering to make useful permanent magnets [25]. The five element system Ce Sm (Co Fe Cu) has many variations of properties and these magnets are often preferred over $SmCo_5$ due to improved cost/performance, particularly in the Japanese market. However, the properties do not allow substitution for $SmCo_5$ in some situations.

An important class of rare-earth cobalt permanent magnet has the general composition $R(Co, M)_z$ where R stands for at least one rare-earth element, M for a combination of transition metals and/or copper and z is between 6 and 8.5. These multicomponent magnet alloys are generally

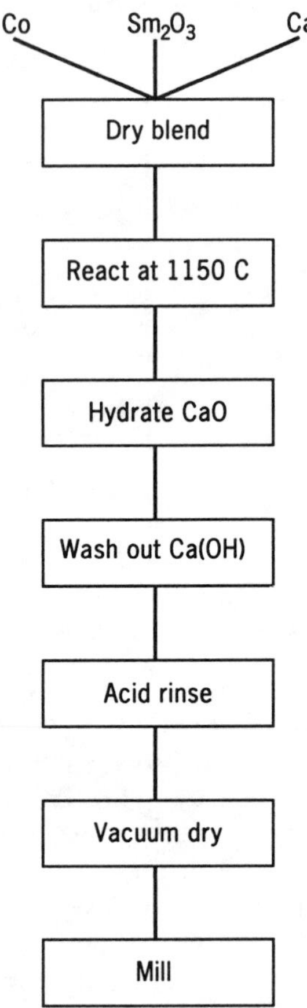

Figure 4.26 Reduction/diffusion process.

termed the 2:17 system. They have the advantage of higher magnetization and use less Sm and Co. The processing steps are very similar to those for $SmCO_5$, except the heat treatment is longer and may involve step aging. In early investigations of these compounds, Strnat et al. [26] found that generally, the 2:17 compounds had an easy plane of magnetization and only $SmCo_{17}$ had easy axis magnetization necessary for coercivity. By replacing some of the cobalt with iron, the anisotropy field was increased to useful levels. Additional improvements in this system came from adding small amounts of Cr, Mn, Ti, Hf, and Zr. Michra et al. [27] described a 30 MGOe magnet having an H_{ci} of 13 kOe with a composition $Sm(Co_{0.63}Fe_{0.28}Cu_{0.05}Zr_{0.02})_{7.7}$. The commercial properties available generally involve two distinct levels of coercive force. For low values of z, H_{ci} is about 6 kOe and for higher z values with higher Fe content H_{ci} levels of 12.5 kOe are offered. In these materials the coercive force mechanism is domain wall pinning at grain boundaries.

MMPA Brief Designation	IEC Code Reference	Chemical Composition[a]		Magnetic Properties (nominal)[b]			
		Alloys	Possible Elements	Maximum Energy Product $(BH)_{max}$ (MGOe)	Residual Flux Density B_r (G)	Coercive Force H_c (Oe)	Intrinsic Coercive Force H_{ci} (Oe)
5/16	R5-1	RE Co_5	RE = Sm, Nd, MM	5	4700	4500	16,000
14/14	R5-1	RE Co_5	RE = Sm, MM	14	7500	7000	14,000
16/18	R5-1	RE Co_5	RE = Sm, Nd	16	8300	7500	18,000
18/20	R5-1	RE Co_5	RE = Sm, Pr, Nd	18	8700	8000	20,000
20/15	R5-1	RE Co_5	RE = Sm, Pr, Nd	20	9000	8500	15,000
22/15	R5-1	RE C_{05}	RE = Sm, Pr, Nd	22	9500	9000	15,000
22/12	R5-2	RE_2TM_{17}	RE = Sm, Ce TM = Fe, Cu, Co, Zr, Hf	22	9600	8400	12,000
24/7	R5-2	RE_2TM_{17}	RE = Sm, Ce TM = Fe, Cu, Co, Zr, Hf	24	10,000	6000	7000
24/18	R5-2	RE_2TM_{17}	RE = Sm TM = Fe, Cu, Co, Zr, Hf	24	10,200	9200	18,000
26/11	R5-2	RE_2TM_{17}	RE = Sm TM = Fe, Cu, Co, Zr, Hf	26	10,500	9000	11,000
28/7	R5-2	RE_2TM_{17}	RE = Sm TM = Fe, Cu, Co, Zr, Hf	28	10,900	6500	7000
26/20	R5-3	$RE_2TM_{14}B$	RE = Nd, Pr, Dy, Tb TM = Fe, Co	26	10,400	9900	20,000
27/11	R5-3	$RE_2TM_{14}B$	RE = Nd, Pr, Dy, Tb TM = Fe, Co	27	10,800	9300	11,000
30/18	R5-3	$RE_2TM_{14}B$	RE = Nd, Pr, Dy, Tb TM = Fe, Co	30	11,000	10,000	18,000
33/11	R5-3	$RE_2TM_{14}B$	RE = Nd, Pr, Dy, Tb TM = Fe, Co	33	11,800	10,800	11,000

[a] Temperature compensated materials and materials with maximum energy products of 40 MGOe are available from various manufacturers. To achieve the properties shown in this table, care must be taken to magnetize to technical saturation.
Source: from MMPA Std.

Table 4.3 shows the typical unit properties available for the now numerous types of rare-earth cobalt magnets, in addition to rare-earth iron properties.

4.4.5 NdFeB (Sintered)

In June 1983, Sumitomo Special Metals of Japan publicly announced the development of a new rare-earth iron permanent magnet having a $(BH)_{max}$ of 35 MGOe. Sagawa et al. [28] later described compositional and processing details. The new magnet involved essentially the same powdered metallurgy approach as used for making $SmCo_5$. In a later paper, Sagawa et al. [29] described refinements and the record achievement of a 50.6 MGOe magnet made under laboratory conditions. During the past 5 years there has been a tremendous development effort on sintered NdFeB magnets. Today a wide variety of commercial magnets are offered with properties from

Table 4.4 Comparison of the magnetic and physical propeties of NdFeB and SmCo magnets

	SmCo Magnet	NdFeB Magnet
Elements	Sm, Co, Fe, Cu additives	Nd, Fe, B additives
Process	Sintering	Sintering/melt spinning
Magnetic characteristics		
$(BH)_{max}$ (MGOe)	16–32	24–37
B_r (kG)	8.2–11.6	10.2–12.6
H_{ci} (kOe)	6.2–20.0	9.6–20.0
μ_r	1.05	1.05
Temperature coefficients		
B_r (%/°C)	−0.03 to −0.04	−0.12 to −0.15
H_{ci} (%/°C)	−0.14 to −0.40	−0.40 to −0.70
Reversible temperature coefficient (%/°C)	0.045	0.13
Curie temperature (°C)	800	320
Density (g/cm^3)	8.2–8.4	7.3–7.5
Thermal expansion coefficient		
$C_{\parallel}$(0–100°C) (/°C)	8×10^{-6}	4.2×10^{-6}
$C_{\top}$(0–100°C) (/°C)	11×10^{-6}	2.8×10^{-6}
Bending stength (kg/mm^2)	15	25
Compressive strength (kg/mm^2)	82	75–110
Tensile strength (kg/mm^2)	3.6	7.5
Vickers hardness	500–550	550–600
Specific resistivity (cm)	86×10^{-6}	137×10^{-6}
Minimum magnetizing field (kOe)	$SmCo_5$ 15 Sm_2Co_{17} 25 (high H_c type)	18–20

27–35 MGOe. Table 4.3 gives the approximate range of properties. Some producers offer a "H" grade in which compositional additions to the normal NdFeB magnet have been made to enhance H_{ci} and hence improve the high temperature stability. Dy, Al, and Tb have been used [30] to produce the "H" grade.

The process flow chart is very similar to that outlined for $SmCo_5$. Again there are two routes to alloy preparation. Either the vacuum melting of the elements or the calciothermic oxide reduction process can be used. Table 4.4 shows an interesting comparison between sintered rare-earth cobalt and sintered rare-earth iron magnets. The mechanical properties of NdFeB are superior to those of $SmCo_5$. The principle disadvantage of NdFeB is the large change in magnetic properties with temperature.

4.4.6 NdFeB (Melt Spun)

Croat et al. [31] in 1984 described work at General Motors on the melt spinning approach to making a rare-earth iron magnet. Isotropic properties of 14 MGOe were achieved. Some what later Lee [32] described a hot deformation technique for orientation of the melt spun material giving anisotropic properties of 40 MGOe. In this melt spinning process, the molten alloy is forced under pressure through a small orifice onto the surface of a water-cooled revolving wheel as shown in Figure 4.27. The alloy

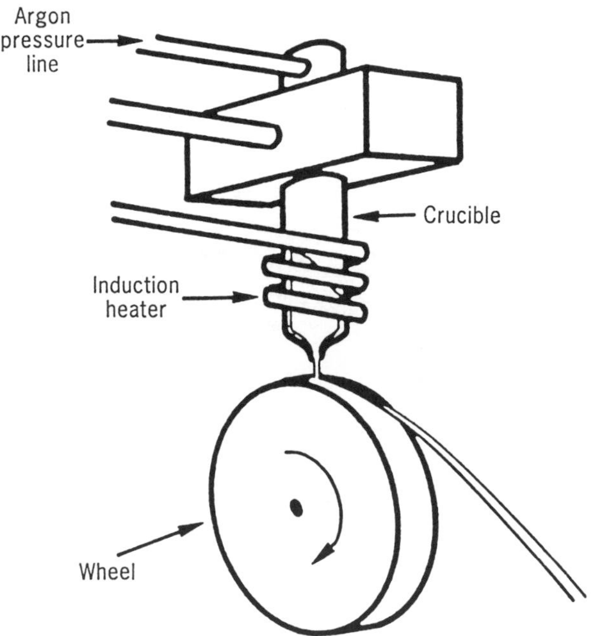

Figure 4.27 Rapid solidification process.

is quenched rapidly and a ribbon is formed with a very fine microcrystalline structure having high crystalline anisotropy. By control of the solidification rate the microstructure is optimized. The NdFeB melt is under an argon atmosphere. The ribbon fragments from the wheel are the basic building blocks for several unit property demagnetization characteristics having different physical and magnetic parameters. General Motors calls this process Magnequench. The process flow chart and the end processing variations leading to MQ I, MQ II, and MQ III grades are shown in Figure 4.28. MQ I is a matrix or bonded version and is isotropic. MQ II is a fully dense magnet achieved by hot pressing and is also isotropic. MQ III magnets are anisotropic and are made by taking the fully dense, hot MQ II magnet and hot deforming the magnet so that its area is increased and its dimension parallel to pressing is decreased. This hot deformation involves a transverse plastic flow. There is some thinning of the ribbon fragments, resulting in a magnetic alignment parallel to the pressing direction. Figures 4.29–4.35 describe the processing events in making Magnequench magnets. General Motors starts the process with an oxide reduction step to produce the alloy (Figure 4.29). The second step is the vacuum alloy furnace (Figure 4.30). The jet casting operation is shown in Figure 4.31 with MQ I, MQ II, and MQ III forming events shown in Figures 4.32–4.35. Table 4.5 compares the magnetic and physical properties of Magnequench magnets.

The Magnequench crushed ribbon is relatively large (200 μm) and can be exposed to air without appreciable oxidation. This new processing approach

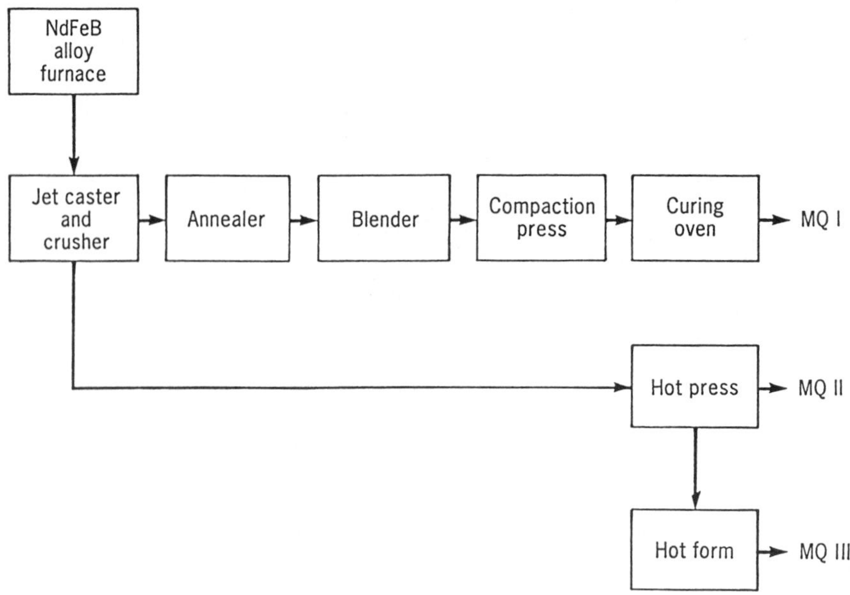

Figure 4.28 Magnequench process.

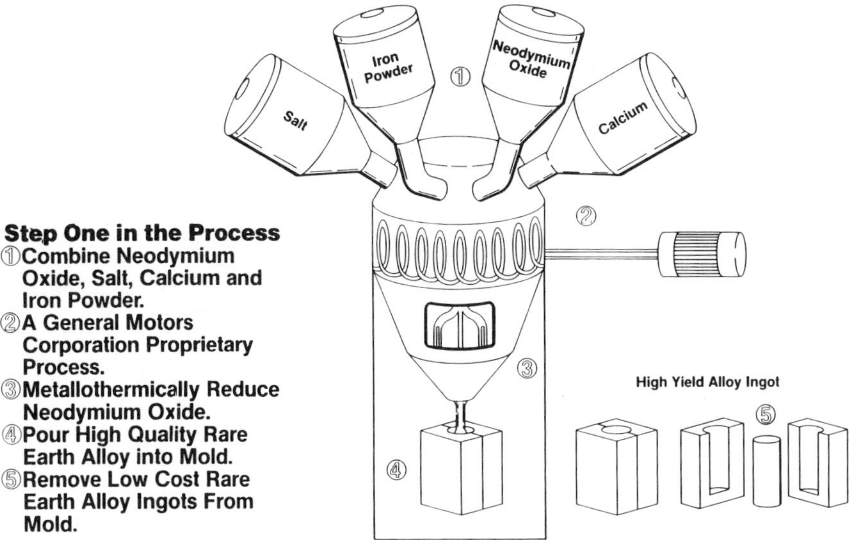

Figure 4.29 Metal making by Neochem (GM patented process).

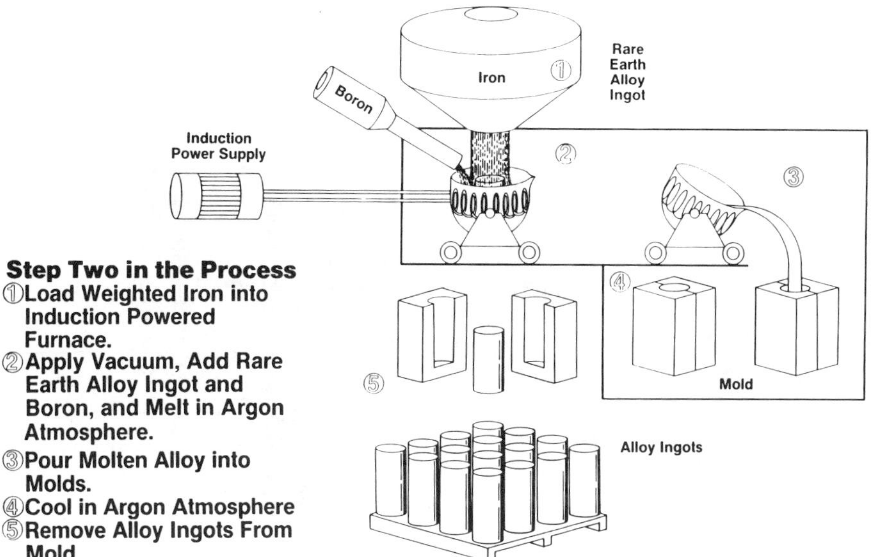

Figure 4.30 Vacuum alloying furnace.

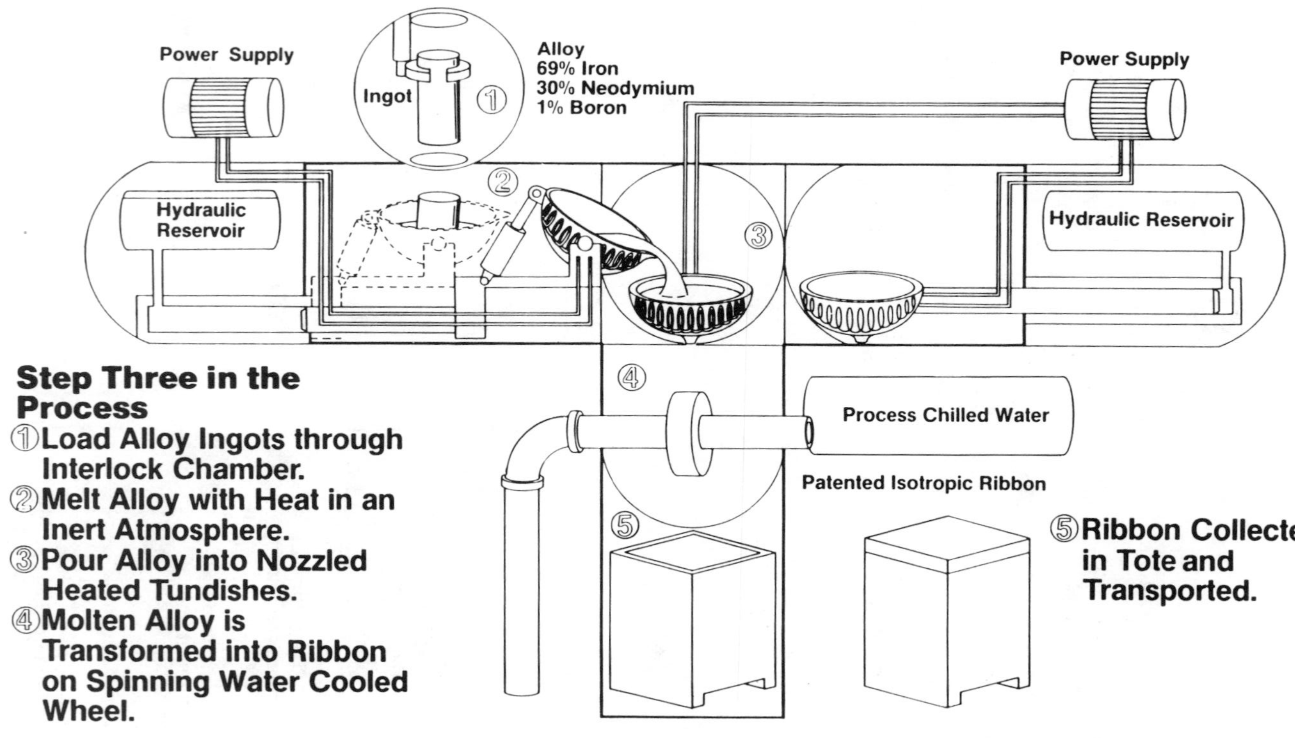

Figure 4.31 Jet casting.

Step Six in the Process

① Magnequench Powder is Weighed.

② Proper Amounts of Epoxy and Die Lubricant are Added.

③ Homogenize Magnequench Powder, Epoxy and Die Lubricant.

④ Store Processed Powder in Tote.

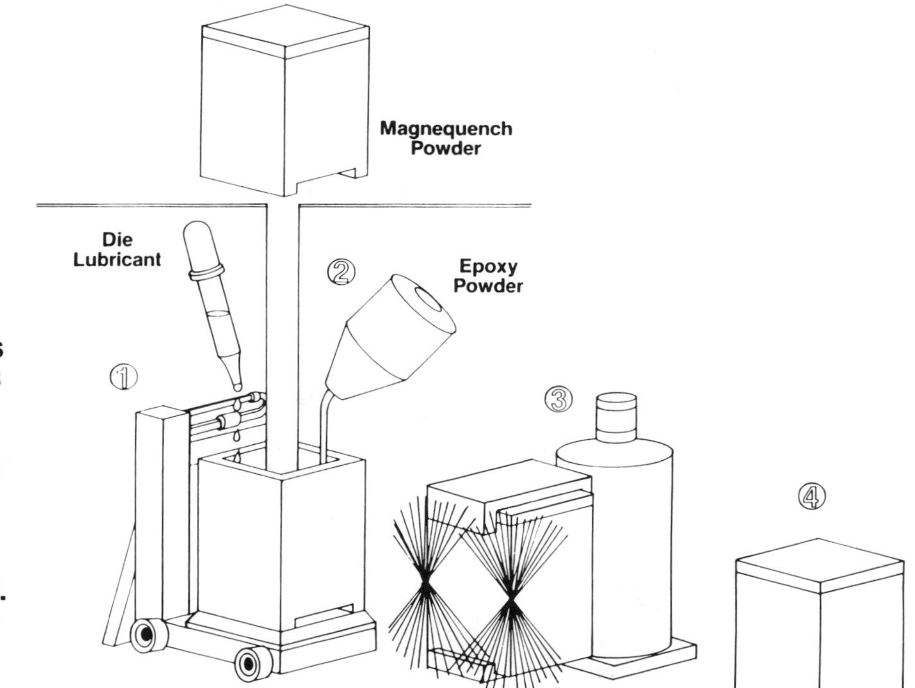

Figure 4.32 MQ I blending.

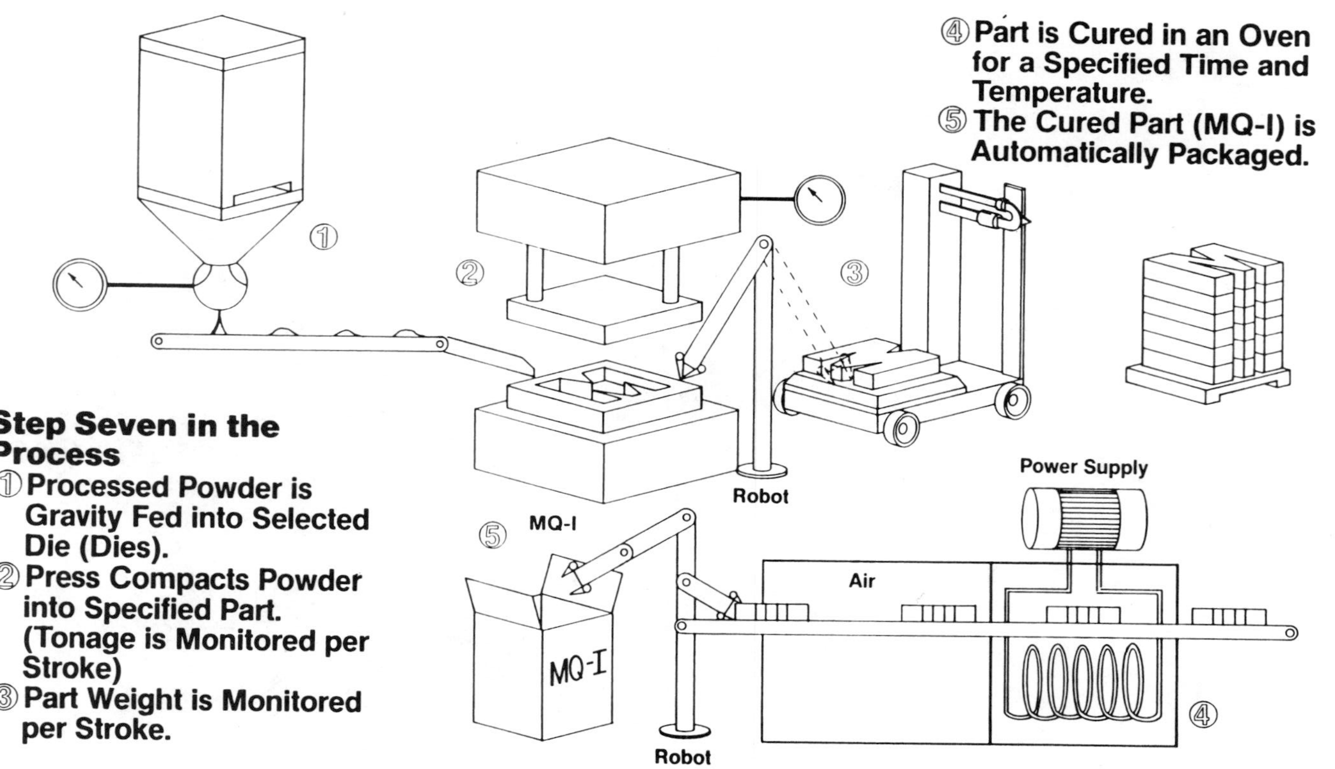

Figure 4.33 MQ I compaction and curing.

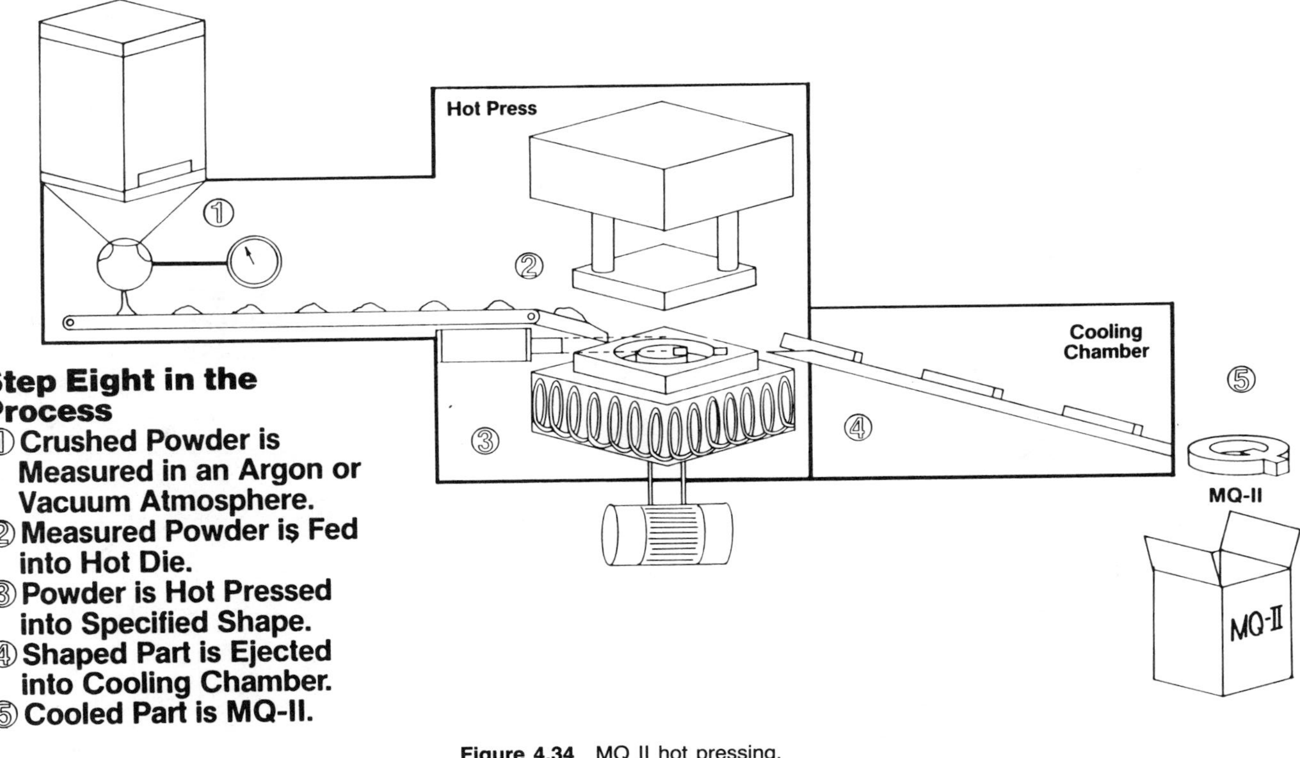

Figure 4.34 MQ II hot pressing.

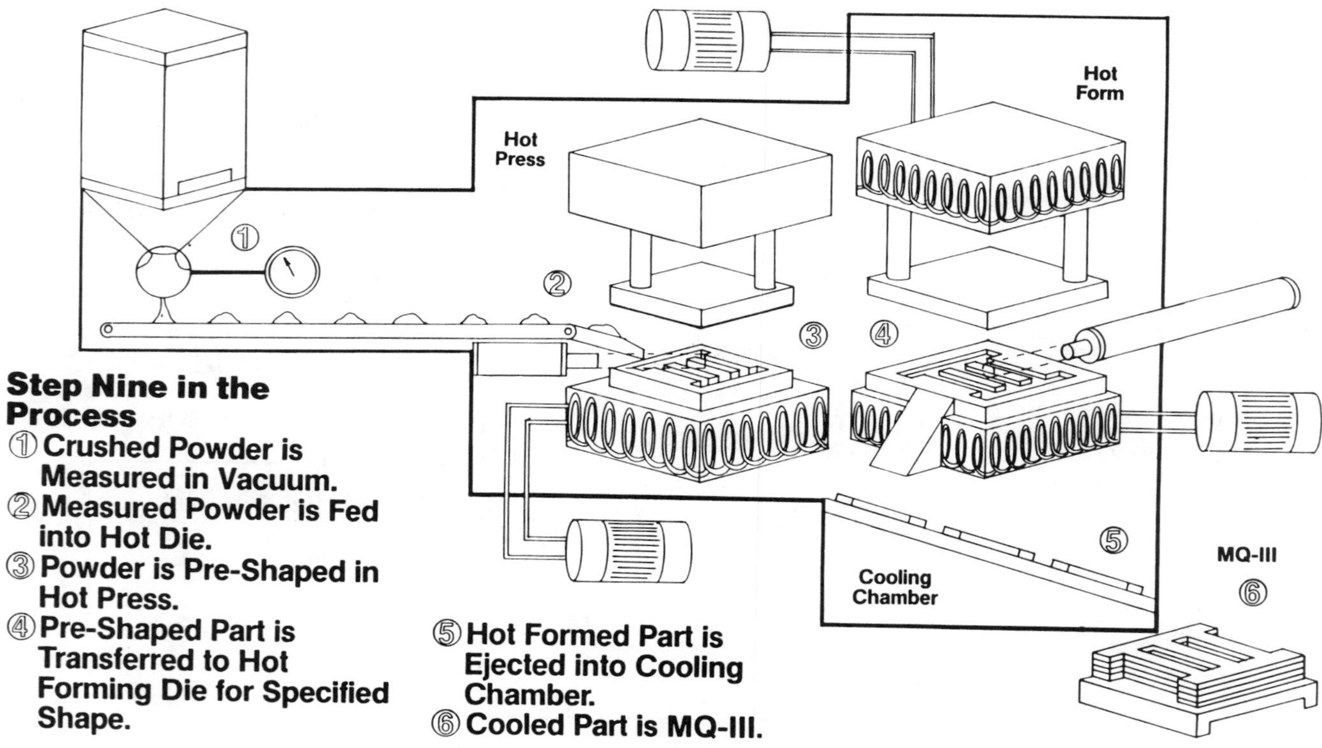

Figure 4.35 MQ III hot forming.

Table 4.5 Magnequench properties and characteristics

	MQ I	MQ II	MQ III	Units
Magnetic characteristics				
Maximum energy product $(BH)_{max}$	8.0	13.0	32	MGOe
Residual induction (B_r)	6.1	8.0	11.75	kG
Coercive force (H_C)	5.3	6.5	10.50	kOe
Intrinsic coercive force (H_{ci})	15.0	16.0	13.0	kOe
Magnetizing force (H_s)	35.0	35.0	25.0	kOe
Recoil permeability (μ_r)	1.15	1.15	1.05	G/Oe
Physical properties				
Density	0.217	0.271	0.271	lb/in^3
	6.0	7.5	7.5	g/cm^3
Curie temperature	312.0	312.0	312.0	°C
Maximum operating temperature	125.0	150.0	150.0	°C
Reversible temperature coefficient at a load line of 1	−0.192	−0.157	−0.157	%/°C
			−0.157	%/°C
Coefficient of thermal expansion	3.8	Nonlinear	Anisotropic	in/in × 10^{-6}°C
Electrical resistivity at 25°C	18000.0	160.0	160.0	μohm cm
Hardness	36–38	60	60	Rockwell-C

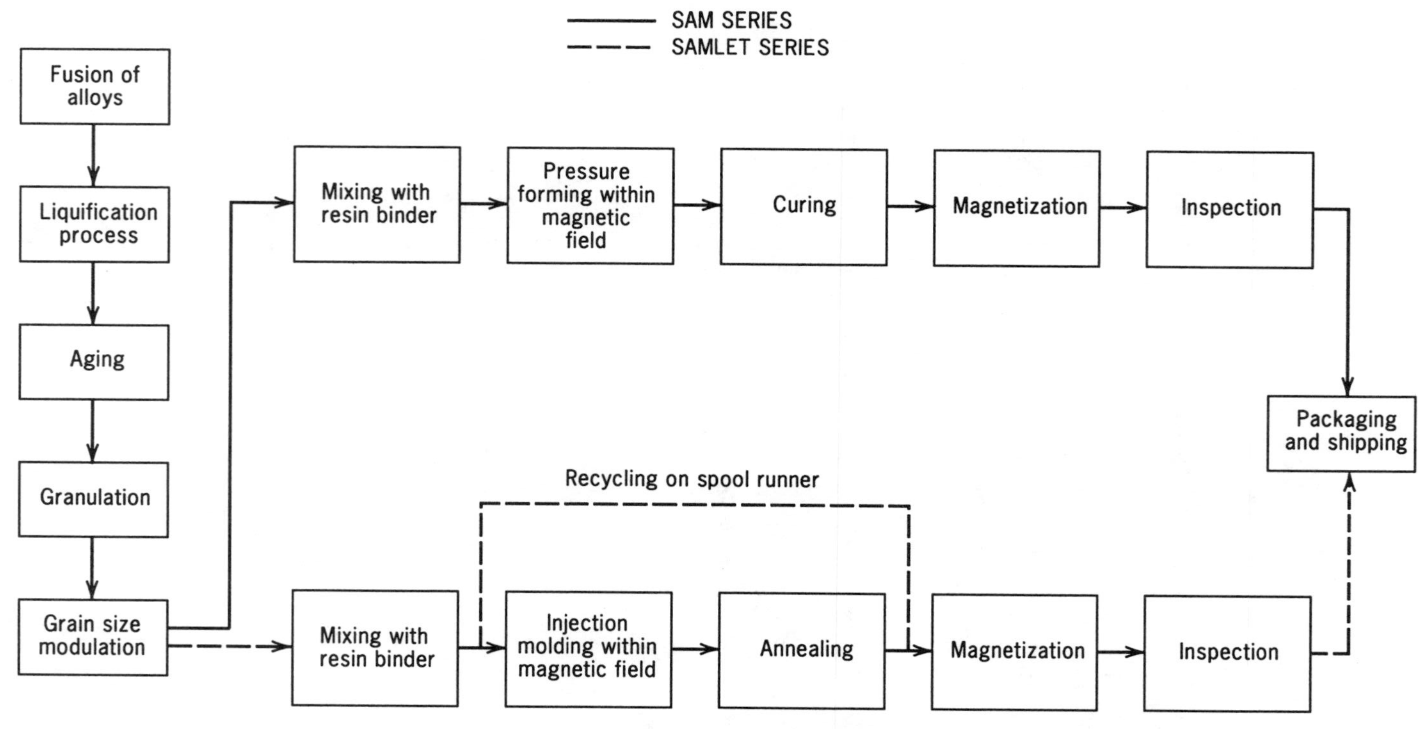

Figure 4.36 Production process for bonded $Sm_2(CoFe)_{17}$ magnets. From SUWA Seikosha Co., Ltd.

moves magnet making from the traditional batch processing approach toward the automated factory with major advantages such as improved quality, cost effectiveness, and versatility. This forming approach opens the way for making integrated permanent magnet circuits. For example, iron powder and magnet material may be pressed together in one die system to form magnet and return path circuit.

4.5 MATRIX OR BONDED MAGNETS

A major class of permanent magnets has evolved from ferrite and rare-earth powders imbedded in rubber, resins, and plastics. Depending on the magnetic material loading these products may either be flexible or of rigid form. These permanent magnets are offered in a wide range of properties, from 0.5 to near 20 MGOe. They may be formed by rolling, extrusion, compression, and injection molding. The magnet components can be formed in complex shapes with close mechanical tolerances. Bonded magnets, of course, have lower levels of B compared to their fully densified counterparts. Both isotropic and anisotropic products are offered. The ferrite bonded magnets have good high temperature stability, but stability is a problem in the rare-earth bonded magnets. The best stability is achieved with metal bonded magnets described by Strnat et al. [33]. Resin bonded magnets are limited to approximately 100°C at this time. Suwa-Seikosha in Japan, offers a wide range of properties and Figure 4.36 shows the process steps involved in making both compression and injection molded magnets.

Magnequench powders are also being used in injection molded magnet products with 6 MGOe isotropic properties attainable. Magnequench MQ III powders have recently been made into bonded magnets exhibiting properties up to 16 MGOe with H_{ci} levels of 17 kOe [34]. The coercive force levels were achieved with Ga additive (NdFeGaB). The stability of these magnets appears to be excellent up to 150°C.

4.6 SEMI-HARD MAGNETS (HYSTERESIS ALLOYS)

Several permanent magnet materials described in this chapter find wide use in hysteresis motors, brakes, and torque drives. The maximum torque which a hysteresis device can develop is a function of E_h, the loss per cycle or the area of the hysteresis loop of the rotating material. There are actually two fundamentally different hysteresis effects. We tend to be most familiar with alternating hysteresis in which B and H vectors are parallel in space and the magnitude of H is changed. The other type of hysteresis loss involves rotating a volume of magnetic material in a field of constant magnitude, but varying in direction. This is known as rotational hysteresis loss. Figure 4.37 illustrates domain wall movement in each hysteresis relationship. In some

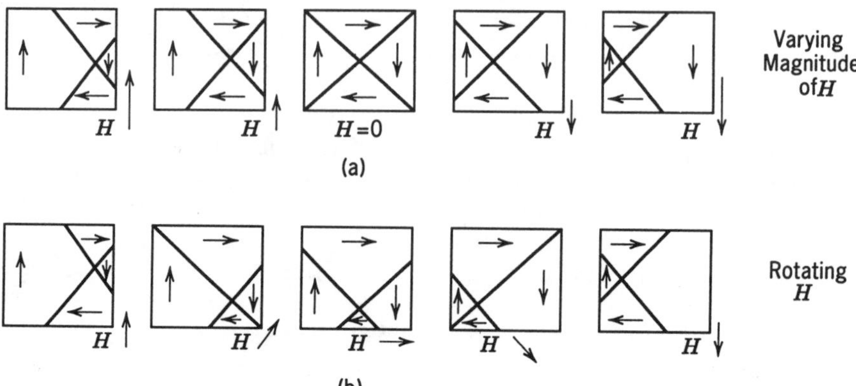

Figure 4.37 Domain wall motion in hysteresis devices.

hysteresis devices the loss and torque developed is largely due to rotational effects, while in others both effects are present. Figure 4.38 shows how each effect changes with material flux density B. The convenient way to measure rotational hysteresis loss is to draw torque curves using a torque magnetometer. A torque curve is drawn and then redrawn, turning the specimen in the opposite direction. The two curves will be displaced relative to each other and the area between them will be proportional to the rotational hysteresis. Owing to the difficulty in measuring and evaluating the rotational influence, hysteresis devices tend to be analyzed and property comparison made on the basis of alternating hysteresis.

A general figure of merit for a hysteresis device is the ratio E_h/H_p where H_p is the peak magnetizing force developed in the device. Figure 4.39

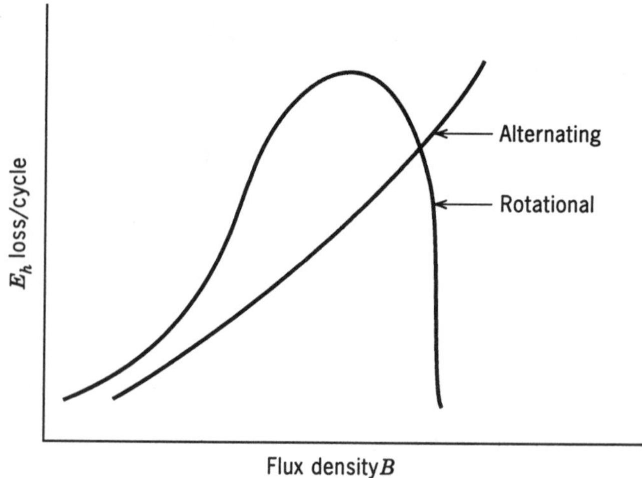

Figure 4.38 Hysteresis relationships.

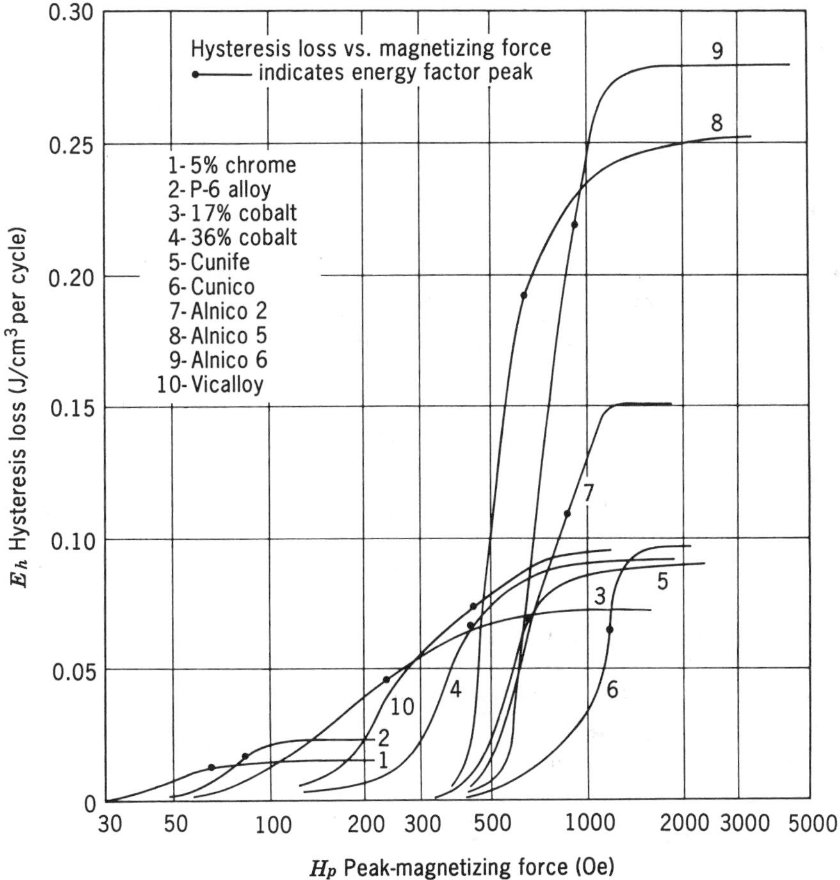

Figure 4.39 Alternating hysteresis loss vs. magnetizing force.

compares several materials by plotting E_h as a function of H_p. A dot on each curve indicates the most efficient operating point in terms of maximizing E_h/H_p. Due to the complex relationships involving hysteresis loss, field magnitude, and field direction selecting properties and optimizing, a hysteresis device tends to be a rather empirical process.

REFERENCES

[1] K. Honda, U.S. Patent 1,338,132–134, 1920.

[2] T. Mishima, U.S. Patent 2,027994-5-6-7-8-9.

[3] G. B. Jonas, U.S. Patent 2,295,082.

[4] J. Cahn, J. Appl. Phys. 34 (1963) 3581.

[5] M. McCaig, J. Appl. Phys. 35 (1964) 958.
[6] C. Marks, Transactions of the First European Conference on Magnetism, Paper 2.9, Vienna, 1965.
[7] G. Howe, U.S. Patent 2,192743-44.
[8] L. I. Mendelsohn, F. E. Luborsky and T. O. Paine, J. Appl. Phys. 26 (1955) 1274.
[9] L. I. Mendelsohn, F. E. Luborsky and T. O. Paine, J. Appl. Phys. 28 (1957) 344.
[10] H. Kaneko, M. Homma and K. Nakamura, AIP Conf. Proc. No. 5 (1971) 1088.
[11] S. Jim, N. V. Gayle and J. Bernadini, IEEE Trans. Magn. 16 (1980) 1050.
[12] J. J. Went, G. W. Rathenau, E. W. Gorter and G. W. Van Oosterhout, Philips Tech. Rev. 13 (1952) 194.
[13] K. Fries, First European Conference on Magnetism, Vienna, September, 1965.
[14] A. J. J. Koch, J. Appl. Phys. 315 (1960) 755.
[15] Kubo, U.S. Patent 3,976,519.
[16] T. Ohtani, S. Kato, S. Kojima, K. Kojima, Y. Sakamoto, I. Konno, M. Tsukahara and T. Kubo, IEEE Trans. Magn. 13 (1977) 1329.
[17] K. J. Strnat and G. Hoffer, USAF Materials Lab. Report AFML TR-65-446, 1966.
[18] K. H. J. Buschow, R. A. Naastepad and F. F. Westendorp, J. Appl. Phys. 40 (1969) 4029.
[19] D. K. Das, IEEE Trans. Magn. (1969) 214.
[20] D. L. Martin and M. G. Benz, Cobalt No. 50 (1971) 11.
[21] R. E. Cech, J. Metals 26 (1974).
[22] C. Herget and H. G. Domazer, Goldschmidt Inf. 35. (1975) 3.
[23] E. A. Nesbitt, R. H. Willens, R. C. Sherwood, E. Buehler and J. H. Wernick, Appl. Phys. Lett. 12 (1968) 361.
[24] Y. Tawara and H. Senno, Jpn. J. Appl. Phys. 7 (1968) 966.
[25] Y. Tawara and H. Senno, IEEE Trans. Magn. 8 (1972) 560.
[26] K. J. Strnat and A. R. Ray, Goldschmidt Inf. 35 (1975) 47.
[27] R. K. Mishra, G. Thomas, T. Yoneyama, A. Fukumo and T. Ojima, J. Appl. Phys. 52 (1981) 2517.
[28] M. Sagawa, S. Fujimura, N. Togawa, H. Yamamoto and Y. Matsuura, J. Appl. Phys. 55 (1984) 2083.
[29] M. Sagawa, S. Hirosawa, H. Yamamoto, Y. Matsuura, S. Fujimura, H. Tokuhara and K. Hiraga, IEEE Trans. Magn. 22 (1986) 910.
[30] M. Tokunaga, N. Meguro, M. Endoh and S. Manigawa, IEEE Trans. Magn. 21 (1985) 1964.
[31] J. Croat, J. F. Herbst, R. W. Lee and F. E. Pinkerton, J. Appl. Phys. 55 (1984) 2078.
[32] R. W. Lee, Appl. Phys. Lett. 46 (1985) 790.
[33] R. M. W. Strnat, S. Liu and K. J. Strnat, Proceedings of the 5th International Workshop on Rare-Earth Cobalt Permanent Magnets, 1987.
[34] T. Tadatoshi, Gorham Conference, 1987.

5

PERMANENT MAGNET STABILITY

5.1 Introduction
5.2 Classification of Magnetization Changes
 5.2.1 Reversible Changes
 5.2.2 Irreversible Changes
 5.2.3 Structural Changes
 5.2.4 Displaying Reversible and Irreversible Loss Data
5.3 Theoretical Considerations
5.4 Temperature Effects
 5.4.1 Effects of Temperature on Magnetization and Coercive Force
 5.4.2 Complete Demagnetization Curves at Various Temperatures
 5.4.3 Time Effects at Constant Temperature
 5.4.4 Examples of Thermal Cycling Behavior
 5.4.5 Modeling and Predicting Change Due to Temperature and Time
5.5 Magnetic Field Effects
5.6 Temperature Compensation
 5.6.1 Temperature Compensation in a Device
 5.6.2 Compensation by Compositional Change
5.7 Mechanical Energy Input and Stability
5.8 Corrosion and Surface Oxidation
5.9 Nuclear Radiation
5.10 Enhancing Stability
5.11 Stabilization Techniques
5.12 Conclusions and Comparison of Materials

5.1 INTRODUCTION

A compelling reason to use permanent magnets in many devices and systems is the magnet's ability to maintain a constant flux output over a very long period of time. Nearly all applications subject the magnet to influences that tend to alter the magnet flux. If one knows the magnitude and nature of these influences, it is possible to predict the flux change. It is also possible, by exposing the magnet to these influences in advance, to render the magnet insensitive to subsequent changes in service. For many years, permanent magnets have exhibited long-term stability in meters and instruments of the order of one part in 10^3. More recently, investigations in conjunction with inertial guidance systems for space vehicles have shown that long-term stability of one part in 10^5 or 10^6 can be achieved. To attempt to achieve this order of stability with regulated electromagnets would be prohibitively complex and costly if, indeed, possible at all.

Today's stability achievement with modern permanent magnets is in sharp contrast to very early permanent magnets in which both structural and magnetic changes caused a significant loss of magnetization with time.

5.2 CLASSIFICATION OF MAGNETIZATION CHANGES

Magnet stability is a sophisticated subject and experience has shown it advantageous to classify these changes as to their nature and cause.

5.2.1 Reversible Changes

The reversible changes in the magnetic properties of permanent magnet materials as a function of temperature originate in the change in spontaneous magnetization. These changes tend to obey the same temperature law as does saturation magnetization. Reversible changes are functions of temperature or circuit loading and are in no way time dependent. They disappear completely without need for remagnetization when the permanent magnet is returned to its initial temperature.

5.2.2 Irreversible changes

In this type of change, after the removal of the disturbing influence the magnetization does not return to its original value. Examples of such changes are:

1. ambient temperature changes;
2. statistical local temperature fluctuations leading to viscosity effects, also known as after-effects;
3. magnetic field-induced changes such as a change in magnetic circuit or exposure to an external field.

In this kind of change, the magnetization may be fully restored by remagnetization.

5.2.3 Structural changes

Changes resulting from a permanent change in the structural or metallurgical state are generally time-temperature dependent. Examples of such changes are:

1. growth of precipitate phase;
2. oxidation;
3. annealing effects;
4. increase in proportion of an ordered phase;
5. radiation damage.

Remagnetization does not restore the original state of magnetization after this kind of change. In the literature this type of change is often referred to as aging since there is time dependence. The temperature at which changes in properties first become noticeable corresponds to the beginning of structural changes and corresponds closely to the maximum service temperature.

The distinction between reversible and irreversible changes is shown in Figure 5.1. If A is the level of magnetization at a reference temperature, e.g. room temperature T_r, then AC is the irreversible loss after exposure to temperature T with measurement made after the specimen has been returned to T_r. Irreversible loss is usually expressed as a percentage of the

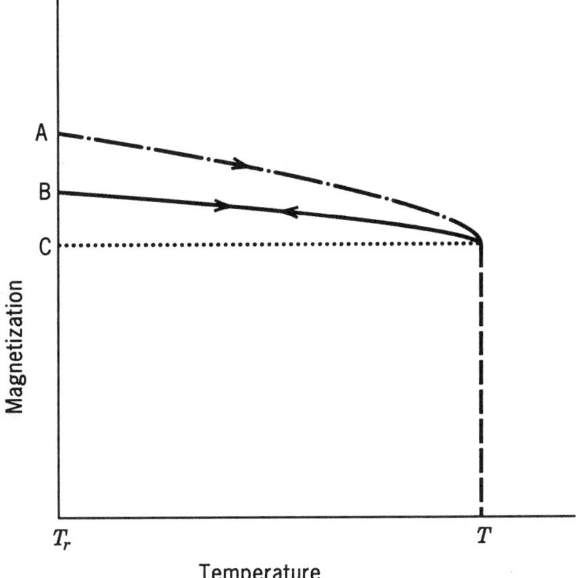

Figure 5.1 Irreversible and reversible changes.

reference temperature level. The change CB represents the reversible change and is usually expressed as an average coefficient over a limited temperature range.

5.2.4 Displaying Reversible and Irreversible Loss Data

In applying the permanent magnet one is interested in knowing the reduction of B_d the flux density at a specific point on the demagnetizing curve caused by a short-term temperature excursion from room temperature. This change in B_d is often expressed as a percentage of the initial room temperature value.

More usual than the numerical data in table form, are curves of B_d versus temperature for various B/H unit permanence values showing one or more full heating-cooling cycles. Figure 5.2 shows a display of loss data in such form after Tenzer [1].

In Figure 5.2 the induction B changes reversibly along a line representing the B-coordinates or intercepts of a unit permeance line with the demagnetizing curve at various temperatures. If one has a set of such demagnetization curves at a temperature, the B_d versus temperature curves can be constructed for various B/H values. B_d (25°C) − $B_d(t)$ is the reversible loss. This loss is fully recoverable by returning the magnet to room temperature.

In Figure 5.2b, a larger deviation from room temperature is shown. In this case, only a part of the B_d change will be recovered by cooling to room temperature $B'_d - B'_d(t)$. The other part, $B_d - B'_d$ is an irreversible loss and represents a permanent magnetic change, only recoverable by remagnetizing at room temperature.

In Figure 5.2c, we encounter structural changes caused by much higher temperatures for longer periods of time. For each material there will be a time temperature product at which this kind of change will be encountered. The structural change is $B_d - B''_d$. Structural change is a permanent change which cannot be recovered by remagnetization. In some instances it may be recovered by reprocessing the magnet.

On repeated cycling, small additional irreversible losses generally occur in all materials. If stable repeatable flux values are required in a high precision application then the magnet should be subjected to the highest anticipated temperature and cycled several times. A reverse field will also serve to stabilize the magnet. However, experience has shown that a field will not completely stabilize against change due to temperature excursions. For some materials, a field will remove most of the change.

5.3 THEORETICAL CONSIDERATIONS

One must remember that there is no energy required to maintain a magnetic field. Energy is only required to change a field. The magnetization in a

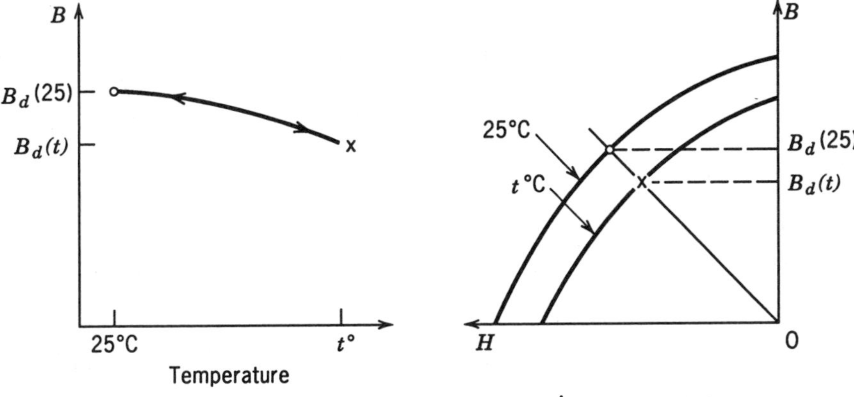

(a) Reversible effect: $B_d''(25) = B_d'(25) = B_d(25)$

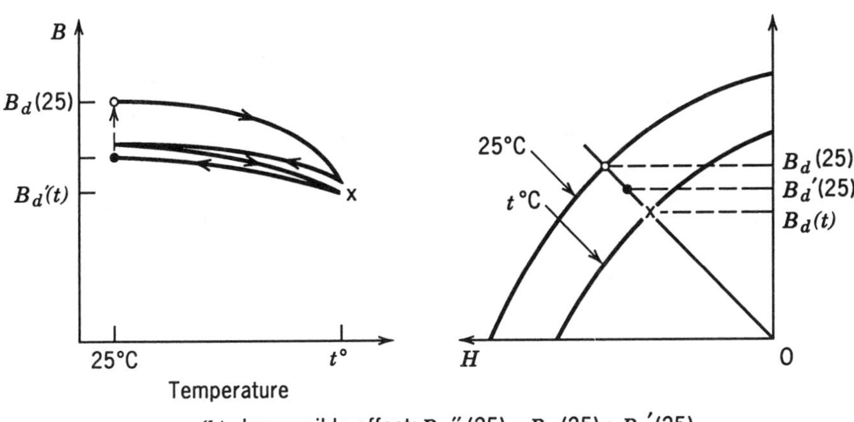

(b) Irreversible effect: $B_d''(25) = B_d(25) > B_d'(25)$

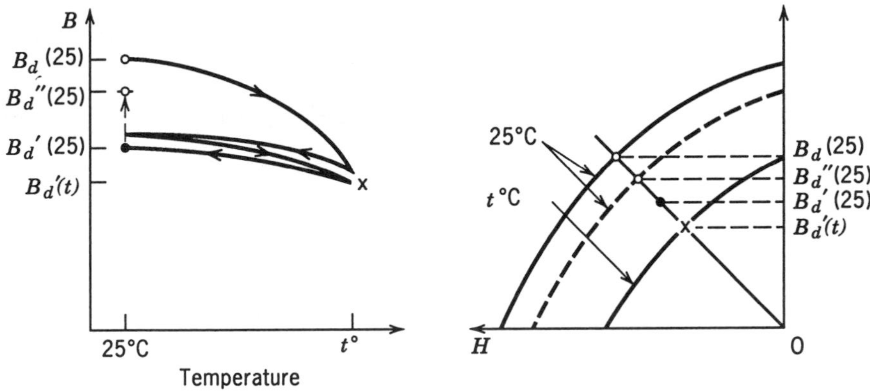

(c) Material effect: $B_d''(25) \neq B_d(25)$

Figure 5.2 Schematic diagrams showing temperature effects.

permanent magnet is held by a net internal field $(H_{ci} - H_d)$ (see Figure 5.3). The dynamic energy balance may be expressed as:

$$KT = \tfrac{1}{2}B_i(H_{ci} - H_d)V_d \tag{5.1}$$

where K is Boltzman's constant, T is temperature, B_i, the magnetization of the specimen and V_d, the volume of material in which a magnetization reversal may take place, such as domain. Irreversible loss is a result of the rearrangement of domain regions due to either a change of coercivity with temperature, thermal after-effects or a change in the magnetic loading causing a change in H_d.

In a typical device a multitude of adverse influences may be encountered acting together. For example, the magnet may be exposed to elevated temperature, self-demagnetization, external fields and mechanical shock. For ease in analysis and to allow some order, one influence is isolated and experimental data collected. Very little information is available on stability as a function of combined influences. In a demanding stability problem, the designer and material scientist must become knowledgeable about the subject. They very often will find themselves in the position of adding to the body of information on stability.

5.4 TEMPERATURE EFFECTS

5.4.1 Effects of Temperature on Magnetization and Coercive Force

Temperature changes bring about changes in a magnet's properties. These changes are often complex and difficult to control. The different property systems give very different responses to temperature change. Thermal hysteresis effects as a result of temperature cycling also have to be controlled.

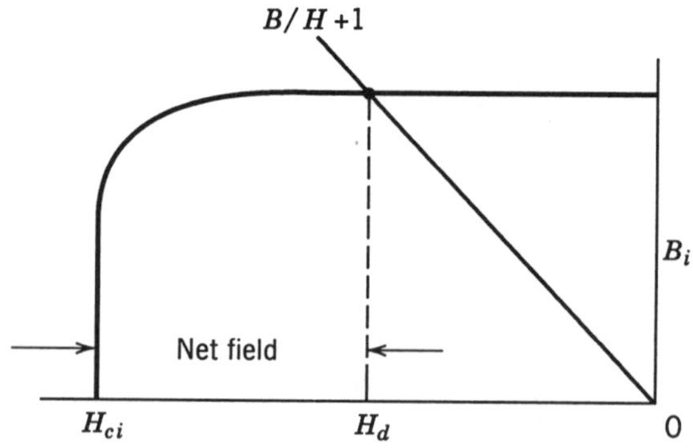

Figure 5.3 Permanent magnet internal field balance.

Table 5.1 Magnetization losses on heating above room temperature (20°C)

Material	Dimensional Ratio (l/d)	Temperature (°C) 100 I[a]	100 II[b]	200 I	200 II	300 I	300 II	400 I	400 II	500 I	500 II
Alnico 2	8.00	2.0	98	3.1	94	4.2	90	6.1	86	8.2	80
	3.62	3.1	98	4.0	92	6.9	88	8.6	84	12.0	78
	2.00	3.5	97	4.7	91	7.4	89	10.7	85	13.1	81
Alnico 5	8.00	0.1	99.9	0.2	96	0.4	93.6	0.7	91.2	1.2	88.0
	4.68	0.4	99.6	0.8	96.3	1.1	93.8	1.7	91.1	2.0	88.2
	2.00	0.5	99.4	1.7	96.6	2.1	94.1	2.6	92.2	3.0	88.6
Alnico 6	20.00	0.1	98.2	0.2	95.6	0.4	93.0	0.8	89.7	1.8	86.5
	4.12	0.5	98.7	0.9	95.6	1.2	92.7	2.0	89.4	3.0	85.2
	2.00	0.7	99.1	1.2	97.2	1.5	94.2	2.1	90.5	3.3	86.0
Alnico 8	5.62	0.7	98.8								
	2.85	0.7	99.0								
	1.91	0.9	99.4								
	1.01	1.0	99.8								
Barium ferrite (all grades)	All	0	85	0	68	0	50				
Platinum cobalt	31.67	0	97.9								
	16.02	0	98.5								
	10.63	0	98.8								
	15.56	0	97.9								

[a] Percent irreversible remanence loss at room temperature after heating to indicated temperature.
[b] Percent of initial room temperature remanence found stable at indicated temperature.

Table 5.2 Magnetization changes on cooling below room temperature (20°C)

Material	Dimensional Ratio (l/d)	% Irreversible Loss at Room Temp after Exposure to −190°C	−60°C	Reversible Temperature Coefficient (% remanence change per °C)
Alnico 2	5.29	0	0	−0.025
	3.69	0	0	−0.021
	2.66	0	0	−0.018
	1.77	0	0	−0.009
	0.94	0	0	−0.014
Alnico 5	8.00	0	0	−0.022
	5.36	4.6	1.4	−0.012
	3.63	9.0	2.5	−0.002
	2.72	6.2	3.6	+0.010
	1.84	7.9	2.1	+0.016
	0.94	8.5	3.4	+0.007
Alnico 6	8.00	0	0	−0.045
	6.03	1.8	0.4	−0.020
	3.57	8.5	1.3	−0.007
	2.70	10.1	4.1	+0.007

	1.78	10.5	4.2	+0.022
	0.89	7.9	3.1	+0.046
Alnico 8	5.62	0	0	−0.013
	2.85	0.5	0.1	+0.003
	1.91	0.7	0.3	+0.015
	1.01	1.3	0.5	+0.033
Barium ferrite	0.50	4.0	0[a]	−0.19
(isotropic)	0.28	–	1.3	−0.19
	0.09	–	2.4	−0.19
Barium ferrite	1.20	–	0[a]	−0.19
($(BH)_{max}$ anisotropic)	0.50	–	–	−0.19
Barium ferrite	0.40	–	0[a]	−0.19
((H_c) max anisotropic)				
Platinum cobalt	31.67	0	0	−0.015
	16.02	0.3	0	−0.015
	10.63	0.2	0	−0.015
	5.56	0.2	0	−0.015

[a] In the case of the low temperature irreversible loss occurring in oriented barium ferrite, only the smallest dimension ratio resulting in no irreversible loss at $-60°C$ is shown.

Source: *R. Parker and R. Studders, Permanent Magnets and Their Application, (John Wiley, New York, 1959), pp. 345–348.

The relationship between intrinsic magnetization and temperature is predictable, in some cases the shape of the B_i versus T curves may be calculated, provided there are details of the crystalline and magnetic structure of the phases present. In most cases such information is not available and instead, direct measurements of B_i versus T are made. Figure 5.4 shows the temperature magnetization relationship for several permanent magnet materials. The change of B_i with T depends on the extensive properties of the system, that is, particle size, internal field, and domain structure. These changes with temperature generally must be determined experimentally for specific properties and geometries. Magnetization changes as a function of temperature for several early permanent magnets are shown in Tables 5.1 and 5.2. For each material several values of length/diameter ratios are given so that the influence of B/H on the internal field can be observed.

The change of coercive force with temperature is a very important characteristic which is not well understood. The change of H_{ci} with temperature may be predicted in some materials if detailed knowledge of the anisotropy temperature relationship is known. However, for most materials several anisotropies contribute to H_{ci} and the resulting complexity leads to the need for direct measurements of H_{ci} versus T.

A summary and comparison of the temperature coefficients of B_r and H_{ci} for the major property systems is shown in Table 5.3. The Curie temperature and maximum service or operating temperature are also indicated.

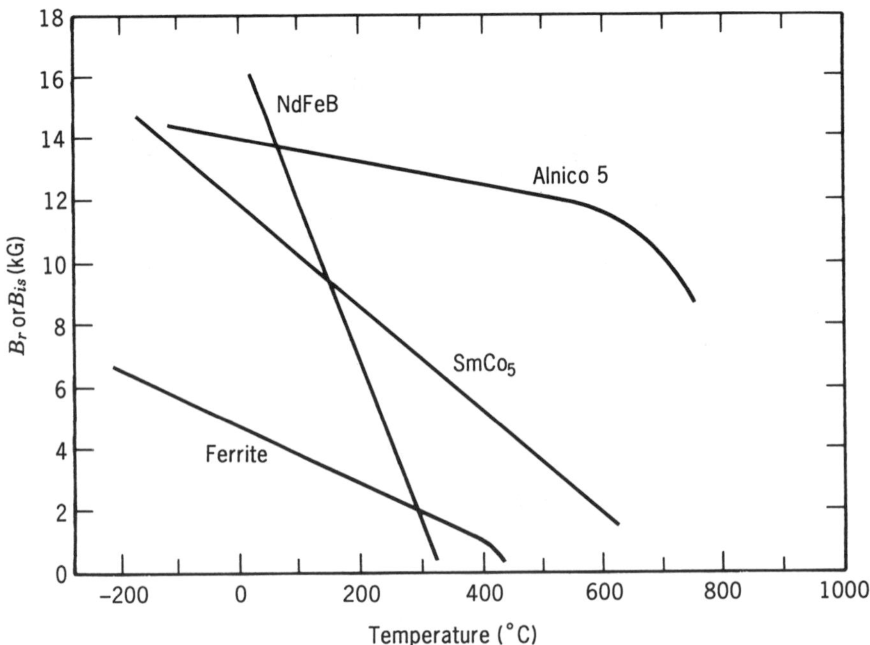

Figure 5.4 Temperature dependence of saturation magnetization.

Table 5.3 Temperature characteristics comparison

Magnet Material	Curie Temperature (°C)	B_r Reversible Coefficient (%/°C)	H_{ci} Reversible Coefficient (%/°C)	Max. Service Temperature (°C)
Alnico 5	720	−0.02	−0.03	520
Ferrite	450	−0.20	+0.40	400
$SmCo_5$	725	−0.04	−0.30	250
$Sm(Co, Cu, Fe, Zy)_{7.5}$	825	−0.035	−0.30	300
NdFeB	310	−0.12	−0.60	150

Table 5.4 compares the stability achievement of the major property systems in terms of irreversible loss at both low and high temperatures. Note that the comparison is for B/H near $(BH)_{max}$, for each material.

5.4.2 Complete Demagnetization Curves at Various Temperatures

The temperature coefficients of magnetization and coercive force in many instances do not give enough information about how a magnet will respond to temperature change. In many magnets the demagnetization curve is not well defined by B_r and H_{ci}. Changes in the curve shape and the intersection with load lines can only be seen from a complete set of demagnetization curves measured at several different temperatures over the temperature range of interest. Figure 5.5 shows families of demagnetization curves at various temperatures for three different permanent magnet materials. These curves show a wide range of response to temperature change. For type I magnets where $H_{ci} \ll B_r$ and a knee is exhibited in the second quadrant for all temperatures, the normal demagnetizaiton curves are all that the designer needs. For type II magnets with $H_{ci} \geq B_r$ both intrinsic and normal curves are required. In Figure 5.5 note that the coefficients for B_{is} and H_{ci} are negative for the materials shown except that for ferrite the coefficient of

Table 5.4 Comparison of irreversible loss[a]

Material	−60°C	+100°C	+200°C	+300°C	+400°C	+500°C
Alnico 5	−1.4%	0.4	0.8	1.1	1.7	−2.1%
Ferrite (C-5)	−27.1%	0	0	0		
$SmCo_5$	0	0.5	1.7			
NdFeB	0	6.5	65.0			

[a] This table indicates loss of remenance measured at 20°C after exposure to the indicated temperature. B/H is near $(BH)_{max}$ for each material

Source: from MMPA Guide Lines 1988.

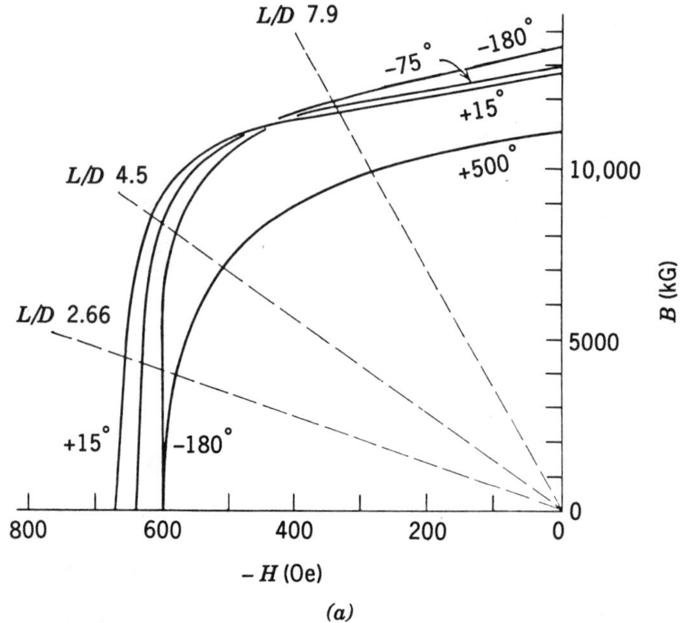

(a)

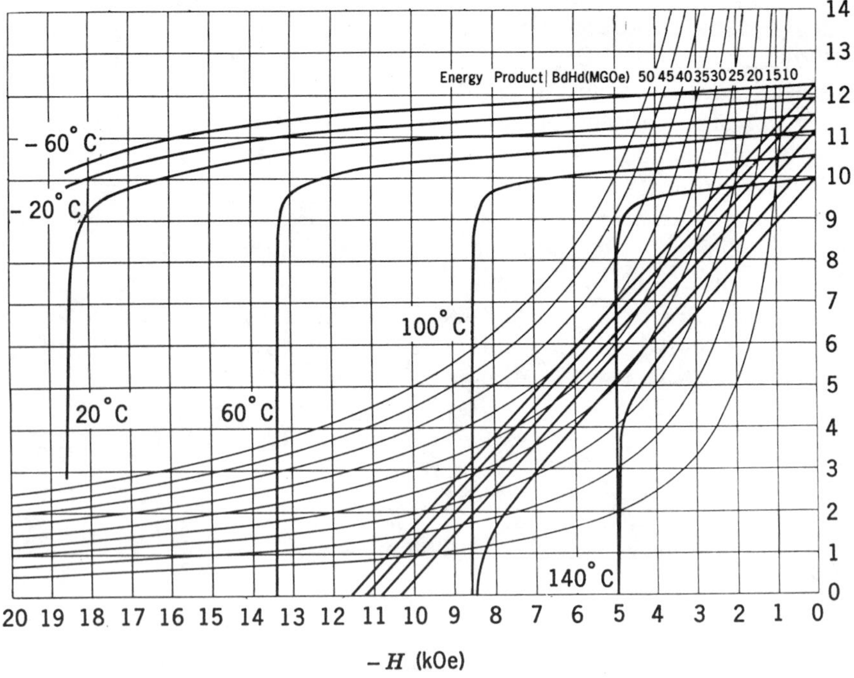

(b)

Figure 5.5 Demagnetization at different temperatures for three different magnets: (a) Alnico 5; (b) NdFeB; (c) Ceramic 5.

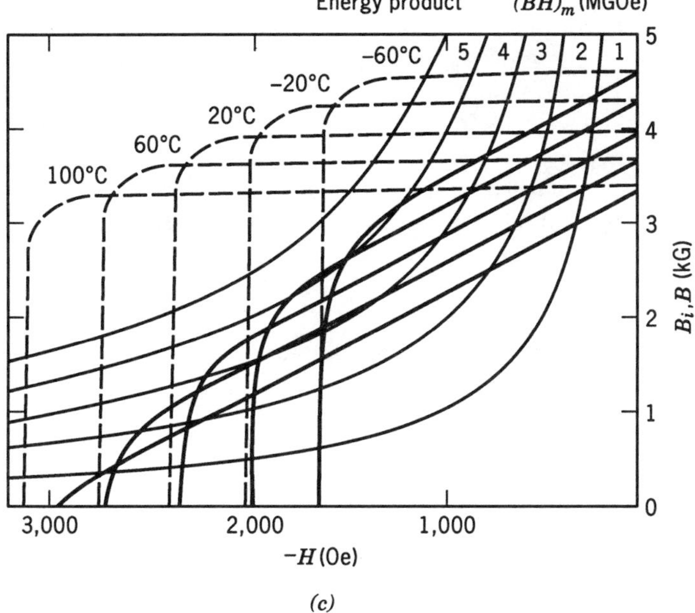

(c)

Figure 5.5. (*Continued*)

H_{ci} is positive. A complete set of demagnetization curves as a function of temperature are given in Appendix 3 for the most widely used magnets. This display has load lines indicated to aid in graphical design analysis.

5.4.3 Time Effects at Constant Temperature

The time adjustment of magnets at constant temperature is generally referred to in the literature as magnetic viscosity. The terminology must be used in the broad context that viscosity describes a material property which yields steadily before a constant stress. The domain regions of magnetization are in a static self-imposed demagnetizing field. Additionally, they are in a field which fluctuates in time. The time dependence of this superimposed field was suggested by Neel [2]. This field is dependent on the logarithm of time. At a domain site the fluctuating field could arise from such external ambient conditions as stray magnetic fields and temperature changes, and from internal temperature fluctuations. The temperature fluctuations are seen as field changes because of the intrinsic magnetization temperature dependence. A freshly magnetized magnet has a certain number of domains whose magnetization vectors are in positions that might be called "metastable." An energy input can initiate a transition to a more stable position. These events would involve a discontinuous orientation change. The adjustment of magnetization is essentially an activation process. Because this adjustment is thermally activated, it is accelerated at increased temperatures, so that quick stabilization can be achieved by raising a magnet's temperature for a short time. Since the changes result from magnetization reversals, they can be anticipated and stabilization effected by subjecting the magnet to alternating or steady state fields sufficient to demagnetize to the extent of the loss which would occur during the time period of interest. Street and Wooley [3] have shown that the time temperature dependence is given by

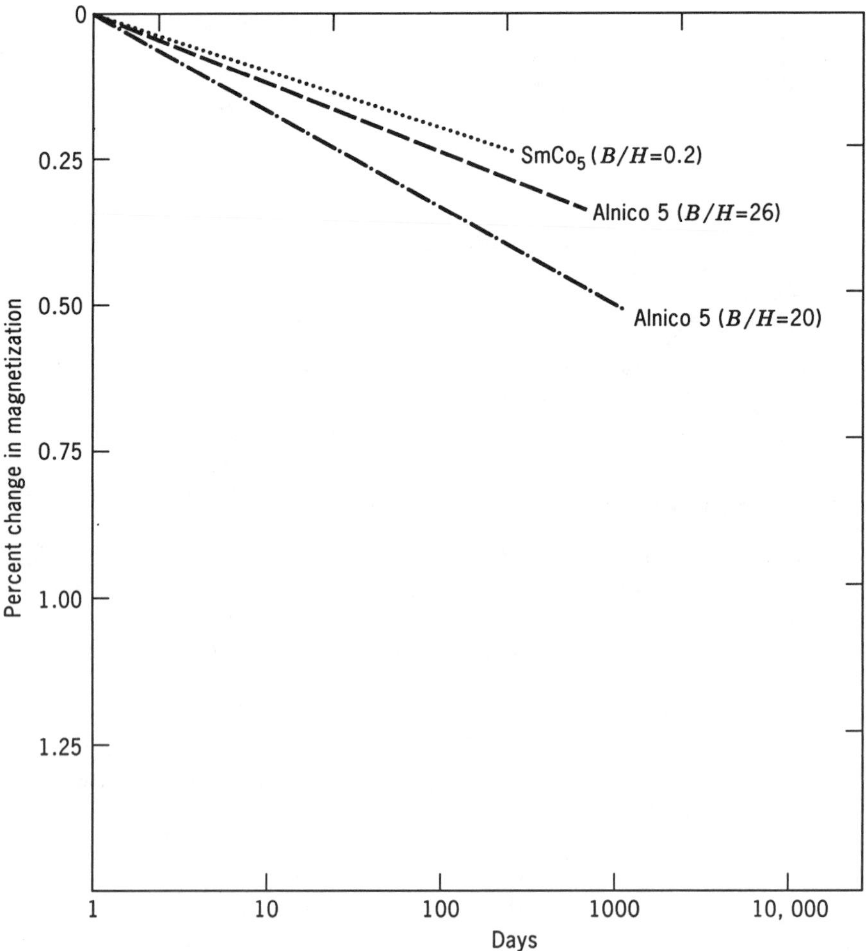

Figure 5.6 Comparison of Alnico 5 and $SmCo_5$ with respect to time adjustment of magnetization.

$$B_i(t) = \text{constant} + \lambda N B_{is} kT \log t$$
$$= \text{constant} + S \log t \qquad (5.2)$$

where $S = \lambda N B_{is} KT$, N is the number of domain regions of magnetization, B_{is} is the saturation magnetization per unit volume, λ is the constant probability energy density, and K is Boltzmann's constant. T is the constant absolute temperature. Since these factors are relatively independent of temperature, except near T_c, S is nearly directly proportional to T. However, N and λ will depend on the level of self-demagnetization of the sample. The results of experiments are in general agreement with (5.2) as shown in Figure 5.6.

5.4.4 Examples of Thermal Cycling Behavior

In using permanent magnets, it is necessary to be able to predict how a particular type will change its magnetization when cycled over a wide temperature range. Different property systems react to cycling in totally different ways. Three examples are shown here to make the point that one must be careful not to allow the experiences with one property system to be translated to all property systems.

Figure 5.7 shows some experimental results of the change in B_d on temperature cycling of Alnico 5. For a long bar (a) with B/H above $(BH)_m$ the change is reversible. In (b), for a short bar with B/H below $(BH)_m$, the first cooling cycle results in a substantial loss. After the initial low temperature exposure, the changes are reversible, but at a level below initial B_d. In (c), the results of an even shorter bar are shown. The interesting result here is that a reversible temperature coefficient, very much lower than normal is obtained. These data suggest that by proper choice of dimension, a coefficient near zero could be achieved over a limited temperature range. For long bars with high B/H values, the flux density increases steadily as the magnet is cooled resulting in a reversible change. This increase in B_d near B_r is typical of all types of magnets. For a short bar, the B/H intercept indicates lower B_d at lower temperature due to the decrease in H_{ci}. This decrease in H_{ci} is associated with a change in anisotropy and microstructure. The loss in B_d indicates reversal of some regions of the bar magnet. These reversals cannot be restored without remagnetizing. This is the reason for the irreversible loss on the first low temperature exposure.

Figure 5.8 shows a family of demagnetization curves for a ferrite magnet material at various temperatures. As an aid in understanding the reversible and irreversible changes, consider the change in curve shape and the two load lines shown in Figure 5.9. Suppose a magnet is designed to operate at point a (load line P_1) at +20°C and the temperature drops to −40°C, then the operating point will shift to point b and an irreversible loss is encountered. The load line remains the same while the demagnetization curves

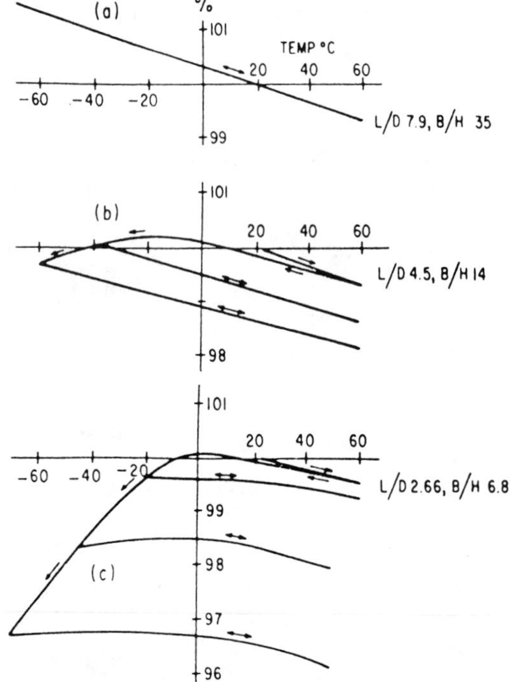

Figure 5.7 Temperature variation of Alnico 5 bars of varying *l/d* dimensional ratio. From A. Clegg, Br. J. Appl. Phys. 6 (1955) 120.

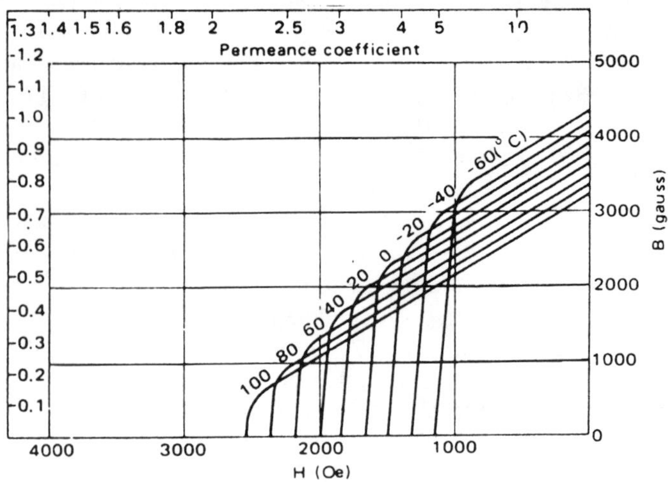

Figure 5.8 Properties of ferrite magnet.

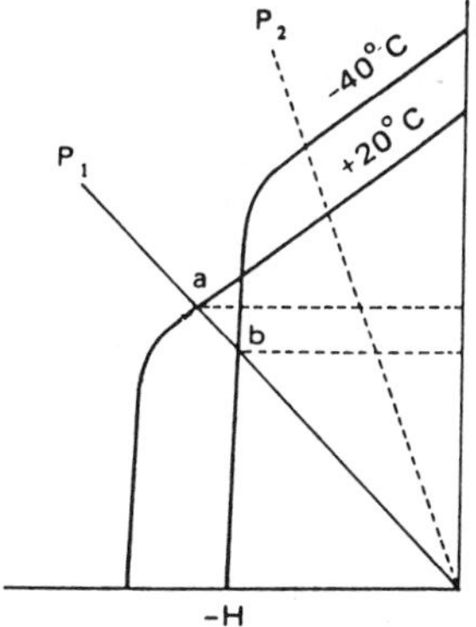

Figure 5.9 Irreversible change for ferrite magnet.

move with the temperature change. To avoid this irreversible loss, the designer could locate the operating point along permeance coefficient P_2. Figure 5.10 shows the variations of B_d versus t for the various combinations of loading and temperature. For high values of B/H the loss is reversible. For B/H values near the knee of the curve, heating will lower the flux but the operation is still on the reversible part of the main hysteresis loop. However, cooling will lead to some irreversible loss. At low B/H levels much higher irreversible loss is encountered (bottom part of Figure 5.10). It should be noted that recooling and further heating events lead to a reversible process as long as the changes follow recoil lines.

In Figure 5.11, NdFeB properties are shown as a function of temperature. We see that for a long magnet (high B/H) the loss is very reversible over the temperature range. For low B/H values there is a very large irreversible loss which is quite limiting in this class of magnets.

5.4.5 Modeling and Predicting Change Due to Temperature and Time

In Section 5.2.2 it was shown that irreversible loss is due to either a loss of coercivity with temperature, or to the viscosity or thermal after-effect. Adler and Marik [4] have shown a simple procedure for predicting these losses based on the idea that the demagnetization curve can be viewed as a super position of the single coercivities of all of the grains in the magnet, and the assumption that all these coercivities depend on temperature in the same way.

The change in magnetization due to a decrease in H_c may be written as $\Delta M_{H_c} = M \cdot \delta_{H_1}$ where δ_{H_1} is the relative demagnetization of a magnet after opposing field of magnitude H_1 has been applied. At room temperature, the field can be determined by the following considerations.

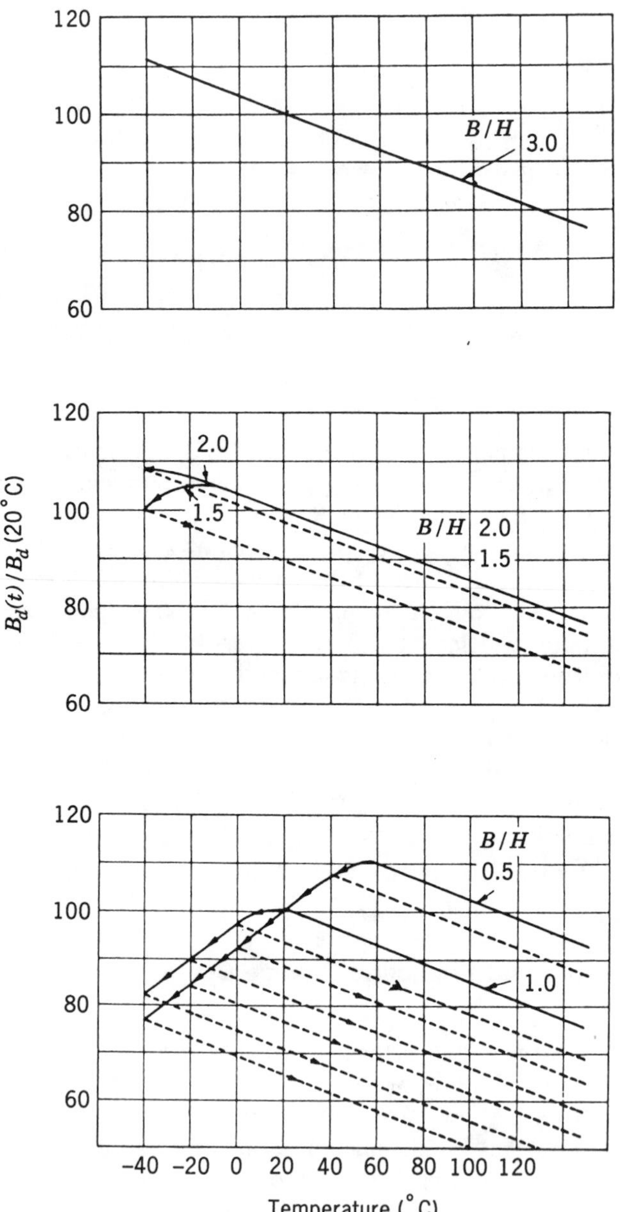

Figure 5.10 Low temperature characteristics of a ferrite magnet. From TDK Electronics Co. DME 794-001B, 1979.

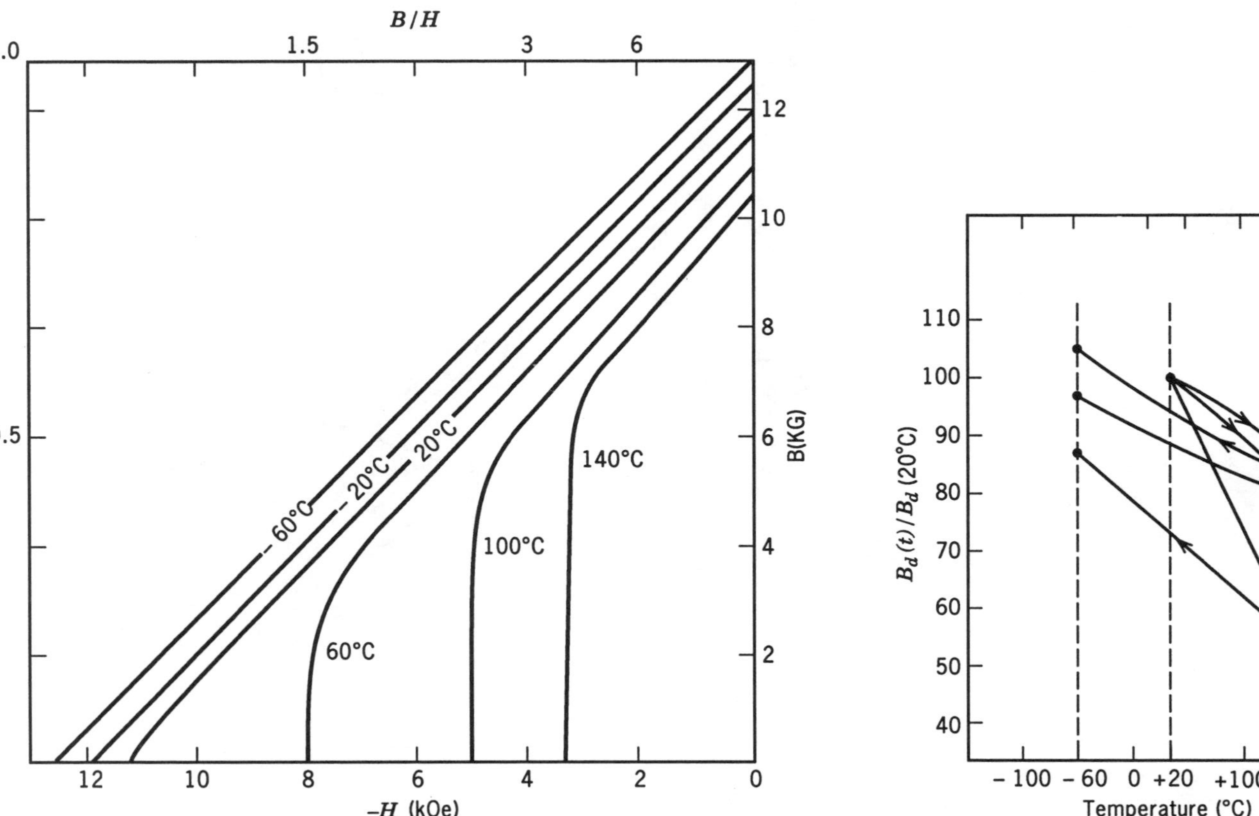

Figure 5.11 Temperature cycling behavior of NdFeB magnet.

In Figure 5.12, a demagnetization curve is shown at room temperature T_0 and at a higher temperature T. After cooling back to T_0 the working point will be at c on the virtual curve $m(-H)_T$. This curve is curve $M(H)_T$ renormalized to the remanence at T_0. One can simulate the irreversible loss at the higher temperature by application of an opposing field H_1 at room temperature. The main idea is that $M(H)_{T_0}$ and $m(H)_T$ have the same dependence on H or

$$m(H)_T = M(\alpha \cdot H)_{T_0} \tag{5.3}$$

with

$$\alpha = \frac{H_c(T_0)}{H_c(T)}$$

since

$$H = -N(M) \quad \text{and} \quad H_1 = -(\alpha - 1)N(M) \tag{5.4}$$

If we know the temperature dependence α of H_c of a given material, we can determine the adverse field to apply H_1. Figure 5.13 illustrates the temperature dependence for $SmCo_5$.

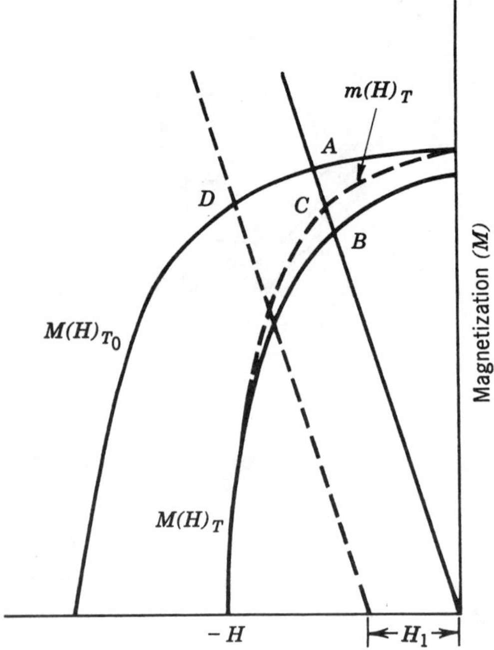

Figure 5.12 Demagnetization curves at T_0 and T where $T > T_0$.

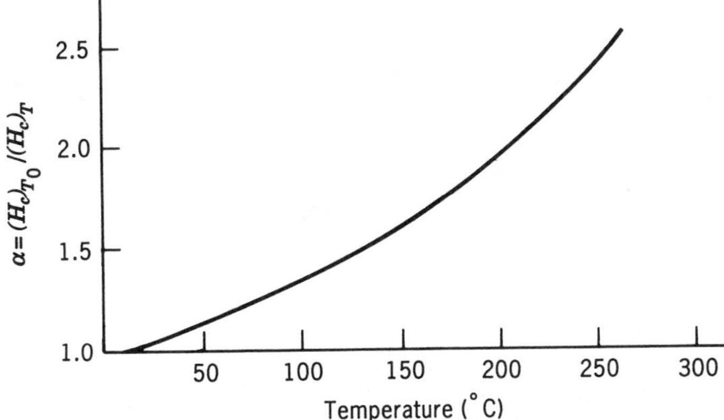

Figure 5.13 Temperature dependence of coercivity for $SmCo_5$.

For modeling and predicting the magnetization loss due to thermal after-effect (ΔM_{th}), we can think in terms of an effective field H_{th} (From Section 5.4.3)

$$H_{th} = S(\ln t) \tag{5.5}$$

The thermal after-effect can be simulated by applying an opposing field at room temperature or

$$M_{th} = M\delta_{H2} \quad \text{or} \quad H_2 = -(\alpha - 1)S \ln \frac{t_a}{t_o} \tag{5.6}$$

where t_a is the estimated time the magnet is to remain in use and t_o is the time the opposing field is applied.

5.5 MAGNETIC FIELD EFFECTS

The level of flux supplied by a permanent magnet can be irreversibly changed by exposing the magnet to a field. The effect of external field while acting on the magnet and the final effects after its removal can be predicted. Intrinsic and normal demagnetization curves are shown in Figure 5.14a. Upon exposure to an external field, ΔH, the operating point moves down the curve and the change in intrinsic properties can be projected to the normal curve which shows a change ΔB occurring while the field is applied. The recovery along an interior loop when the field is removed is shown in Figure 5.14b. ΔB is, of course, smaller than in Figure 5.14a. Subsequent

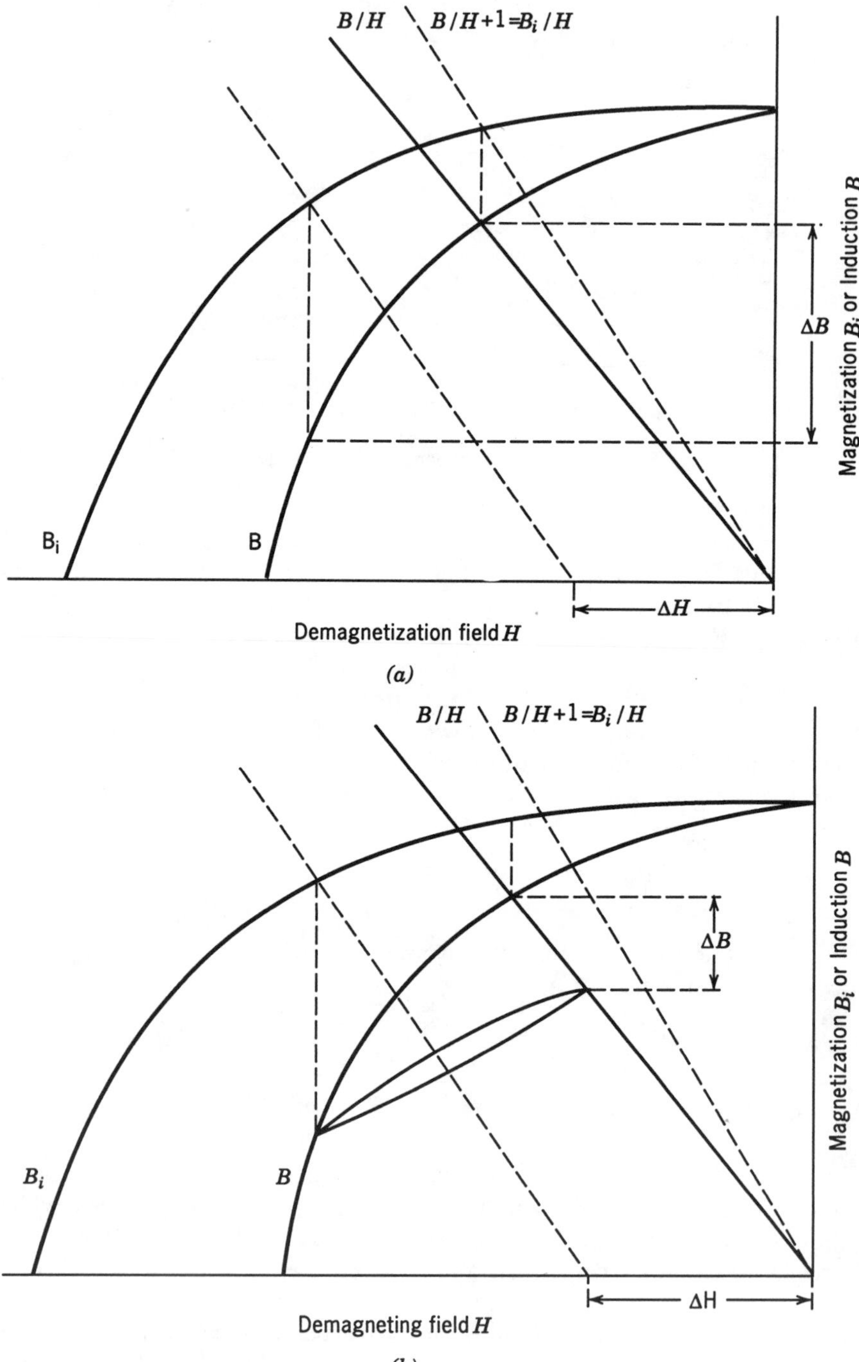

Figure 5.14 Magnetic field effects. (a) Effects of a demagnetization field; (b) recoil of magnetization.

exposure to fields of lesser strength give small changes within the previously established minor loop, and upon removal, no change in magnetization results. A given reverse field ΔH will cause different changes in ΔB depending on the permanence line. The loss will be greater when a magnet is operating in the steepest part of the demagnetization curve. Instead of exposing the permanent magnet to a field ΔH, the same effective stabilization will result if the air gap is increased and decreased so that the same minor hysteresis loop is established. For example, in the removal of a magnet from a magnetizing yoke and its subsequent transfer into an instrument gap and return path, stabilization is achieved.

5.6 TEMPERATURE COMPENSATION

5.6.1 Temperature Compensation in a Device

Performance variation with temperature cannot be tolerated in many magnet applications. If it is impractical to maintain the device at a constant temperature the designer must resort to some type of compensation. Variations with temperature will be due to the magnet and the electrical properties of other materials. It is the practice to compensate the magnet to accommodate the total performance of the device, which generally means overcompensating the reversible change of the magnet. There are two methods available for compensation. The first, and most popular method, requires a temperature-sensitive shunt to be located adjacent and parallel to the permanent magnet. The second method employs a bimetal strip to vary the air-gap volume but is generally avoided because of serious regulation problems.

The parallel shunt system is quite simple in theory. At low temperatures the shunt permeability increases, enabling the shunt to carry or divert more flux from the air gap. At higher temperatures the reverse condition exists; the shunt permeability decreases and less air-gap flux is diverted. Special temperature-sensitive magnet alloys are needed for this purpose. The most widely used materials are nickel-iron alloys, sometimes referred to as Curie alloys. Permeability versus temperature curves for a commercial compensating alloy are shown in Figure 5.15. This curve represents permeability at $H = 46$ Oe, approaching saturation, and is used in compensating permanent magnets. Compensation is usually attempted only over a limited temperature. A typical range is $-60°C$ to $+150°C$.

The permeability versus temperature curves are used to calculate and dimension compensator shunts. The total flux carried by the shunt at a given temperature, T, is given by

$$\phi = 46\ \mu a \tag{5.7}$$

where ϕ is shunt flux at temperature T, μ is material permeability at

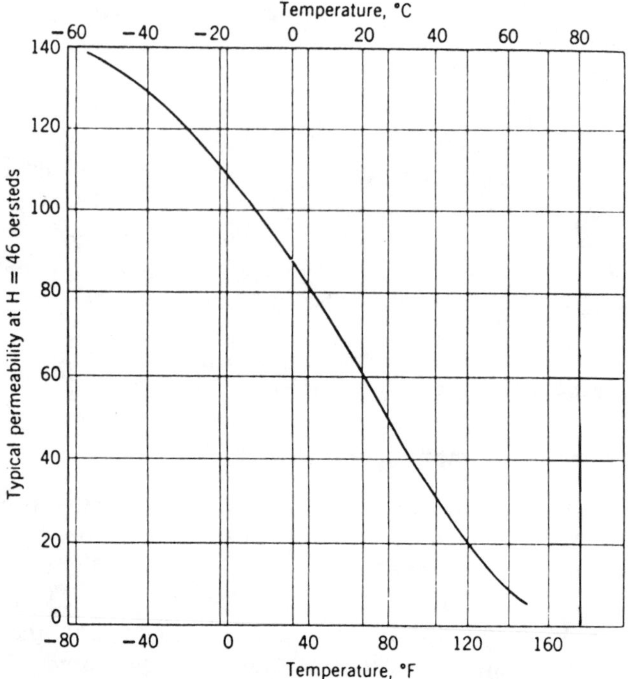

Figure 5.15 Typical temperature vs. permeability curves for type 2 iron-nickel alloy (Carpenter compensator alloy 30).

temperature T from curve, a is the shunt cross-sectional area in cm^2, and 46 is the constant field strength H for permeability curves.

In the event that the flux change to be compensated is known the shunt cross-section may be calculated for a reasonable fit. If the degree of total compensation required in the device is unknown, a trial with a shunt of known dimension will establish the parameters involved. The automotive eddy current speedometer shown in Figure 5.16 is a good example of device temperature compensation. Figure 5.17 shows the benefit of the compensated system over the uncompensated system.

5.6.2 Compensation by Compositional Change

For many devices it is not feasible to compensate by shunting the flux. In such cases, compensation at the atomic level has led to permanent magnets with near zero reversible temperature coefficients. With $SmCo_5$ and other light rare-earth compounds, the magnetic moment has a strong temperature dependence in the useful temperature range. There are rare-earth compounds containing Gd, Tb, Dy, Ho, and Er which exhibit a different temperature dependence because of a different mode of coupling between

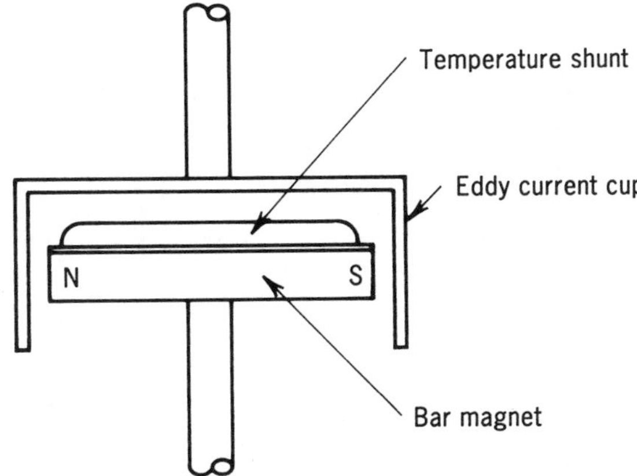

Figure 5.16 Elements of temperature compensated eddy current speedometer.

magnetic atoms. In Figure 5.18, the variation of magnetization with temperature is shown for $SmCo_5$, $ErCo_5$, and $GdCo_5$.

For the temperature range of interest (−40°C to +150°C) we see that the light rare-earth compound has a negative temperature slope while the heavy rare-earth compounds have a positive slope. By combining both light and heavy rare-earth compounds, one can produce a near zero reversible temperature coefficient over a limited temperature range. However, the overall moment and resulting properties tend to be disappointing. For example, $Sm_{0.7}Gd_{0.3}Co_5$ results in only a 10 MGOe magnet. Of particular interest is work to investigate temperature compensation in the

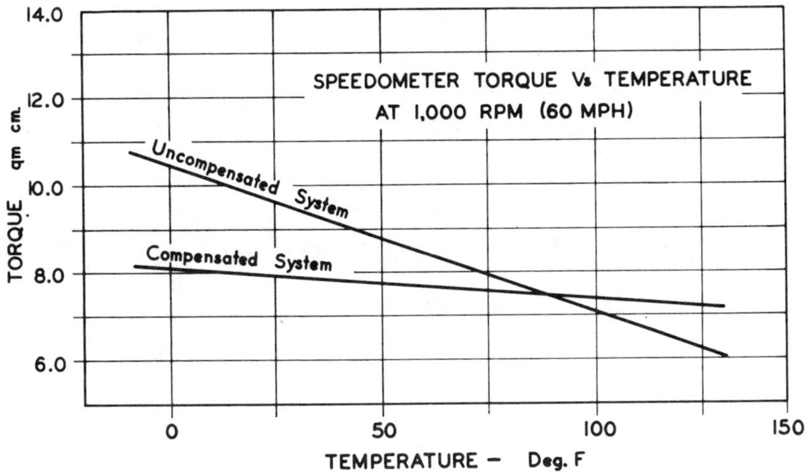

Figure 5.17 Speedometer performance with and without temperature compensation.

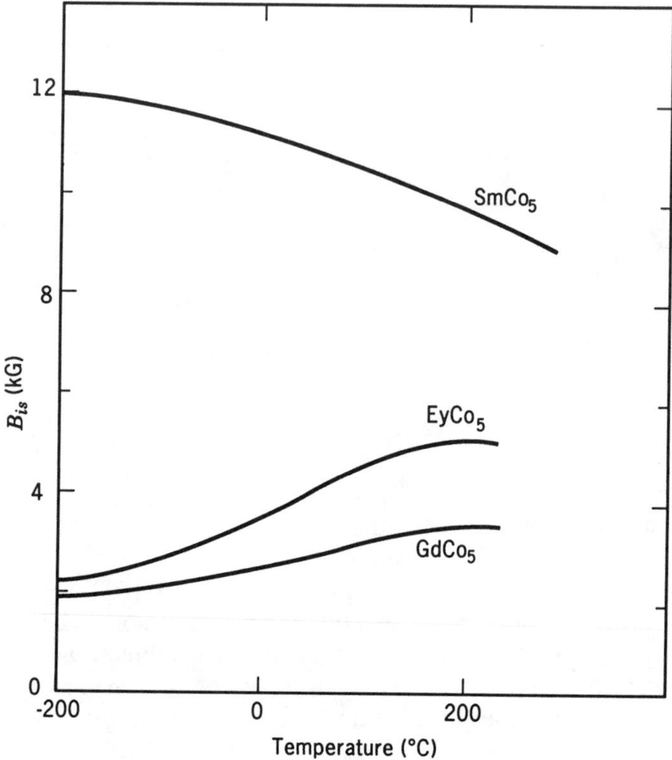

Figure 5.18 Comparison of magnetization changes for light and heavy rare-earth compositions.

$Sm_2(CoFe)_{17}$ alloy system. Leupold et al. [5] have reported that an alloy of composition $Sm_{0.6}Er_{0.4}(CoFeCuZr)_{7.22}$ gave these magnetic properties B_r = 9.3 kG, H_{ci} = 9.4 kOe, $(BH)_{max}$ = 16.5 MGOe, α = 0.004%/°C. There appears to be most interest in compensation in the highest property materials although a substantial drop in properties is inherently the price paid for atomic alloy temperature compensation.

5.7 MECHANICAL ENERGY INPUT AND STABILITY

Mechanical shock and vibration add energy to a permanent magnet to decrease the magnetization in much the same manner as discussed for the case of the thermal after-effect. One difference is the energy imparted thermally to the magnet is precisely KT while the energy imparted mechanically is difficult to know with precision. Repetitive shock or vibration should reduce magnetization with the same logarithmic relationship as was the case

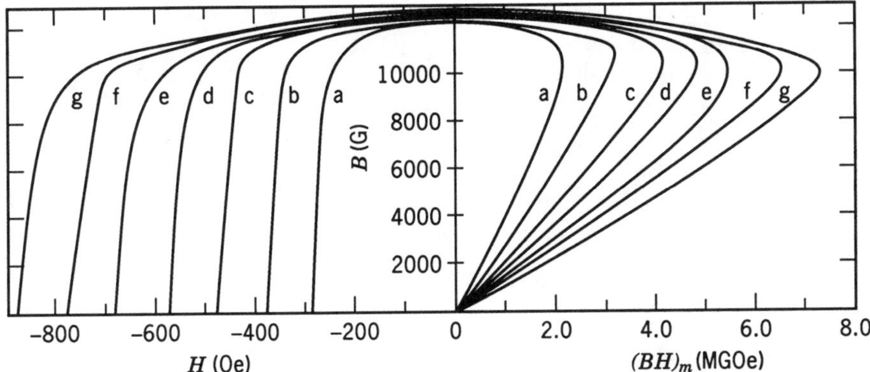

Figure 5.19 Effect of stress on Vicalloy. Applied stress of (a) 0; (b) 51; (c) 101; (d) 152; (e) 203; (f) 254; and (g) 305 kg/mm^2. After Shur, Luzhinskaya, and Shubina, Tr. Inst. Fiz. Metal., Akad. Nauk SSSR, 20 (1958) 11 and Fiz. Metal i Metalloved. 4 (1957) 54.

for temperature. The time variable would be replaced with the number of impacts. Figure 5.19 shows some results for Alnico 5 and earlier type magnets.

There has been little study regarding stabilization due to mechanical energy because with today's high energy density magnets there is little need for concern. There is limited information which suggests that both thermal

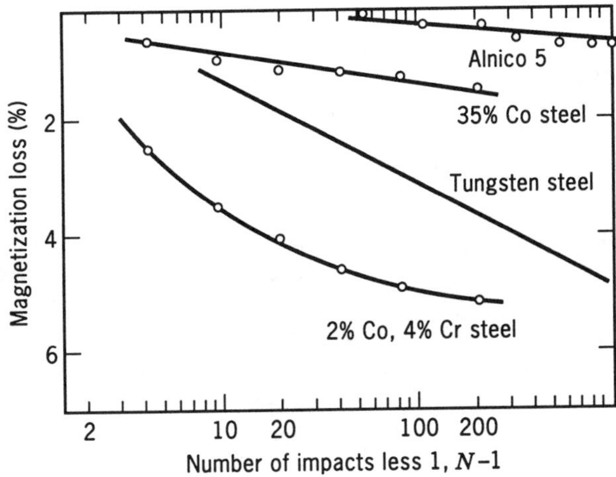

Figure 5.20 Flux loss in bars dropped 1 m onto hardwood floors working at their $(BH)_m$. After D. Hadfield, ed., Permanent Magnets and Magnetism (John Wiley, New York, 1962), Chapter 10, and R. Parker and R. Studders, *Permanent Magnets* (John Wiley, New York, 1962).

and field exposure will minimize but not entirely eliminate changes in magnetization due to shock.

Some magnet materials change properties when subjected to tension or compression. Vicalloy is one such material as shown in Figure 5.20. These changes are probably due to the contribution strain makes to the total anisotropy of the system as described in Section 2.1.

While man's earliest magnets were magnetized and demagnetized by shocks in the earth's field, today's high coercive force magnets require very different energy input levels to change magnetization.

5.8 CORROSION AND SURFACE OXIDATION

Alnico magnets are very resistant to corrosion, nearly as resistant as stainless steel. Only a slight discoloration of the surface is encountered when a magnet is heated in air to 450°C.

Ferrite magnet surfaces are stable and not subject to oxidation. Rare-earth magnets have oxidation problems and this limits their maximum operating temperatures. Surface oxidation will lead to a reduction in magnetic output. Adler and Marik [4] have studied this problem in depth and conclude that there are really three processes to distinguish: (i) selective surface oxidation; (ii) easy nucleation due to the deterioration of a surface layer with a thickness of a grain diameter and (iii) interior structural oxidation. In actual use, for $SmCo_5$ up to 200°C, one has to be concerned mostly with surface problems since interior oxidation will occur only at higher temperatures. As time and temperatures increase, oxygen diffuses into the material causing an oxidized layer to form. The thickness d_{s0} grows according to the relationship

$$d_{s0} = K(T)\sqrt{t} \tag{5.8}$$

This outer layer has a changed composition and a much lower coercive force so that the internal field of the magnet can reverse the magnetization of the outer layer. This leads to compounding of the flux loss of the magnet since the outer layer represents a loss of flux because of the volume which is not effective and it also acts as a shunt in reducing the output of the interior magnet. $SmCo_5$ magnets need some kind of protection above 200°C if structural change is to be avoided. The amount of protection needed will depend on the magnet's surface to volume ratio.

One effective form of protection is a heavy nickel plate. Kawashima et al. [6] have shown this form of protection effective for a traveling wave tube up to 300°C (Figure 5.21).

Oxidation and corrosion are major problems in non-sintered rare-earth

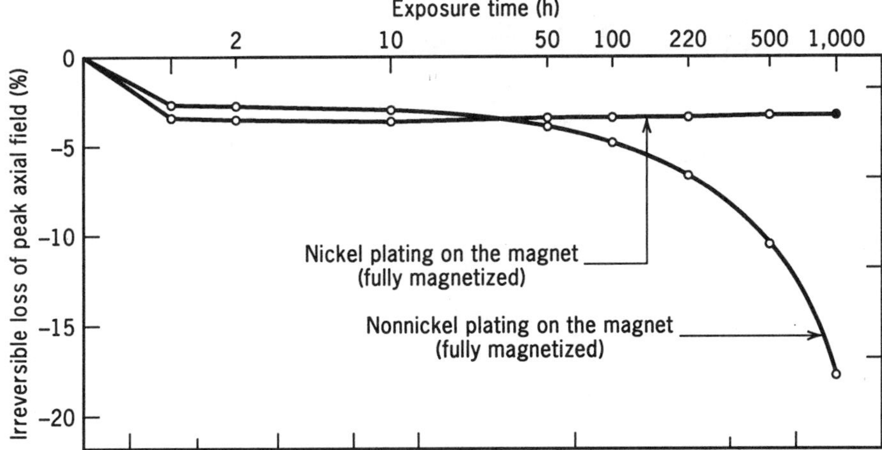

Figure 5.21 Exposure time dependence of peak axial field at 300°C. Changes are compared for a nickel-plated surface and an untreated surface [6].

magnets. In a magnet with porosity, we have interior oxidation at much lower temperatures. To protect the fine particles is difficult and the form of protection occupies volume and hence severe property reduction is experienced.

5.9 NUCLEAR RADIATION

There are a growing number of applications of magnets in nuclear environments. There are reported tests on both hard and soft magnetic materials. For soft magnetic materials, the tests are quite conclusive in showing increased coercivity. For permanent magnets there is little evidence that the loss in flux output is anything beyond what can be attributed to temperature effects. The work reported to date does not clearly separate irreversible loss from structural loss. Figure 5.22 shows flux loss for $SmCo_5$ and $Sm_2(CoFe)_{17}$ magnets as a function of level of proton irradiation [7]. Preliminary tests on NdFeB indicate they are inferior to the rare-earth cobalt magnets in terms of flux stability.

5.10 ENHANCING STABILITY

Permanent magnets have established a remarkable record with respect to maintaining constant flux over very long periods. There is still much further

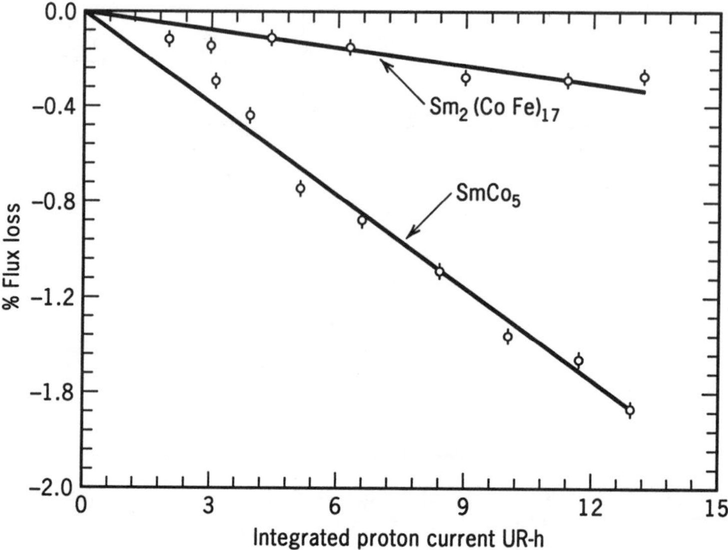

Figure 5.22 Demagnetization of REC magnets due to proton irradiation.

to go in terms of optimizing processing to focus on stability rather than property level. It appears that in the rare-earth cobalt magnets at least, and probably also with NdFeB, the most important properties to control are H_{ci} and H_k to minimize short-term irreversible losses. The order is important and square intrinsic demagnetization curves are desired. Another index is

$$\Delta_s = H_A(T) - H_k(T) \tag{5.9}$$

The larger one can make Δ_s by control of process and by magnetic circuit design, the less will be the irreversible loss.

Das et al. [8] have extensively studied $SmCo_5$ processing and conclude that instability in this material results from the formation of Sm_2Co_{17} and Sm_2Co_7 phases. If the magnet is composed of stoichiometric $SmCo_5$ with low oxygen content that remains in solution at the operating temperature, then $SmCo_5$ can offer much improved stability. Coercivity, composition, and structure all appear to impart stability and high stability may only be achievable with some specific processing detail.

5.11 STABILIZATION TECHNIQUES

When a magnet is newly magnetized and its flux level is observed over a long time period, one sees the relationship with respect to time as shown in

Figure 5.23. There is initially an appreciable loss in a short time. This loss is irreversible and is followed by a relatively long period of near constant flux called the plateau. In this region the loss per logarithmic cycle is constant. After long periods of time all materials at some higher temperature will exhibit a rapid flux decline. This is the region of structural change, perhaps oxidation or phase transformation. There is a temperature at which any magnet will undergo this kind of change. It may be polymer bonded $SmCo_5$ at 75°C or Alnico 5 at 650°C. The example in Figure 5.22 is for $SmCo_5$ at 250°C. The initial loss can be anticipated and stabilized against by heating to a temperature just above the expected operating temperature.

One can pre-stabilize by applying an a.c. or d.c. field. Most, but not all, of the irreversible loss due to temperature will be eliminated. This field might also be combined with a temperture level above room temperature. In stabilizing against the after-effect, an a.c. field is most desirable. The alternating field is gradually reduced to zero or the magnet is slowly drawn from a field of constant magnitude in order to leave the magnet in a symmetrical state. As an alternative, the after-effect can be anticipated by holding at a temperature higher than T_a for a short time period.

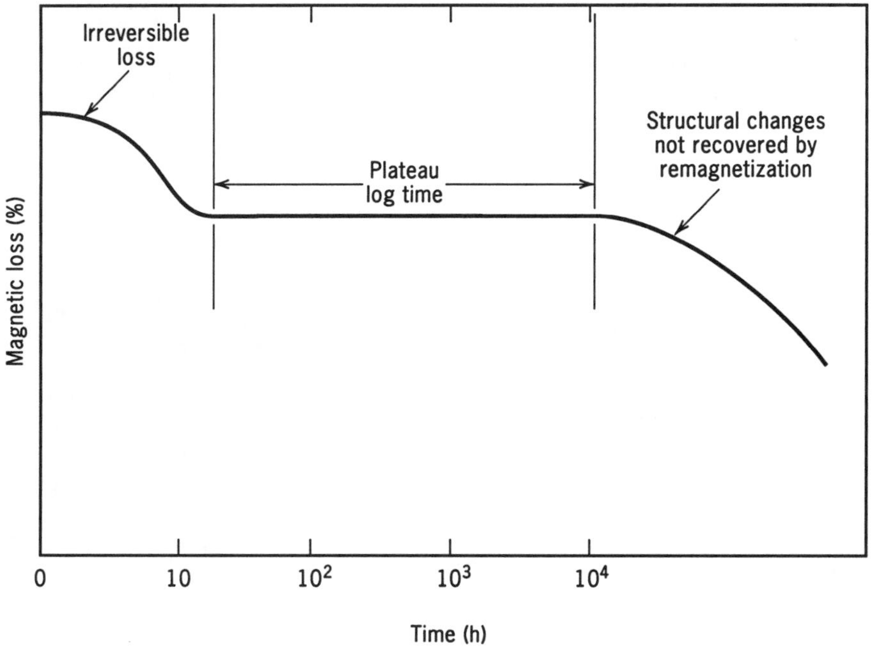

$SmCo_5 B/H = -1$ 250°C

Figure 5.23 Magnetic changes with time.

Magnetic stabilization tests for a cylinder of $SmCo_5$ having $B/H = 1.6$ are shown in Figure 5.24. The pre-stability conditions are described in the left position of the figure. One surprising result is the increase in flux resulting when heating to 100°C and 200°C after a high level of field demagnetization. This thermal magnetization apparently results from a wide range of coercive force over the magnet volume. A near perfect crystal could have an anisotropy field of perhaps 100 kOe surrounded by imperfect crystals having a much lower anisotropy field. When the temperature is raised the less perfect crystals are remagnetized by the near perfect crystal region resulting in an overall increase of magnetization in the specimen. Note that magnetic contact test results in small changes which can be stabilized for by either field or temperature exposure. For most test conditions either thermal or field energy can be used to pre-stabilize. Or they can be used in combination. With high energy density magnets like rare-earth cobalt or NdFeB, it is sometimes advantageous to combine the energy sources so that modest levels of each may be used in both magnetizing, demagnetizing, and stabilizing.

	Prestability
A	Full Magnetized
B	5–10% d.c. Demag
C	25–30% d.c. Demag
D	5–10% Thermal Demag
E	25–30% Thermal Demag

		Irreversible Loss (%)				
Test	Condition	A	B	C	D	E
	(1) −40°C	0	0	0	0	0
Thermal	(2) −20°C	−0.6	0	0	0	0
Exposures[a]	(3) RT	−0.7	0	0	0	0
	(4) +100°C	−4.2	−0.3	+0.7	0	0
	(5) +200°C	−6.2	−1.3	+2.4	−0.3	0
Magnet	(1) Magnet-Magnet (Repulsive)	−2.3	0	0	−0.1	−0.3
Contact[b]	(2) Magnet-Magnet (Attractive)	−0.7	0	+0.4	0	+0.1
Impact	150 G, 3.5 ms	0	0	0	0	0
Vibration	0.5–30 G, 5–2000 Hz	−0.5	0	0	0	0

[a] After 10,000 h exposures. [b] After 100 times.
Test Sample: $B/H = 1.6$

Figure 5.24 Magnetic stability of $SmCo_5$.

5.12 CONCLUSIONS AND COMPARISON OF MATERIALS

Choosing the magnet material that allows for a stable design in a particular environment involves many considerations. In this section, the intent is to pull together the many issues that have been discussed in prior sections of the stability chapter.

Alnico magnets and CrCoFe magnets are in general the most stable permanent magnets. Their only disadvantage is lower coercive force and resistance to adverse fields. They can be used up to 450°C without structural change. They have actually been applied for limited time devices in the 600°C range. By careful choice of operating permeance, near zero coefficients can be obtained.

Ferrite type magnets have excellent structural stability up to their Curie temperature. However, their high reversible coefficient, along with their limited low temperature H_{ci} severely limits their use. At 150°C their energy product is half of the room temperature value. This is tolerable because of their economic leverage in many devices. MnAl magnets like ferrite have properties at 150°C which reduce the energy product to about half that exhibited at room temperature. In this case the temperature coefficient of coercivity is negative and leads to very high irreversible loss.

$SmCo_5$ magnets are the most stable with respect to adverse fields. Regarding thermal stability, they are second only to Alnico magnets. Although their irreversible loss is appreciable, it can be anticipated and stabilized against. In the design, the choice of load line can minimize irreversible loss. Surface oxidation is a problem above 200°C and some surface protection is required. It is difficult, if not impossible, to use $SmCo_5$ in calibrated devices above 250°C for any length of time.

Copper containing $Sm_2(CoFe)_{17}$ type magnets of the high coercive force type, have improved stability compared to $SmCo_5$. Lower coefficient of coercivity and somewhat less surface oxidation make this material the best performer, regarding stability, of all of the rare-earth, high energy density magnets.

NdFeB magnets are still evolving at this time. It is clear that they are quite limited by a very high coercivity temperature coefficient. Corrosion and oxidation problems are also more severe than those of rare-earth cobalt magnets. In the writer's opinion, the properties are so high and the potential costs so low, that NdFeB will be widely accepted and the stability problems will be tolerated as they were with ferrite. In the designs, the stability shortcomings can be compensated for. If the economic leverage is present, most devices will be able to use NdFeB to advantage.

REFERENCES

[1] Indiana General Corp., Temperature Effects on Permanent Magnets—Applied Magnetics, Vol. 16 (Indiana General Corp., 1969).

[2] L. Neel, Ann. Univ. Grenoble 22 (1946) 299.

[3] R. Street and J. Wooley, Proc. Phys. Soc. London A62 (1949) 562.

[4] E. Adler and H. J. Marik, Fifth International Workshop on Rare-Earth Cobalt Magnets, 1981, p. 335.

[5] H. Leupold, E. Potenzianti, J. P. Clark and A. Tauber, IEEE Trans Magn. 20 (1984).

[6] F. Kawashima, H. Haroda, S. Takimoto and T. Mizuhara, Hitachi Rev. 25 (1976).

[7] E. W. Blackmore, TRIUMF Vancouver, B. C., Canada, unpublished data.

[8] D. Das, K. Fumar, and C. Dauwalter, IEEE Magn. 19 (1983).

6

DESIGN RELATIONSHIPS AND UNIT PROPERTY SELECTION

6.1 Introduction
6.2 Relating Unit Magnetic Properties to Magnet Volume, Magnet Geometry, and Device Parameters
  6.2.1 The Static Gap Design Problem
  6.2.2 Dynamic Recoil Design Problems
6.3 Determination of Permeance
  6.3.1 Introduction and Techniques
  6.3.2 Permeance Evaluation by Flux Plotting
  6.3.3 Permeance Evaluation by Formula
  6.3.4 Open Circuit Magnet Permeance
  6.3.5 Air-Gap Permeance and Flux Distribution
  6.3.6 Some Examples of Permeance Calculations
6.4 Magnetic and Electrical Circuit Analogy
6.5 Use of High Permeability Materials in Permanent Magnet Circuits
6.6 Economic Considerations in Design and Property Selection
6.7 Permanent Magnets and the Laws of Electromagnetic Scaling

6.1 INTRODUCTION

This chapter is concerned with selecting unit properties, arriving at a magnet volume, geometry, and magnetic circuit environment, to yield a specific solution to a permanent magnet problem. We are confronted with a tremendous array of magnetic properties, circuit configurations, device parameters, and economic issues. We want to achieve that one unique solution which is both functional and economically correct.

We are involved in the solution of difficult field problems and the need to determine rather accurately the total flux and its distribution in space around the magnet structure. Many simplifying assumptions are made and experienced is necessary to determine how appropriate these assumptions are in a particular situation.

Over the years, several analytical techniques have been used to give approximate solutions in magnetic design problems. In the simplest cases, the analytical solution can be exact, but in general for any practical problem, analytical solutions lead to large errors. In practice lumped circuit constants and electrical analog methods are often used. Electrostatic analog techniques, such as flux plotting of two-dimensional fields on conducting paper, are also used. For three-dimensional problems, styrofoam and aluminum wraping may be used to model and then the capacitance can be measured and converted to permeance. Today, most designs are optimized by measurement and extrapolation. The experience of working with a particular magnetic circuit is important.

The availability of digital computers has led to quite accurate modeling of circuits, devices, and systems using numerical methods. This is fortunate timing because rare-earth magnets are expensive, and tend to be used in more sophisticated and costly apparatus. Accurate solutions are imperative if one is to fully exploit the potential of these new magnets and to achieve cost-effective designs. We need to be systems thinkers because the permanent magnet is a very interactive high leverage component. We need to explore many alternatives in order to minimize total material and manufacturing cost. The computer allows us to explore the components, the device, and the system.

The value of the device or system often tends to influence our approach to problem solving. If one is working with a simple device that is low in cost, some trial and error modeling, building, and measuring to optimize is still an acceptable approach. However, if the device is large and costly to build and modify, one cannot afford the trial and error approach and an investment in numerical methods is required. At this time, three-dimensional programs are available for engineering work stations. Two-dimensional programs are widely used on personal computers. There are many devices which cannot be represented analytically in only two dimensions, however, if the device is axisymmetric, it can be worked in two dimensions on a personal computer. By geometric modification, it has been shown possible to compute field distribution for devices that are not truly axisymmetric with only small error [1].

As computer costs decrease and as the techniques for getting the problem onto the computer improve, numerical methods will gain further acceptance.

There has been rebirth of interest in electrical analogies since the development of type II magnets [2, 3]. The constant magnetization and linear demagnetization curves really mean we have constant magnetomotive force which is independent of the circuit used with the magnet. We can easily calculate the flux in every leakage path by using the magnetic equivalent of Ohm's law, along with Kirchoff's laws which govern current distribution. Ferrite and rare-earth magnets are very easy to evaluate compared to earlier type I magnets.

6.2 RELATING UNIT MAGNETIC PROPERTIES TO MAGNET VOLUME, MAGNET GEOMETRY, AND DEVICE PARAMETERS

6.2.1 The Static Gap Design Problem

The simplest design consideration is for the permanent magnet used to establish field energy in an air gap of fixed dimensions where the air gap is the only demagnetizing influence encountered by the permanent magnet. To establish the relationships between the properties of the material and the magnet dimension, let us consider the ring-shaped magnet and its demagnetization curve as shown in Figure 6.1. The ring magnet when magnetized to saturation and with no demagnetizing field action retains a residual flux density B_r as represented by ordinate Oa. The introduction of an air gap in the ring creates poles and self-inflicted demagnetizing. A certain amount of flux will exist in the air gap, and the magnetomotive force to sustain it will be supplied by the permanent magnet. By applying the well-known line integral theorem to the circular magnet and gap, we may write a basic equation involving magnetomotive forces in the magnet and its air gap.

$$L_m H_d - L_g H_g = 0 \tag{6.1}$$

where L_m is magnet length, H_d is the magnetizing potential of magnet per

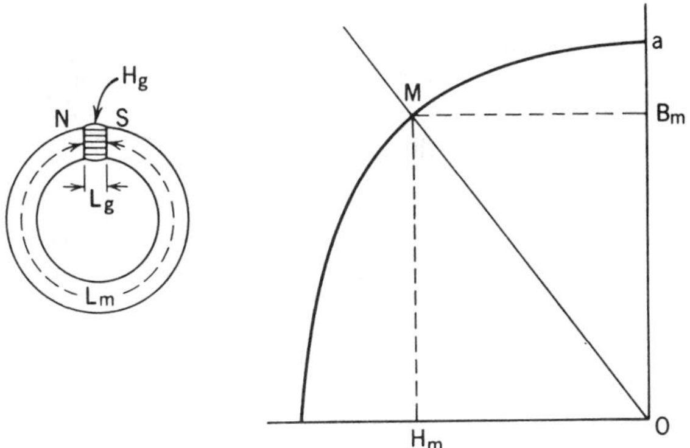

Figure 6.1 Relationships between unit properties and geometry.

unit length, L_g is the length of the air gap and H_g is air-gap unit potential or numerical gap density since $B = H$ in air.

Another basic equation involving magnet dimension and unit material properties comes from the premise that flux lines are continuous and may be equated at any two points in a magnetic circuit. If the air gap and magnet-neutral section (point the in circuit where maximum induction exists) are equated and for the moment the invalid assumption is made that all lines reach the air gap we have

$$A_m B_d = A_g B_g \tag{6.2}$$

where A_m is magnet area, B_d is the magnet unit density, A_g is the gap area and B_g is gap flux density. From (6.1) and (6.2)

$$L_m = \frac{L_g H_g}{H_d}, \qquad A_m = \frac{A_g B_g}{B_d}$$

$$V_m = \frac{L_g H_g^2 A_g}{B_d H_d}, \qquad \begin{array}{l} B_g = H_g \text{ numerically} \\ V_m = \text{Magnet volume} \end{array}$$

$$\frac{B_d}{H_d} = \frac{L_m A_g}{A_m L_g} \tag{6.3}$$

This equation is in terms of B and H and can be plotted as a straight line with a negative slope on a demagnetization curve. The intersection of this line with the demagnetization curve represents the operating point of the magnet. The flux density in the magnet has moved from B_r to B_d at which point the negative rise of magnetomotive force $L_m H_d$ developed around the magnet is equal to the magnetic potential drop across the air gap.

The reciprocal of reluctance $1/R$ is defined as permeance (P), and for ease in calculation of permanent magnet circuits the expression permeance is widely used. The concept of permeance reduced to a unit basis, P_u, as seen from a unit volume of magnet material is simply B_d/H_d. The foregoing relationships show how the magnet dimensions alter the operating point of the magnet on its demagnetization curve. Many factors will influence where the designer will want the operating point to fall. Perhaps the most widely used consideration is that of volumetric efficiency

$$L_m A_m = V_m = \frac{L_g H_g^2 A_g}{B_d H_d}$$

and (6.4)

$$L_m = \frac{L_g H_g^2 A_g}{B_d H_d A_m}, \qquad A_m = \frac{L_g H_g^2 A_g}{L_m B_d H_d}$$

This expression has great significance in the economics of magnet design and for a given gap volume A_gL_g permeated by a field intensity H_g, the volume of magnet required will be minimum when B_dH_d is a maximum. Consequently, dimensions of the magnet configuration are varied so that B_d and H_d will be used in the correct ratio (a critical value of unit permeance). The use of materials having high B_dH_d products (maximum available energy) with the magnet geometry proportioned to operate at the point of maximum available energy will give designs of greatest volumetric efficiency. Dimensional limitation or stability considerations may dictate that the magnet operate at some point other than maximum energy or on an interior recoil line if the application makes this desirable or necessary.

In Appendix 3, demagnetization curves for the most widely used magnet materials are plotted for use in design analysis. Energy contours as well as B/H values are shown. By drawing a line through the origin to a particular B/H reference slope, one can obtain a convenient graphical solution. Rapid exploration of the design parameters is possible.

The foregoing analysis, linking unit properties to magnet geometry and volume, requires some correction in a real problem. The most important correction has to do with the existence of leakage flux. Not all of the flux passes through the useful air gap. It is common practice to introduce a factor K_1, which is the ratio of total flux in the circuit to useful flux in the gap, or

$$K_1 = \frac{\text{Useful flux} + \text{leakage flux}}{\text{Useful flux}} \tag{6.5}$$

To determine K_1, the gap permeance and all of the leakage permeances must be determined; in a sense this is equivalent to an increase in air gap area. In some design situations the useful flux may be somewhat different than the gap flux. In the loudspeaker, for example, we find that the useful flux is greater than the gap flux because the voice coil extends on either side of the pole plate. In other devices with long air gaps, only a part of the total gap flux is useful flux. Often the gap area must be increased so that a region of uniform flux density will exist. One must define what useful flux is and adjust the values in (6.5) accordingly.

In our consideration of equating the rise of magnetic potential in the magnet to the drop of magnetic potential across the gap (6.1), another correction must be considered. Referring to Figure 6.2 and the distribution of magnetic potential, it becomes apparent that our previous assumption of constant potential per unit length in a magnetic circuit is in error. This is particularly true in using long limb lengths of type I magnet material. In working with gaps of appreciable length, some knowledge regarding the variation of H in the gap is necessary if a correct magnet length is to result. To be correct, the $\int H\,dl$ expression must be evaluated along a single flux line taking into account the cosine of the angle between H and dl. Theoretically, between points A and B in the air gap, the density close to a

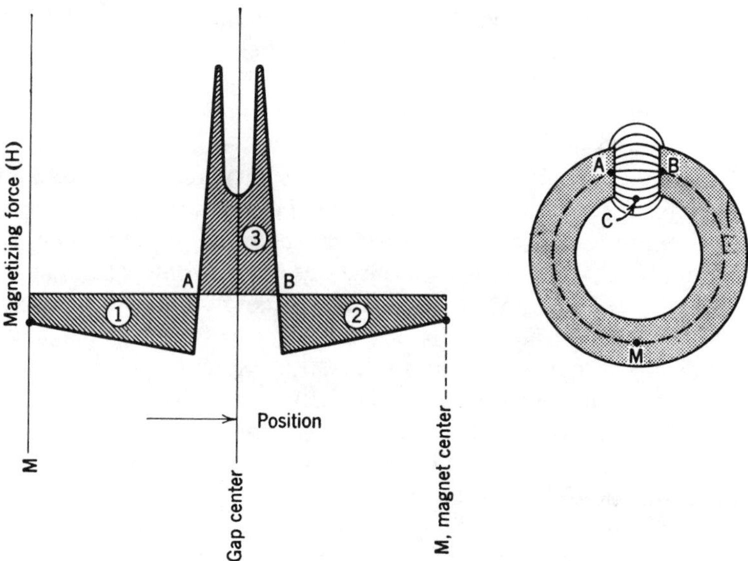

Figure 6.2 Variation of H in a permanent magnet circuit.

central line of flux is the same as in the magnet material since the lines are continuous. However, this theoretical density is of little interest since the flux plot indicates that the density will be much less at point C, and any average measurement of gap density will be less depending on the gap geometry and the curvature of the lines of force. However, in practical magnet design work, a rough approximation of the variation of H around the path is adequate in correcting the relationship $L_m H_d = L_g H_g$ and thus arriving at a fairly accurate magnet length. The H furnished per unit of magnet length varies along the magnet limb due to leakage in type I magnets and the consequent change of operating point on the demagnetization curve. Tapering of the magnet limb is used rather universally in an attempt to compensate for leakage and keep the various sections of the magnet near the point of $(BH)_{\max}$ and thus improve the design efficiency.

To handle the variations in H around a magnetic circuit as well as to acknowledge miscellaneous small drops in potential due to circuit joints and potential loss in soft magnetic circuit elements, it has become common practice to compensate in design by using a reluctance factor K_2 in (6.1)

$$K_2 = F_T / F_g \tag{6.6}$$

where F_T is total magnetomotive force developed in the circuit and F_g is the total magnetomotive force of the gap. We can think of this correction as equivalent to an increase in gap length. In practice K_1 can vary from about 1.5 to perhaps 10 depending on gap permeance. K_2 values would tend to be between 1.1 and 1.5 for most circuits. For a gap of area A_g and length L_g, the corrected permeance would be

$$\frac{A_g}{L_g}\frac{K_1}{K_2} \tag{6.7}$$

6.2.2 Dynamic Recoil Design Problems

In the preceding section the action of the permanent magnet when influenced by dynamic conditions has been considered. At this point, energy considerations and volumetric efficiency in designing magnet structures subjected to dynamic conditions are discussed.

In the fixed-gap problem, the total permeance of the magnet was controlled, and the total available energy was maximized. The leakage permeance paths and the useful permeance of the gap were fixed with respect to each other. In the dynamic application the leakage permeance varies over a fixed cycle, and a portion of the flux change from the leakage paths to the useful path and back again as the useful permeance is changed. As a consequence of this action a corresponding change in the useful and leakage energies takes place. Figure 6.3 illustrate these energy changes.

The minor loop on which the magnet operates is dependent on the slope of line OC, or the total unit permeance of the magnet when subjected to the influence of its free poles (disassembled condition) or maximum reluctance circuit condition. The operating point of the magnet assembled, point K is the intersection of the minor loop with the final unit permeance line OA.

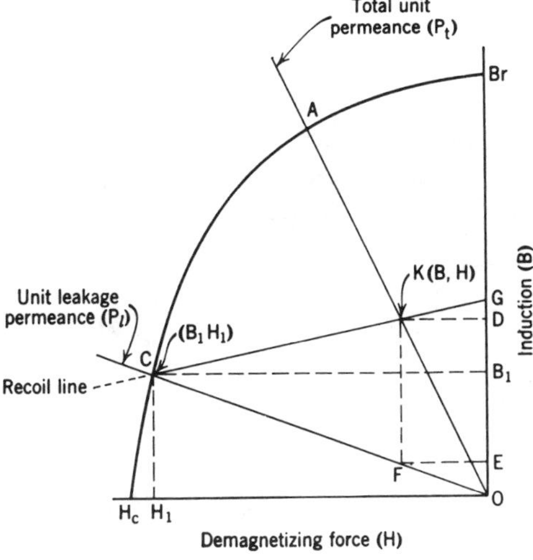

Figure 6.3 Graphical illustration of permanent magnet operating under dynamic load conditions.

Assuming that the leakage permeance line remains constant, a useful flux DE exists, the useful magnetomotive force is EF and the useful energy is represented by area $KDEF$. In a given situation with point C located by the leakage permeance and the recoil line slope established by the nature of the magnet material, for greatest design economy, the total permeance should be set so that point K is midway between C and G. This will maximize $DKEF$ since in a given triangle OCG the inscribed rectangle $DKEF$ is a maximum when $KG = KC$ and $CF = FO$.

From this analysis it is apparent that for economy in the dynamic design problem both the leakage permeance and the total permeance must be controlled. It is possible to select the coordinates B_1 and H_1 of point C by analytical means. Briefly this involves maximizing the area OCG, expressed in terms of the equation of the leakage permeance line, the recoil line, and an imperial equation fitting the demagnetization curve of the material. The resulting equation, after differentiation to maximize the area, is cubic and by trial and error, or equation solving techniques, values of H_1 can be found for maximum energy. The point of maximum recoil energy can then be obtained since it occurs at the midpoint of the recoil line passing point C.

As a matter of convenience it is more feasible to construct recoil energy contours for the various permanent magnet materials and to locate the desired operating point graphically. The following procedure is used to draw recoil energy lines. The equation of the leakage permeance line through point $C(B_1H_1)$ may be expressed as

$$B = \frac{(B_1)}{(H_1)} H \tag{6.8}$$

The recoil line through H_1B_1 with a slope may be written

$$B - B_1 = \mu_r(H - H_1)$$

The useful recoil energy E available by operation at any point (H_1B_1) on a recoil line through H_1B_1 is equal to the product of useful flux density and available H expressed in terms of the equations of the leakage permeance and recoil lines.

$$\begin{aligned} E &= \left[\mu_r(H - H_1) + B_1 - \frac{B_1}{H_1} H\right]H \\ &= H^2\left(\mu_r - \frac{B_1}{H_1}\right) + H(B_1 - \mu_r H_1) \end{aligned} \tag{6.9}$$

For a given value of E this equation gives two values of H corresponding to each point on the demagnetization curve. The two values of H are the abscissas for two points on the recoil line through (H_1B_1). The locus of all such points for a given demagnetization curve forms a closed curve commonly known as an energy contour. As an example, Figure 6.4 shows the

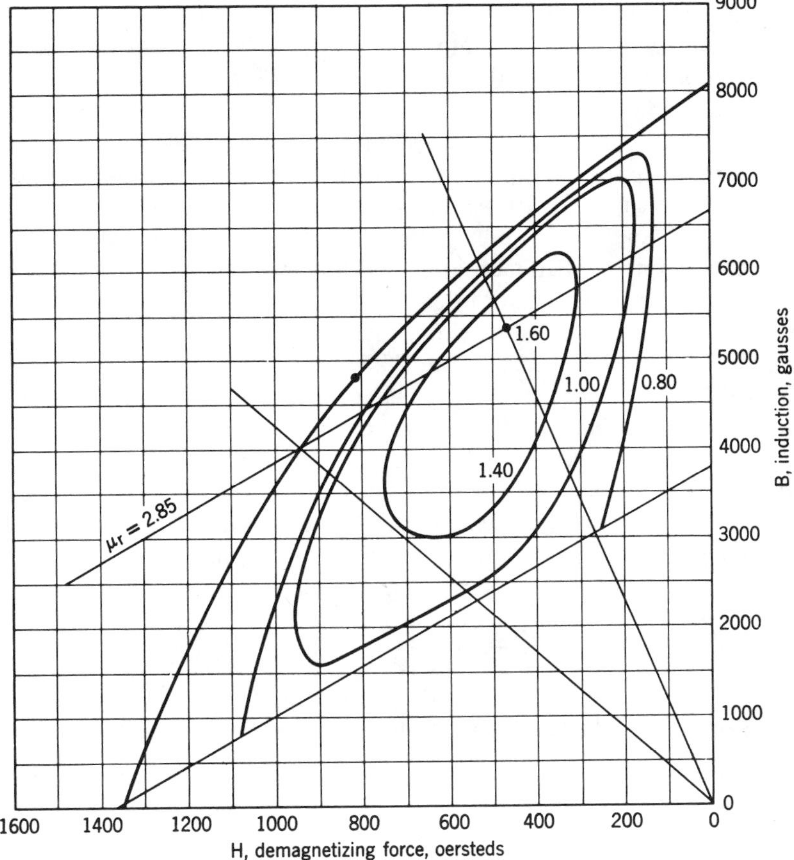

Figure 6.4 Useful recoil energy contours for cast Alnico 8.

energy contours constructed for Alnico 8. For type II material properties where major and minor loops have the same slope, the contours are compressed and one finds that the useful energy approaches the maximum available energy. Again maximum useful energy occurs when the recoil is half way along the interior loop. For efficient designs, the open circuit load line should approach zero. The magnet volume should be in the form of a thin sheet.

6.3 DETERMINATION OF PERMEANCE

6.3.1 Introduction and Techniques

Permeance is the ratio of flux (ϕ) to magnetomotive force (F). It is the reciprocal of reluctance. The proper dimensioning of a permanent magnet depends upon making estimates of the permeance of leakage flux paths. In the region surrounding the magnet the stored field energy may be represented by dividing the space into cells, each of which contains the same quantity of energy. In two dimensions each cell is bounded by two lines having equal potential difference and two lines perpendicular enclosing an equal quantity of magnetic flux. These unit regions or cells arrange themselves so as to make the potential energy of the system a minimum. In all but the most simple cases the arrangement cannot be deduced theoretically. It is well to understand the nature of the problem. We can have a magnetic structure and for given dimensions we can proceed to estimate the permeance paths and the distribution of flux. However, the more general presentation of a permanent magnet problem is to specify an air-gap volume and gap flux density requirement. Now we have to assume a magnet configuration and evaluate the permeance. We have basically a trial and error procedure, since the configuration must be assumed first before permeance can be estimated. In reality, we generally have an extremely difficult three-dimensional field problem to solve. We must consider many techniques and simplifying assumptions for estimating the distribution of flux in a magnetic circuit problem. We will consider flux plotting, evaluation of magnet limb permeance, permeance formulas, single air-gap permeance and flux distribution from measurements. Later in the chapter the use of electrical analogies in problem solving will be considered.

6.3.2 Permeance Evaluation by Flux Plotting

The method of field plotting is most effectively applied to magnetic fields, or portions of magnetic fields which vary in two dimensions but remain essentially constant in the third. The procedure to be followed in field plotting is relatively simple. On a diagram of the magnetic circuit, several equipotential lines are drawn. Then flux lines connecting the surfaces of opposite polarity are added in such a manner so as to fulfill the following requirements:

1. All flux lines and equipotential lines must be mutually perpendicular at each point of intersection.,

2. Each figure bonded by two adjacent flux lines and two adjacent equipotential lines must be a curvilinear square.

When the full plot has been completed, the magnetic permeance can be found by dividing the number of curvilinear squres between any two adjacent equipotential lines, designated as n_e, by the number of curvilinear squares between any two adjacent flux lines, n_f, and multiplying by the length in centimeters (l) of the field perpendicular to the plane of the flux plot. Symbolically, this can be written

$$P = n_e l / n_f \tag{6.10}$$

As an example, consider the C-shaped magnet illustrated in Figure 6.5 for which the permeance of that portion of the field between the pole faces is to be found by the method of flux plotting. The fundamental equipotential line A, which is the intersection of the neutral plane of the magnet with the plane of the flux plot, is drawn first, and then several equipotential lines, such as B and B' and C are sketched in. Starting at the inner edge of one of the pole faces, flux lines are then added to the plot. Though this procedure is largely a matter of trial and error, the patient individual who constantly bears in mind the two requiremenets stated previously will not find the task too difficult. When the plot is completed, the permeance may be found by use of (6.10).

Where any appreciable amount of flux plotting is necessary it is common practice to resort to electrostatic systems which accurately and rapidly allow two-dimensional plots to be drawn. The electrostatic field analogy may be used since the basic field theory and relationships are common in both magnetostatic and electrostatic field systems. One useful technique involves

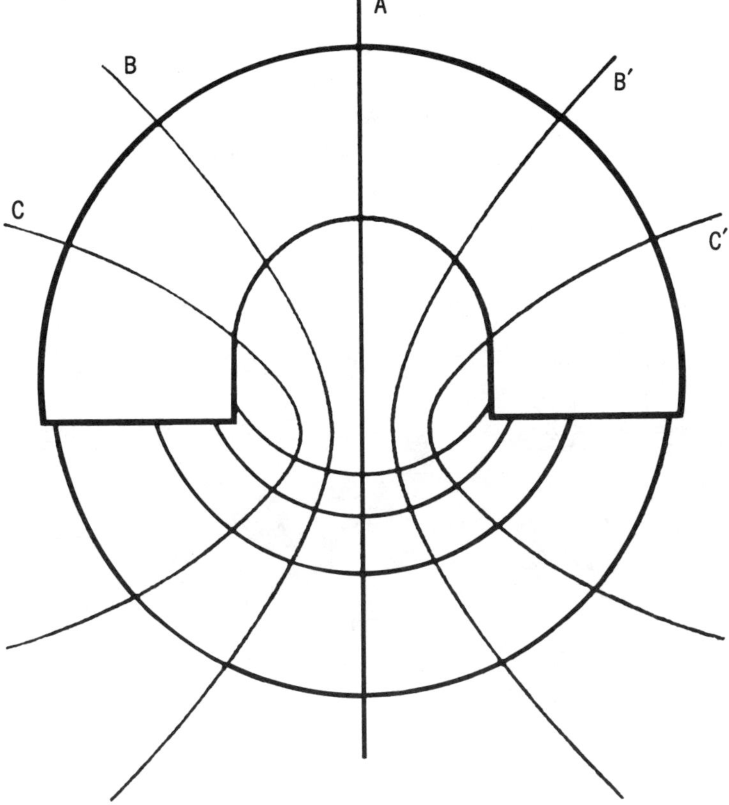

Figure 6.5 Fundamentals of flux plotting.

the use of a thin conducting paper on which the electrodes are painted in the form of silver paint [4]. Figure 6.6 shows the basic elements of such a system. Potential is applied to the electrodes, and the voltage divider allows the spacing of equipotential lines, which are established on a point-by-point basis using the stylus and null detecting instrument. The advantages of this dry method of flux plotting are its low cost and the convenience in set up.

6.3.3 Permeance Evaluation by Formula

Evershed, in 1920, published a series of formulas based on the assumption that magnetic flux will arrange itself in symmetrical fashion for certain simple configurations, and as such the permeance may be expressed by relatively simple mathematical expressions. The following formulas are useful in working with permanent magnets. Additional formulas of interest are developed by Spreadbury [5] and Rotors [6].

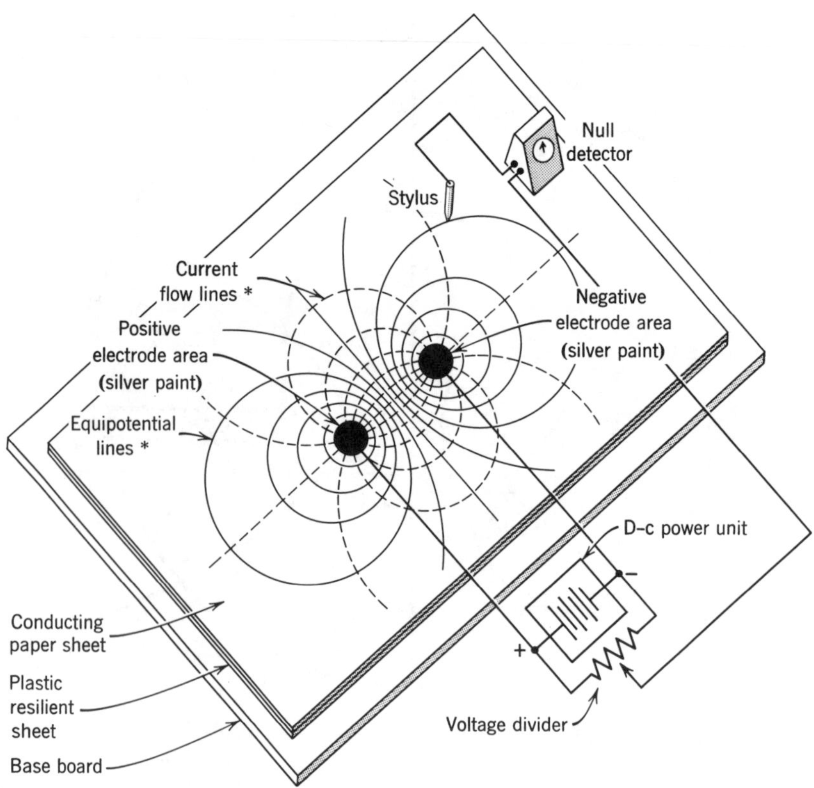

Figure 6.6 Electrostatic field plot and equipment.

Permeance of Air Gap between Rectangular Surfaces (Figure 6.7). When the pole faces are flat rectangular the permeance of the gap is

$$P = \frac{TW}{L_g} \tag{6.11}$$

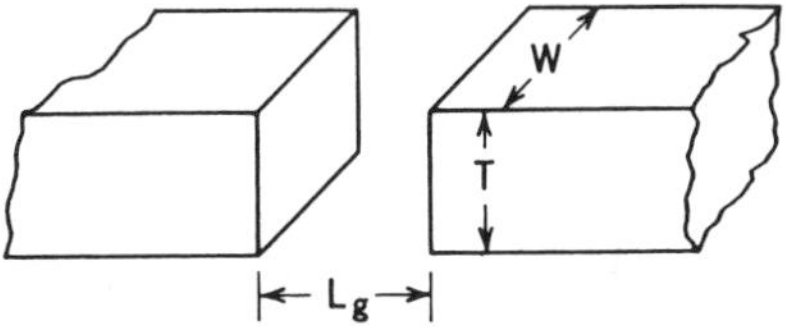

Figure 6.7 Permeance of rectangular air gap.

When L_g is one-fifth of either T or W the permeance will be roughly 4 or 5% greater than that estimated by the formula. This formula does not recognize the existence of fringing flux.

Permeance of Air Gap between Cylindrical Surfaces (Figure 6.8)

$$P = \frac{w\alpha}{\ln_e(1 + (L_g/r))} \quad (\alpha \text{ in radians}) \tag{6.12}$$

When L_g/r is 0.02 or less than the formula reduces to

$$P = \frac{wr\alpha}{L_g}$$

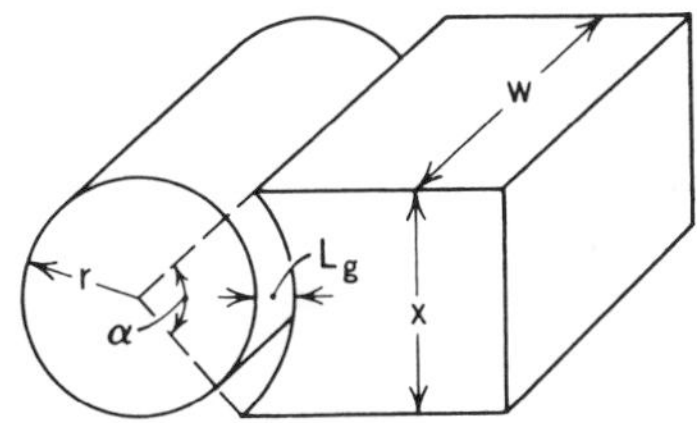

Figure 6.8 Permeance between cylindrical surfaces.

Permeance of Fringe Paths Originating on Flat Surfaces (Figure 6.9). For an elliptical fringe on flat polar surfaces of width, w

$$P = \frac{w}{\pi} \log_e\left[1 + 2\left(\frac{x + (x^2 + xL_g)}{L_g}\right)^{1/2}\right] \tag{6.13}$$

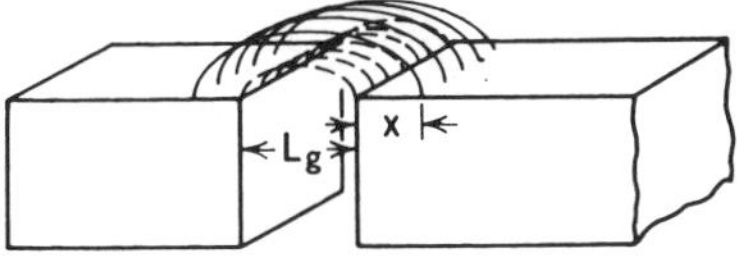

Figure 6.9 Permeance of fringe paths from flat surfaces.

Permeance of Fringe Paths Originating on Cylindrical Surfaces (Figure 6.10). The total permeance of the fringes stated in terms of its extent "x" on each cylinder and a quantity "v" is as follows

$$v = r \log_e \left[1 + 2\left(\frac{x + (x^2 + xL_g)^{1/2}}{L_g}\right)\right]$$

$$P = \frac{\pi(v^2 - x^2)^{1/2}}{\cos^{-1}(x/v)} \qquad \text{when } v > x \tag{6.14}$$

$$P = \pi x \qquad \text{when } v = x \tag{6.15}$$

$$P = \frac{\pi(x^2 - v^2)^{1/2}}{\log_e \left[\frac{x + (x^2 - v^2)^{1/2}}{v}\right]}, \qquad \text{when } v < x \tag{6.16}$$

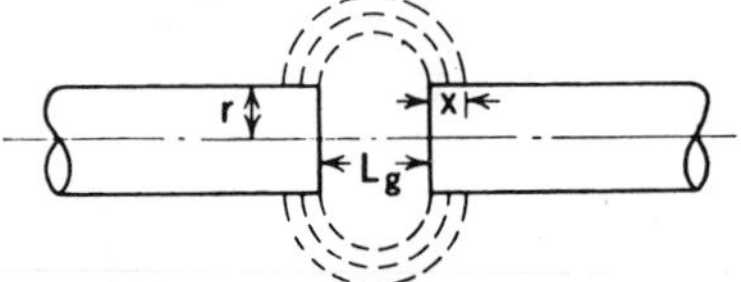

Figure 6.10 Permeance of fringe paths from cylindrical surfaces.

Permeance between Parallel Cylinders (Figure 6.11)

$$P = \frac{\pi}{\log_e \dfrac{L + (L^2 - d^2)^{1/2}}{d}}$$

If $n = L/d$

$$P = \frac{\pi}{\log_e [n + (n^2 - 1)^{1/2}]} \tag{6.17}$$

For rectangular sections

$$d = \frac{\text{perimeter}}{\pi}$$

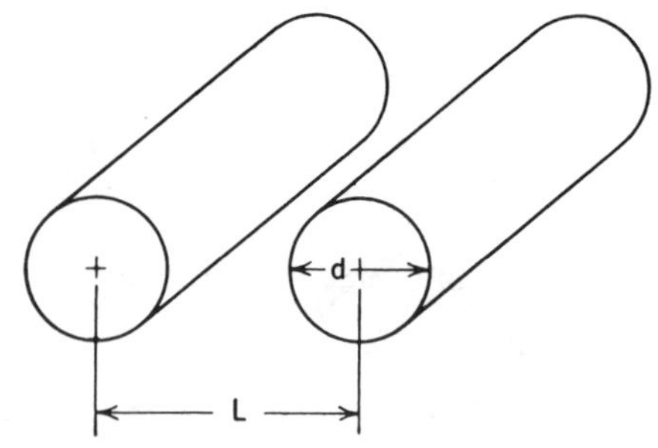

Figure 6.11 Permeance between parallel cylinders.

6.3.4 Open Circuit Magnetic Permeance

In the magnet circuit we have previously found that the load line or coefficient of self-demagnetization may be expressed as

$$\frac{B}{H} = \frac{L_m}{A_m} P$$

The slenderness ratio of the magnet and the total permeance P, determine B/H. For a bar magnet the permeance is the permeance from one limb of the magnet to the other, since there are no pole pieces or well-defined air gap. For the simple bar magnet the operating point on the demagnetization curve is determined only by its geometry and resulting limb permeance.

The B/H value is for the neutral section of the bar, and as we move toward the end of the bar B/H drops due to the fact that flux is leaving the limb. Mathematically it can be shown that if a bar has the outline of an ellipse its flux density B and its potential available for external use H will be constant over the whole bar. The total permeance P may be considered as made up of a unit permeance per centimeter, p, multiplied by the length of the limb. The permeance per centimeter varies slowly being maximum near the neutral section and minimum near the ends of the bar. Evershed [7] in his early work with carbon and tungsten steel bar magnets applied the free polar radiation formula to arrive at values of P with good results. Briefly this formula is developed on the premise that lines of force radiate outward from one polar region to the other. The permeance of the leakage path can be estimated accurately by substituting for the polar area of the magnet limb, a spherical pole of dimensions which make it the equivalent of the actual pole in its leakage effect. The free pole is developed as follows. The permeance of the path extending outward in all directions from the surface of the spherical pole of iron or steel is found as follows. Referring to Figure 6.12 the area of the elementary spherical shell of radius x is $4x^2$ and denoting its thickness by dx the reluctance of the shell is $dx/4x^2$. Hence, the reluctance of the space between the surface of the spherical pole N of radius r, and a spherical boundary of radius nr will be

$$R = \frac{1}{4\pi} \int_r^{nr} \frac{1}{x^2}\, dx = \frac{1}{4\pi r}\left(1 - \frac{1}{n}\right) \tag{6.18}$$

If we suppose the boundary to be at some considerable distance from the pole so that n is a large number, $1/n$ becomes negligible and the reluctance of the path extending outwards from the pole N is simply $1/4\pi r$. Similarly, the reluctance from the distant boundary to a complementary pole of the same radius as N will also be $1/4\pi r$. Hence, the reluctance of the entire path from pole to pole will be $2/4\pi r$, and the permeance from spherical pole to spherical pole will be $P = 2\pi r$.

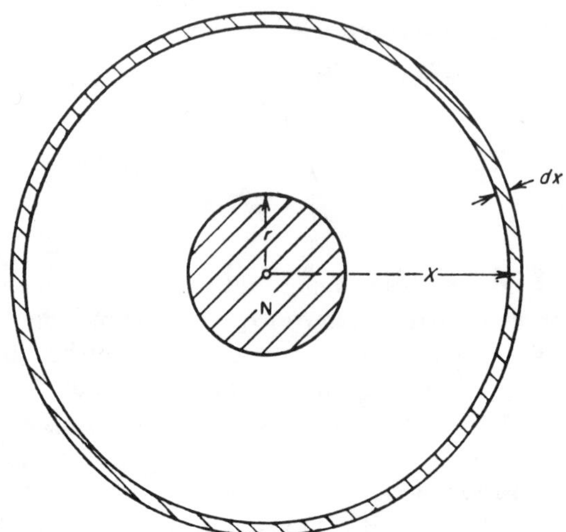

Figure 6.12 Illustration of polar radiation formula.

It is convenient to express the permeance of the leakage path formed in this way in terms of the surface of either pole. Denoting the surface of the pole by s we have $4\pi r^2 = s$, from which $r = (s/4\pi)^{1/2}$, so that the permeance becomes $P = 2\pi(s/4\pi)^{1/2} = (\pi s)^{1/2}$ and since $\sqrt{\pi}$ is 1.77, we have

$$P = 1.77\sqrt{s} \tag{6.19}$$

The use of the spherical pole formula introduced by Evershed and the general premise that real magnet configurations approximate the condition of equipotential surfaces and radiation from these surfaces can be reduced to rather simple terms, is an extremely important consideration; the author has applied this approach in the solution of a wide range of permanent magnet problems.

Many shapes, like bars, rods, and rings, are used to establish field energy in space and are not associated with a well-defined air gap or additional magnet circuit elements. It is common to refer to such magnets as operating under open circuit conditions. In many instances the magnet is subjected to open circuit or air stabilization before being inserted into the magnetic circuit of a device (generator, for example). Under such conditions the origin of the minor hysteresis loop is established by the open circuit permeance, and the degree of accuracy in locating this minor hysteresis loop conditions the accuracy of the whole solution. As mentioned before. Evershed used the free pole formula with early magnet steels, and obviously the extremely long limbs involved represented quite a departure from the assumption of polar radiation and errors were observed. However, with the high coercive force permanent magnet in use, magnet length to cross-section ratios are usually much lower, and 1.77 $\sqrt{s}$ represents the total limb permeance quite accurately. In early permanent magnets the effective spacing of the poles was found to be 0.7 of the total length. Using this effective spacing the basic expression relating magnet flux ϕ, magnetomotive force F, and permeance P becomes

$$\phi = mmf(P)$$

or

$$A_m B_d = 0.7LH_d(1.77\sqrt{s})$$

$$B/H = 1.77(0.7)L_m/A_m\sqrt{s} \tag{6.20}$$

The location of the center of the effective poles is a problem of considerable importance, and the author has studied this with respect to shape for both long and extremely short limbs, and also with respect to the material permeability. Above length to area ratios of 10, the physical departure from a sphere becomes extreme and the formula shows appreciable error, with respect to measurements. Using today's materials this is of little concern, since length to area ratios above 5 are not generally encountered. For ferrites and rare-earth magnets we encounter very low length to area ratios and the flux from the pole cross-section generally outweighs the flux from the periphery of the limb for these low L/A ratios. The free pole formula has been extended to cover a wide range of magnet shapes. B/H values for rings, rectangles, and diametrically magnetized cylinders are plotted as a function of geometry in Figures 6.13–6.16. On each figure the

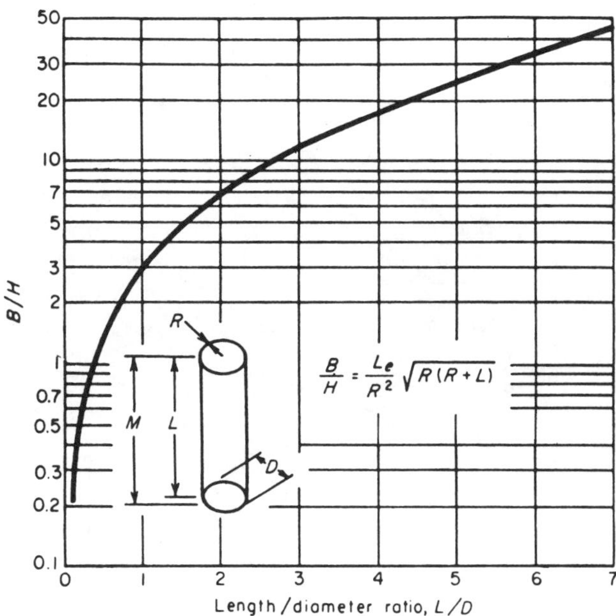

Figure 6.13 Demagnetizing coefficients for cylinders with axial magnetization.

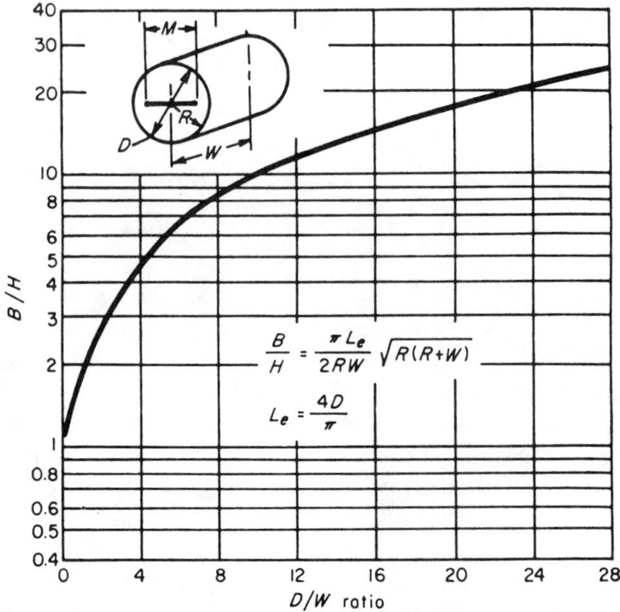

Figure 6.14 Demagnetizing coefficients for cylinders with diametric magnetization.

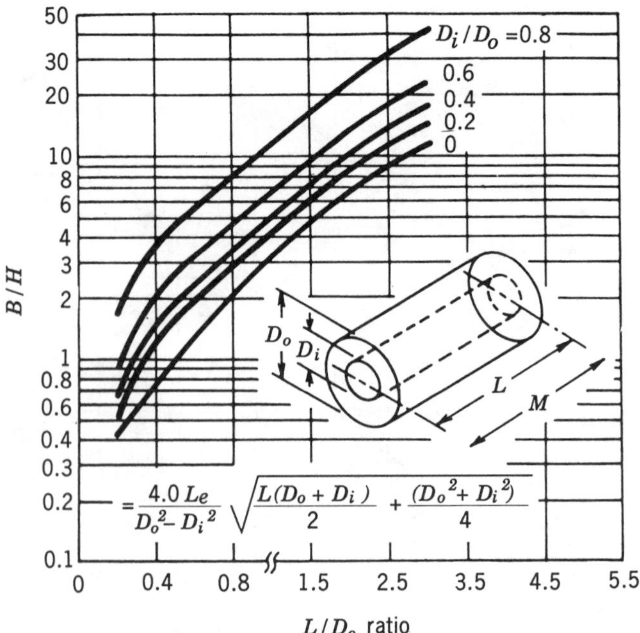

Figure 6.15 Demagnetizing coefficients for tubular magnets with axial magnetization.

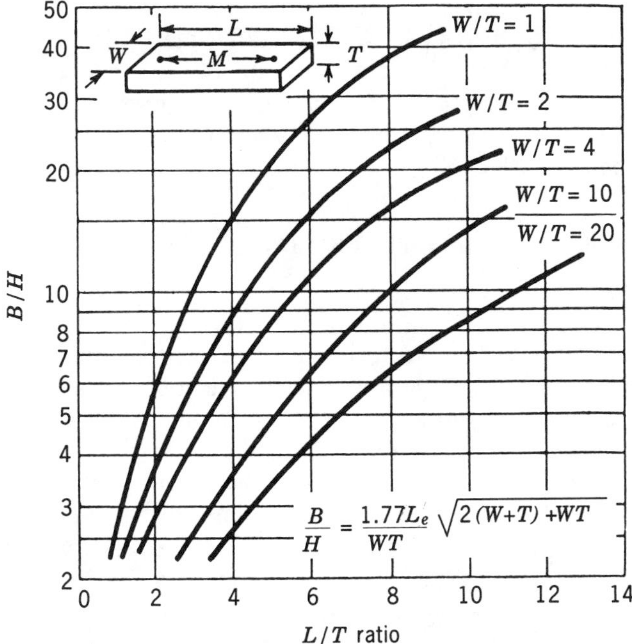

Figure 6.16 Demagnetizing coefficients for rectangular bars.

free pole formula modified for the particular shape is given so that dimensions not plotted can be calculated. In each plot the assumption is made that the effective magnetic length is the true geometric length, hence the plots are for type II magnets having near unit permeability. For earlier magnets such as Alnico 5, the correct B/H would involve using the effective magnetic length L_e of 70% of the full geometric length.

In Figure 6.17 the free pole formula is used to find the neutral section of

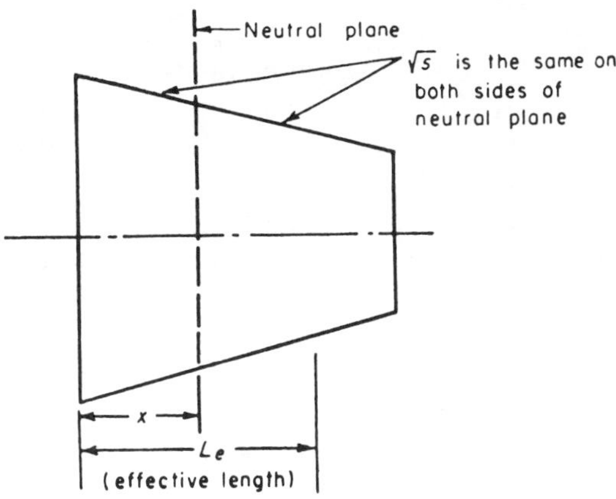

Figure 6.17 Calculation of effective magnetic length of magnet with changing cross-section.

a magnet with a changing cross-section. The distance to the neutral plane, x, can be determined from the consideration that the $\sqrt{s}$ is the same on both sides of the neutral plane. The effective length of this magnet is $2x$ for type II magnets. The polar areas are symmetrical with respect to the neutral section. Since effective length and cross-section are known P and B/H can be determined.

6.3.5 Air-Gap Permeance and Flux Distribution

The permeance of an air gap in its most simple form, Figure 6.18a is $P_g = A_g/L_g$, gap area per gap length. This expression takes no account of fringing flux and yields correct values of permeance only when area to length ratios are very large and all the lines in the gap are parallel. This condition is seldom encountered and even with dimensions T or W five times L_g, the effective gap permeance is of the order of 5% greater than calculated on the basis of the formula. As a rule of thumb in working with high permeance gaps, the flux can be assumed to be uniform in an area $(T - L_g)\cdot(W - L_g)$ (Figure 6.18a), and we have a basis of correcting the formula for permeance. With gap permeance of low values, the problem of arriving at the effective value is extremely difficult since the fringing flux increases the effective area and wide variations in flux density are encountered.

Referring to Figure 6.18a, a brief analysis of the flux distribution resulting from the presence of the various surfaces helps to form perspective concerning the nature of the problem and the approach used in securing useful data. In the case of two thin plates the effective permeance between surfaces $A_g - A_g$ can only be obtained from a consideration of the lines of flux between plates of a condenser using Maxwell's equations relating potential and flux. We can determine that the fringing flux increases the effective pole face by a factor h/π where h is the semi-gap. Consequently if W and T are the dimensions of gap section A_g, the effective permeance becomes $P_g = (W + (h/\pi))(T + (h/\pi))/L_g$. This expression holds until A_g/L_g values as low as 2 are reached, in which case measurements are then necessary. If perpendicular surfaces (b) are present then the whole distribution in the gap is changed and Carter [8] showed that the fringing correction for the case when the perpendicular surface is long relative to h, becomes $0.6h/\pi$, which is a useful limit to know.

It is also of interest to note the influence of surfaces $c - c$. Figure 6.18b shows the permeance per centimeter of width W for three values of H/h plotted against T/h (assuming T to be of appreciable size). Cramp and Calderwood [9] computed this data from Maxwell's equations. An additional air-gap configuration often encountered is shown in Figure 6.18c which shows two parallel surfaces in the same plane. Cramp and Calderwood have by calculation and flux plots arrived at the data shown in Figure 6.18d. Again knowing the limiting values of permeance involved is very useful in approximating the true values involved.

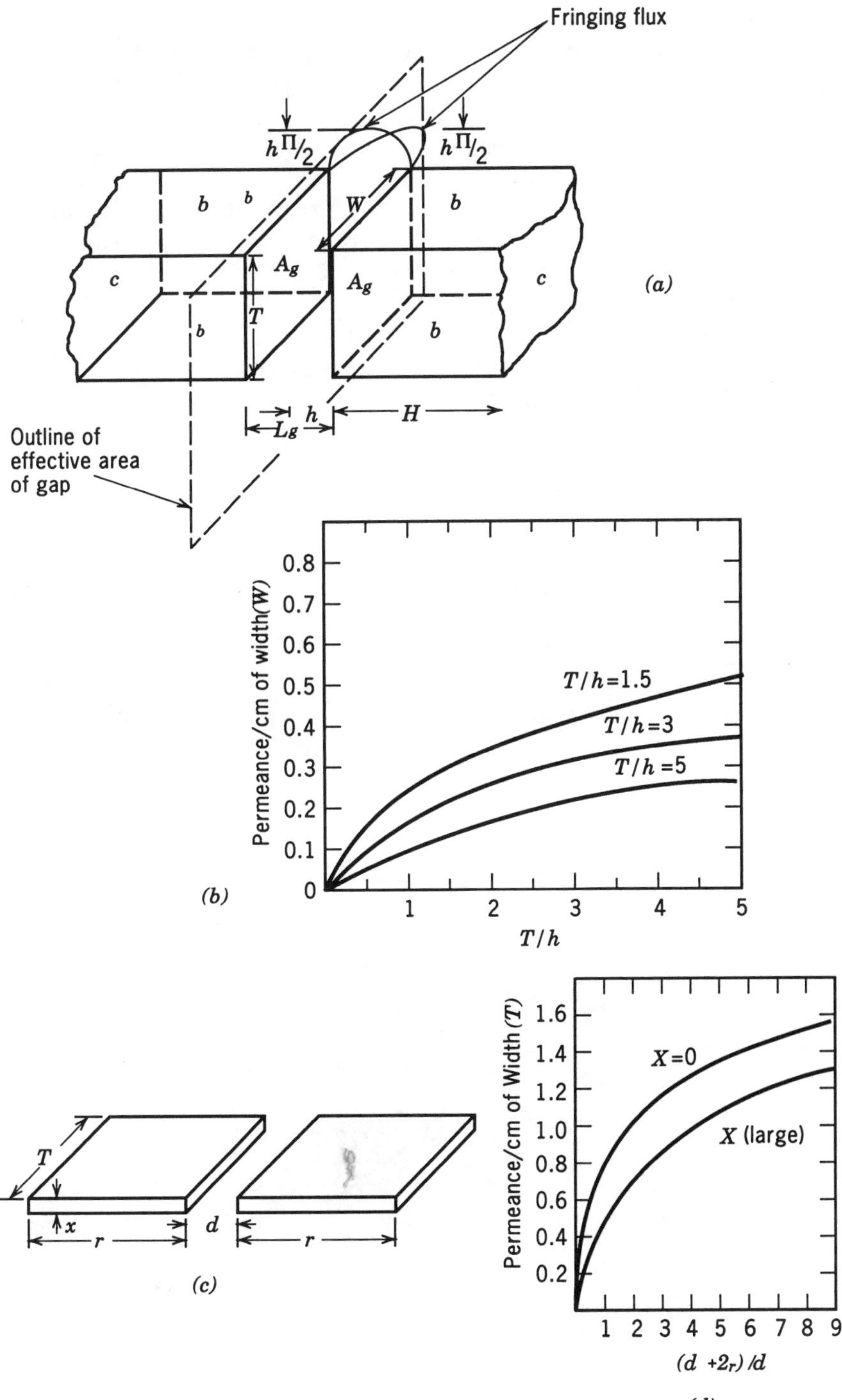

Figure 6.18 Basic considerations involved in evaluating air-gap permeance.

In many permanent magnet arrangements using rectangular pole blocks, the flux is brought to the gap by bringing the permanent magnet either against face c or faces $b - b$. Whatever the arrangement, the basic considerations of Figure 6.18 will aid in estimating the effective permeance.

Figure 6.19 shows a typical gap geometry and how the flux density varies both axially and diametrically on a contour flux plot. Except in very high permeance gaps, denoting air-gap density means little or nothing without specifying where the measurement is to be made and over what area, all of which introduces another variable in magnet design, the necessity to relate flux density at a particular point in the air gap to the total gap flux or flux density existing at the pole of the magnet assembly. In order to approximate the true effective permeance and the distribution of flux in air gaps where the length is appreciable with respect to area, good use can be made of the electrostatic analogy and the premise that the gap faces form the plates of a condenser.

Gap systems in permanent magnets do act for all practical purposes like equipotential systems and we can conveniently measure capacitance and changes in capacitance very accurately with bridge systems specifically designed to give accuracy in the low micromicrofarad range.

The relationship between permeance and capacitance can be obtained as follows

$$P_g = A_g/L_g\,, \qquad C = A_g/4L_g(9 \times 10^{-11})\ \text{F for parallel plates}$$

Therefore

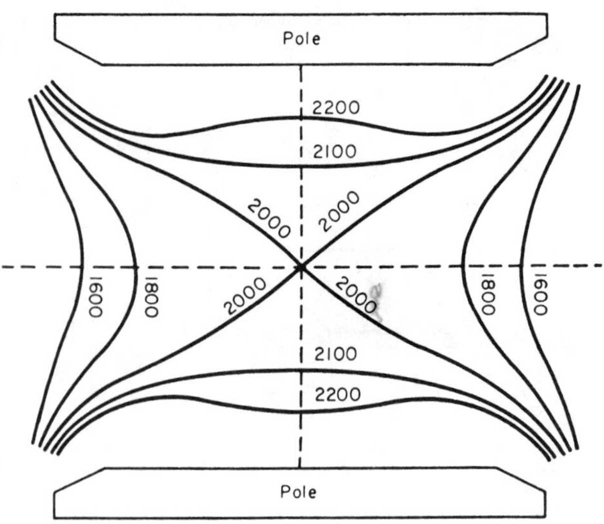

Figure 6.19 Contour flux plot for a specific air-gap geometry.

$$P_g = 11.3C \tag{6.21}$$

when C is in micromicrofarads ($\mu\mu F$).

Figures 6.20 and 6.21 relating true effective permeance to gap geometry were obtained by calculation and measurements in the electrostatic system and checked from measured data in the magnetic system. In using the effective permeance of these basic gap geometries we should keep in mind the principles shown in Figure 6.18. For low permeance gaps it may be necessary to consider the fact that flux radiation from back surfaces is not present due to the presence of the permanent magnet limb.

In order to obtain useful data relating flux density B_g in the center of an air gap to the flux density at the pole face B_p, the ratio B_p/B_g has been found for the most frequently encountered gap geometries

$$B_p = \phi_g/A_g = F_gP_g/A_g$$

where F_g = gap potential

$$B_g = H_g = (F_g/L_g)(F_1) \tag{6.22}$$

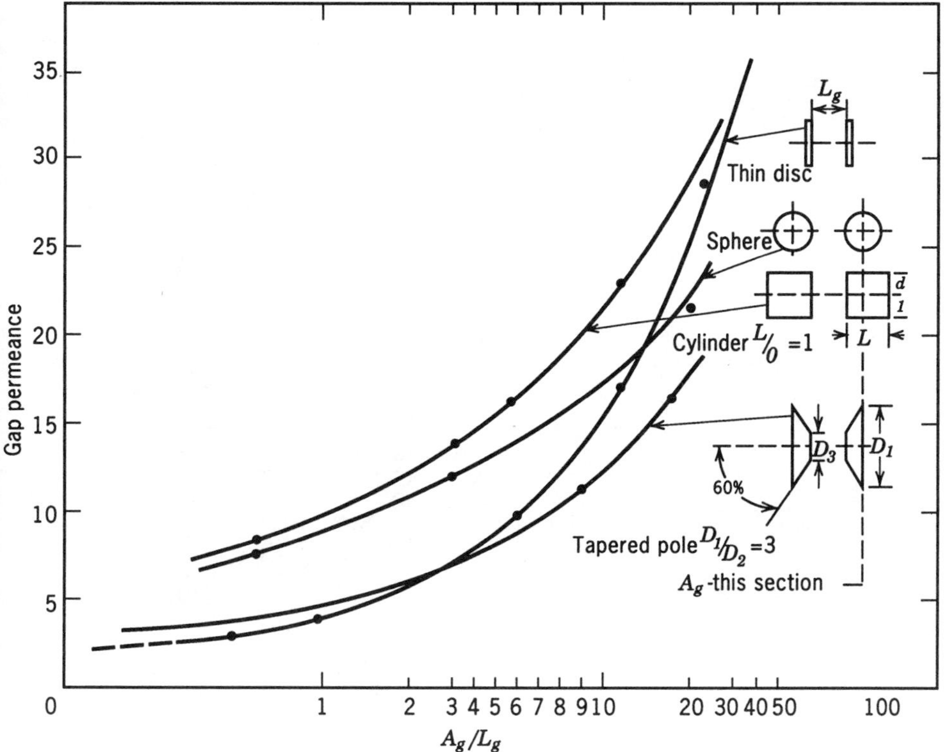

Figure 6.20 Effective gap permeance for air gaps of circular section.

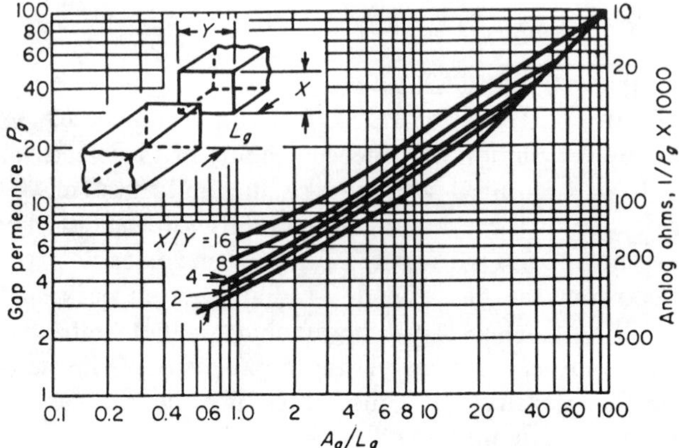

Figure 6.21 Effective gap permeance for rectangular gaps.

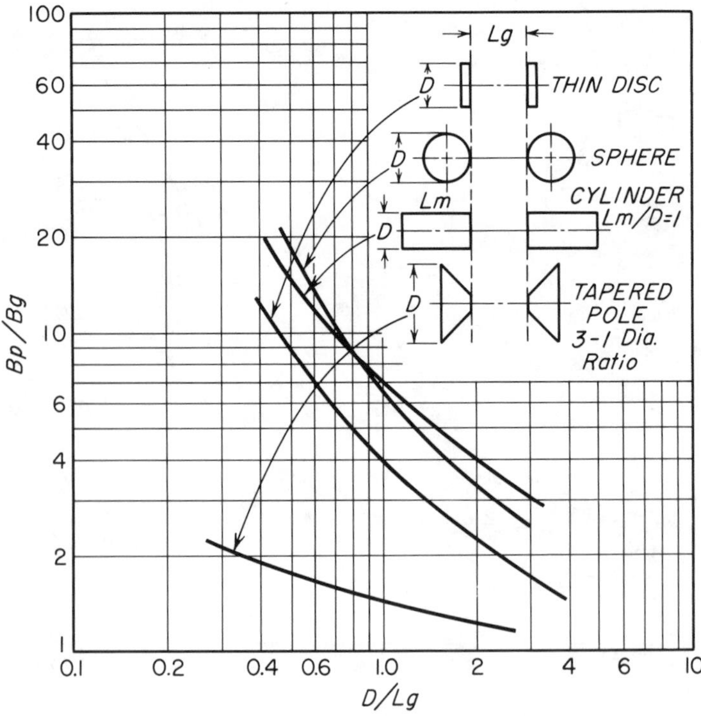

Figure 6.22 Air space flux density relationships for gaps of circular section.

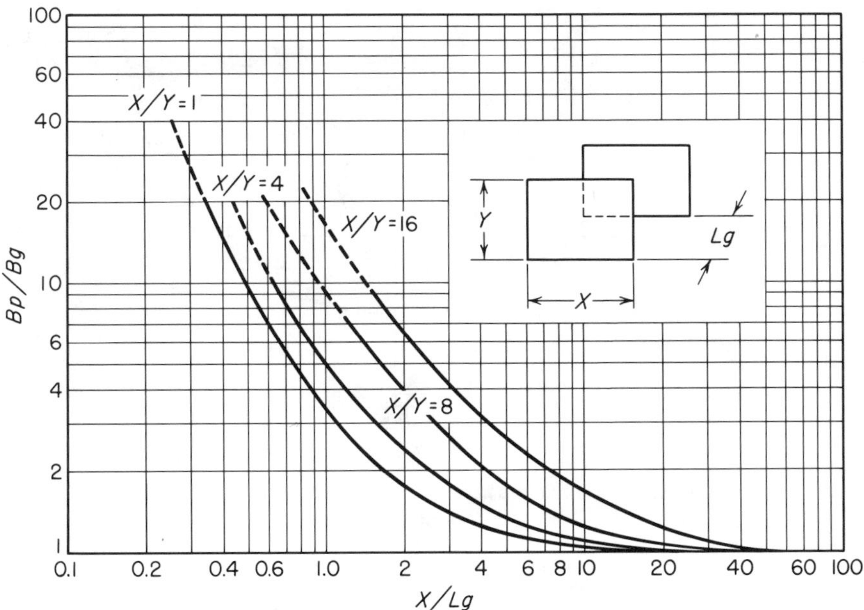

Figure 6.23 Air space flux density relationships for gaps of rectangular section.

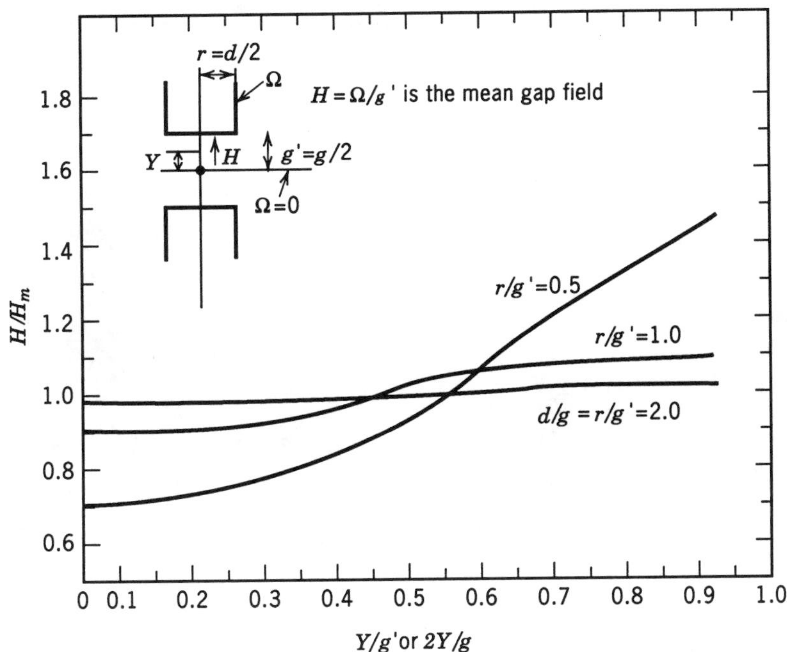

Figure 6.24 Variation of field across the gap at the pole axis.

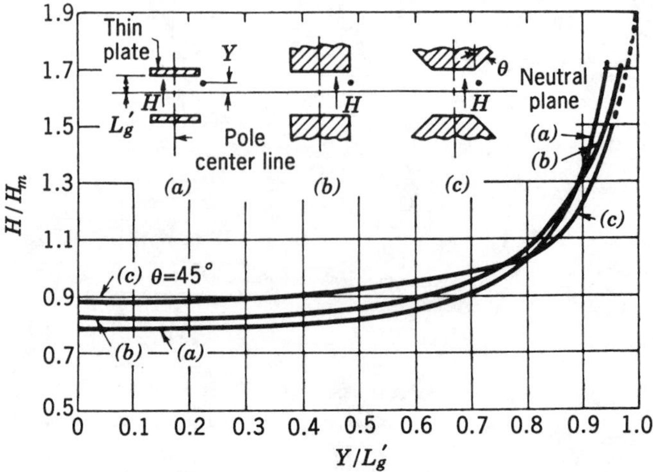

Figure 6.25 Variation of field across the gap at the edge of semi-infinite poles.

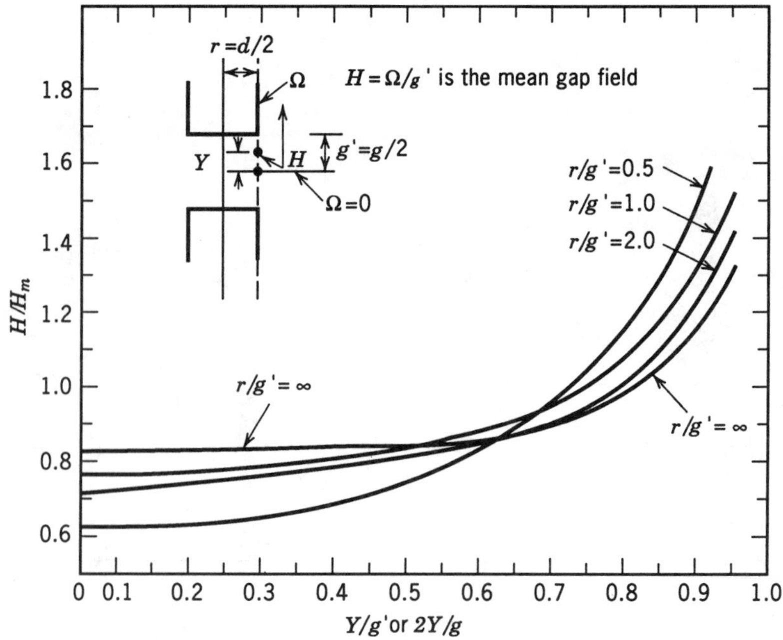

Figure 6.26 Variation of the field across the gap at the edge of similar cylindrical poles.

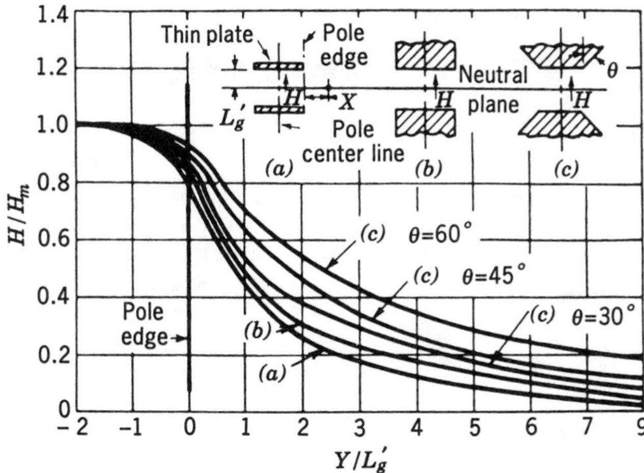

Figure 6.27 Variation of the field midway between similar semi-infinite poles.

where F_1 is a complex function of gap spacing and diameter. B_p/B_g ratios for discs, spheres and rectangular plates (Figures 6.22 and 6.23) of various dimensions have been calculated using electrostatic equations and the flux density relationship measured at several points on the data sheets by actual measurements in the magnetic system. The data shown in Figures 6.24–6.28 show flux distribution data for semi-infinite gap geometries and for finite circular gap geometries. This data was developed by BTH Laboratories [10] for electromagnetic studies, but the data are just as appropriate for permanent magnet problems. In each case the field strength H is expressed as a function of the mean field strength of the gap H_m where H_m is gap potential divided by gap length.

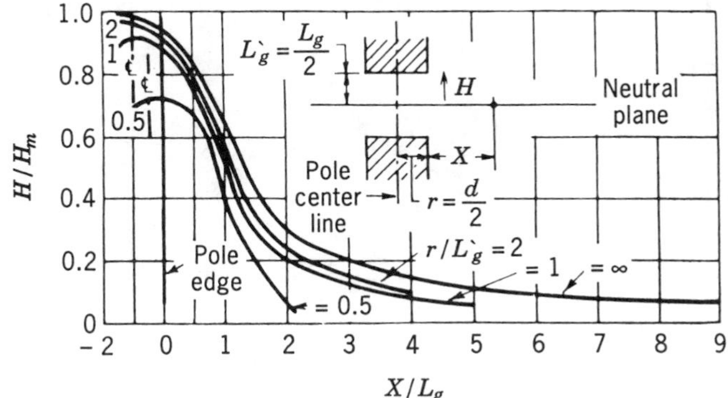

Figure 6.28 Variation of field midway between cylindrical poles.

6.3.6 Some Examples of Permeance Calculations

Problem 1. Given the ferrite loudspeaker structure of Figure 6.29a, the requirement is to calculate the flux density expected in the air gap using ceramic 5 material. We must first calculate the permeance of the leakage paths in Figure 6.29a. The formulas for the estimated paths are shown in Figure 6.30. P_a was derived by Spreadbury; P_b, P_c, P_h, and P_a are from Rotors. P_f, P_k, and P_n were derived from the free pole formula developed in an earlier section. The consideration here is to use the free pole concept on the magnet and pole plates and to assume that the ratio of the permeance of

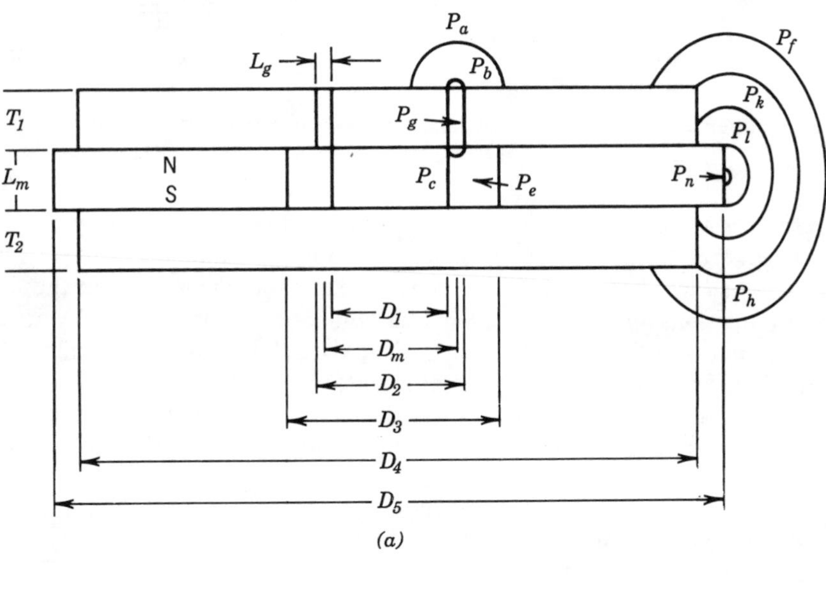

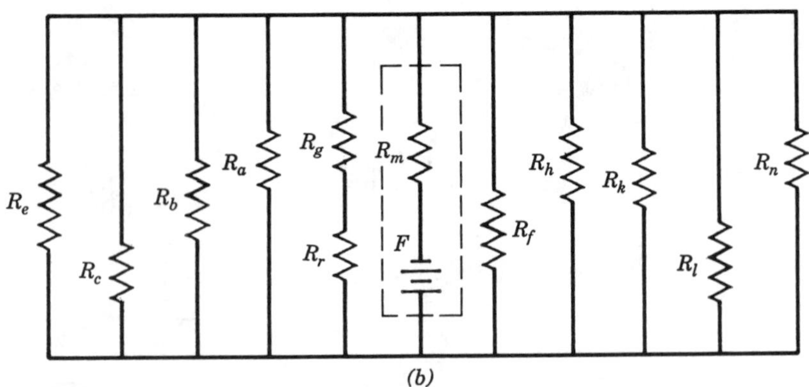

Figure 6.29 (a) Loudspeaker structure. (b) Electrical analog of loudspeaker circuit.

$$P_g = \pi DT_1/L_g = \pi(2.64)(0.64)/0.10 = 53.0$$

$$P_a = D \log_e\left(1 + \frac{D_2}{L_g}\right) 2 \cdot 6 \log_e\left(1 + \frac{2.74}{0.10}\right) = 8.79$$

$$P_b = P_c = \pi D_2/4 = \pi(2.74)/4 = 2.15$$

$$P_e = \frac{2\pi L_m}{\log_e D_3/D_1} = \frac{2\pi(1.25)}{\log_e 1.33} = 16.02$$

$$P_f = \frac{\pi^{3/2}(D_4^2 - D_2^2)}{4S^{1/2}} = \frac{\pi^{3/2}(54.9)}{4(86.6)^{1/2}} = 8.23$$

$$P_h = P_l = \pi D_4/4 = \pi(7.90)/4 = 6.20$$

$$P_k = \pi^{3/2}T_1 D_4/S^{1/2} = \pi^{3/2}(0.64)(7.90)/(86.6)^{1/2} = 3.03$$

$$P_n = \pi^{3/2}\frac{L_m}{2} D_5/S^{1/2} = \pi^{3/2}\frac{(1.25)}{2}(8.60)/(86.6)^{1/2} = 3.22$$

$D_1 = 2.54$ cm $\quad D_2 = 2.74$ cm $\quad D_3 = 3.40$ cm $\quad D_4 = 7.90$ cm $\quad D_5 = 8.6$ cm

$L_g = 0.10 \quad L_m = 1.25$ cm $\quad T_1 = T_2 = 0.64$ cm $\quad D_m = 2.64$ cm

$$S = \pi D_5 L_m/2 + T_1 \pi D_4 + \pi(D_4^2 - D_2^2)/4 + T_1 \pi D_2 + \pi D_2 L_m/2$$

$$S = \pi(8.6)(0.625) + (0.64)\pi(7.90) + \pi(7.90^2 - 2.74^2)/4 + (0.64)\pi(2.74) + \pi(2.74)(0.625) = 86.6 \text{ cm}^2$$

$$P_g = 53.0$$

$$P_t = P_g + P_a + P_b + P_c + P_e + P_f + P_h + P_l + P_k + P_n$$

$$P_t = 53.0 + 8.79 + 2.15 + 2.15 + 16.02 + 8.23 + 6.20 + 6.20 + 3.03 + 3.22$$

$$P_t = 98.9 \quad A_m = \frac{\pi}{4}(D_5^2 - D_3^2) = 48.9 \text{ cm}^2 \quad A_g = \pi(2.64)0.64 = 5.30 \text{ cm}^2$$

$$P_t/P_G = 98.9/53.0 = 1.86$$

$$B/H = \frac{L_m}{A_m}(P_t) = \frac{1.25}{48.9}(98.9) = 2.52$$

$$\phi_g = A_m B_d (P_g/P_t)$$

$$\phi_g = 48.9(2800)\left(\frac{53.0}{98.9}\right) = 73{,}300 \text{ lines}$$

$$B_g = \phi_g/A_g = 73{,}300/5.30 = 13{,}700 \text{ G}$$

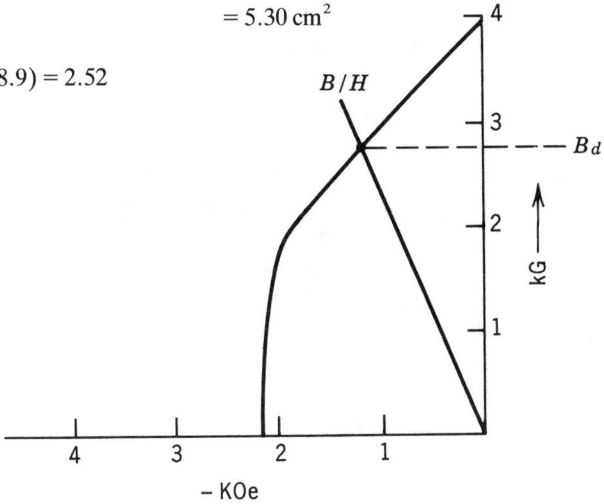

Figure 6.30 Speaker permeance calculations.

each surface element to the total permeance of the pole is simply the ratio of the area of the particular surface element to the total area of the pole. In applying these formulas one is concerned with how to break up the regions and how accuracy will be influenced. Experience seems to suggest, that in general, errors tend to cancel each other. Where one formula will underestimate the permeance, the next may well overestimate the permeance. By dividing the leakage into several paths, reasonably good overall results can be expected.

The electrical analogy of the speaker magnet structure is shown in Figure 6.29b. R_m represents the magnet internal reluctance and is in series with the potential F. Many approximations are involved. For example, the return path reluctance R_r is actually several reluctances. However, experiences have shown that lumping the return path into one reluctance and placing it in series with R_g, the air gap reluctance is a reasonable approximation. Another judgment call is the question of R_e and R_n. The full potential of the magnet is not really acting on these paths and these reluctances could be divided by the estimated fraction of potential that is in effect. This, however, is a minor refinement which has not been used for these calculations. The total permeance is calculated and the load line value can be laid off on the demagnetization curve for ceramic 5 to determine the level of B_d, which is then multiplied by magnet area A_m, to give total flux. The gap flux is determined by multiplying the total flux by the P_g/P_t ratio. B_g, the flux density is then computed by dividing ϕ_g by gap area A_g. With the dimension given, the load line is considerably above the $(BH)_{\max}$ level and one could reduce magnet length L_m to improve efficiency. However, there are some additional considerations. If the magnet is to be exposed to low temperatures, the coercive force will diminish and the operating point must be kept above the knee of the low temperature curve. Additionally, due to the brittle nature of the ferrite material, there is a minimum thickness below which manufacturing losses are excessive. In magnet design this is sometimes the situation; some manufacturing limitation will determine the dimensions rather that the magnetic parameters.

Problem 2. Consider the need to calculate the open circuit load line for three magnet configurations. The magnets all have the same volume, length, and area. The three magnets are dimensioned in centimeters and shown in Figure 6.31. We can compute the load line by applying the free pole permeance formula. The load line is determined by the magnet geometry. The $\sqrt{S}$ is computed for each geometry and (6.20) is then used to compute the load line. The load line could also be estimated by referring to Figures 6.13 and 6.16 where the interrelationships are displayed graphically.

The calculations show that the load line is not only a function of length and area, but of pole surface. The free pole permeance concept allows load lines to be rapidly evaluated.

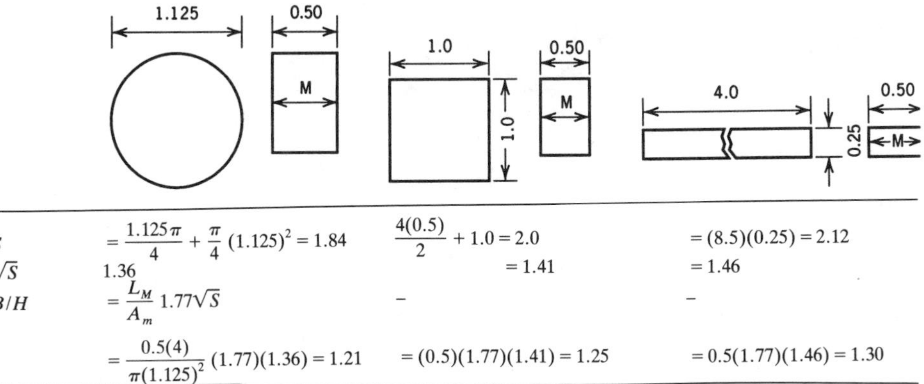

S	$= \frac{1.125\pi}{4} + \frac{\pi}{4}(1.125)^2 = 1.84$	$\frac{4(0.5)}{2} + 1.0 = 2.0$	$= (8.5)(0.25) = 2.12$
$\sqrt{S}$	1.36	$= 1.41$	$= 1.46$
B/H	$= \frac{L_M}{A_m} 1.77\sqrt{S}$	–	–
	$= \frac{0.5(4)}{\pi(1.125)^2}(1.77)(1.36) = 1.21$	$= (0.5)(1.77)(1.41) = 1.25$	$= 0.5(1.77)(1.46) = 1.30$

Figure 6.31 Open circuit permeance calculations.

Problem 3. In Figure 6.32a, a NdFeB sintered magnet with return path and moving armature is shown. This assembly is used in a mechanical work function. The force-distance performance curve is required. In Figure 6.32b, the unit property demagnetization curve for NdFeB is shown. The approach used is to assume three different air gap changes, and to estimate the change in magnetic field energy associated with each gap change. The equivalent mechanical work units are then plotted against air gap values to estimate a force distance relationship as shown in Figure 6.32c. To determine the zero gap force value we may use the expression $F = B^2A/8\pi$. As the armature is moved toward the magnet poles, work will be done. Magnetic field energy will be converted to mechanical energy. The change in field energy expressed in BH units can be converted to ergs (dyne-cm) by dividing by 8π. The force is then determined by dividing the change in energy in ergs by the change in air gap to give a force value in dynes. The blocks of mechanical work are plotted on the force-distance graph and then F_1, F_2, and F_3 values associated with the energy changes are shown. On Figure 6.32b the load lines are estimated from L_m/L_g ratios, since A_g and A_m are the same. At large air gaps this approximation will underestimate the true permeance. Magnet limb permeance and fringing permeance could be added to improve the solution accuracy.

6.4 MAGNETIC AND ELECTRICAL CIRCUIT ANALOGY

In working with magnetic circuits it is often helpful to consider the analogy between electrical and magnetic circuits [11]. The analogy is far from being complete because of inherent differences; however, many magnetic circuit concepts are often more easily understood by a consideration of the electrical equivalent.

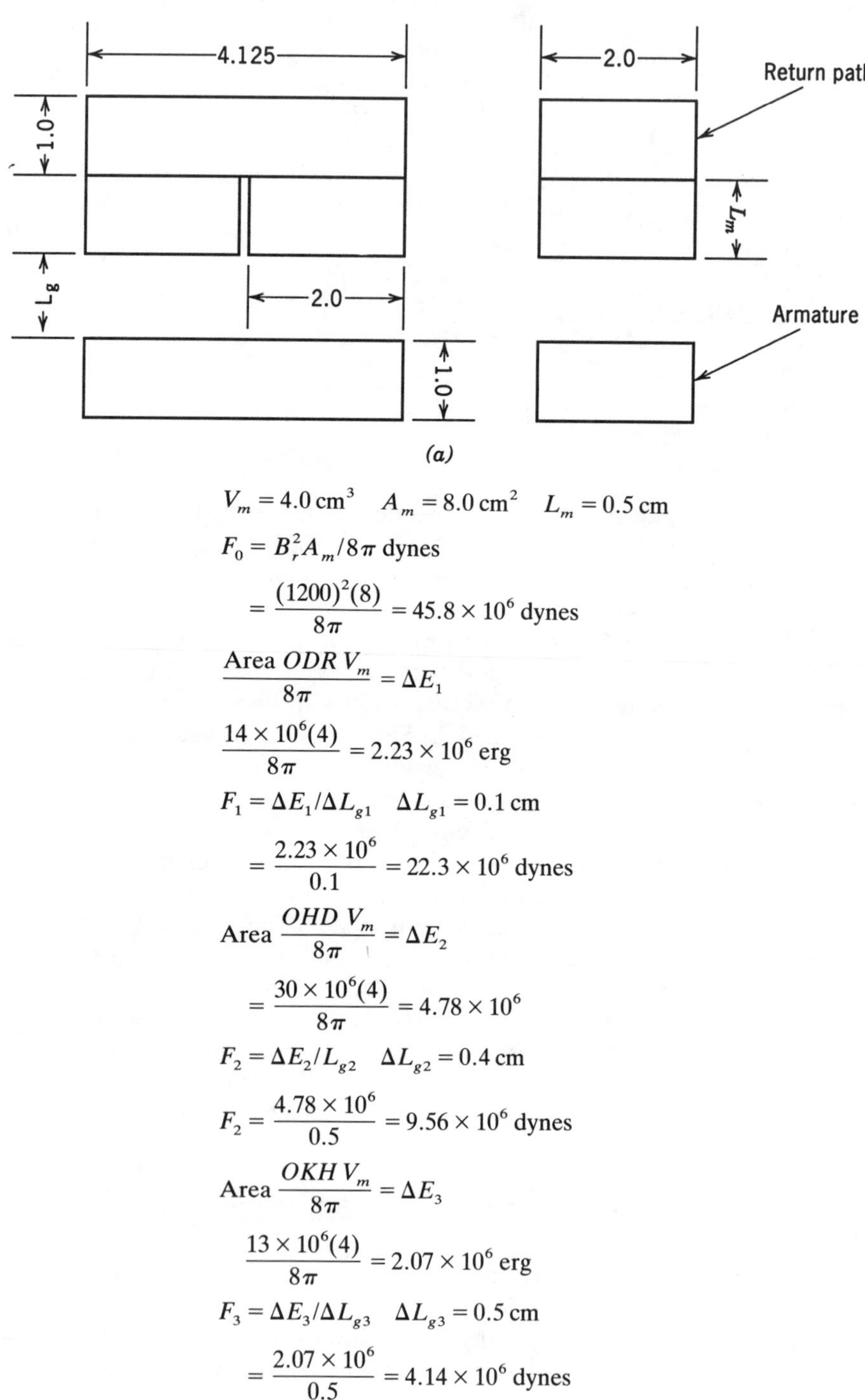

$$V_m = 4.0\ \text{cm}^3 \quad A_m = 8.0\ \text{cm}^2 \quad L_m = 0.5\ \text{cm}$$

$$F_0 = B_r^2 A_m / 8\pi \ \text{dynes}$$

$$= \frac{(1200)^2(8)}{8\pi} = 45.8 \times 10^6 \ \text{dynes}$$

$$\frac{\text{Area } ODR\ V_m}{8\pi} = \Delta E_1$$

$$\frac{14 \times 10^6(4)}{8\pi} = 2.23 \times 10^6 \ \text{erg}$$

$$F_1 = \Delta E_1 / \Delta L_{g1} \quad \Delta L_{g1} = 0.1\ \text{cm}$$

$$= \frac{2.23 \times 10^6}{0.1} = 22.3 \times 10^6 \ \text{dynes}$$

$$\text{Area } \frac{OHD\ V_m}{8\pi} = \Delta E_2$$

$$= \frac{30 \times 10^6(4)}{8\pi} = 4.78 \times 10^6$$

$$F_2 = \Delta E_2 / L_{g2} \quad \Delta L_{g2} = 0.4\ \text{cm}$$

$$F_2 = \frac{4.78 \times 10^6}{0.5} = 9.56 \times 10^6 \ \text{dynes}$$

$$\text{Area } \frac{OKH\ V_m}{8\pi} = \Delta E_3$$

$$\frac{13 \times 10^6(4)}{8\pi} = 2.07 \times 10^6 \ \text{erg}$$

$$F_3 = \Delta E_3 / \Delta L_{g3} \quad \Delta L_{g3} = 0.5\ \text{cm}$$

$$= \frac{2.07 \times 10^6}{0.5} = 4.14 \times 10^6 \ \text{dynes}$$

Figure 6.32(a)(b)(c) Mechanical work problem.

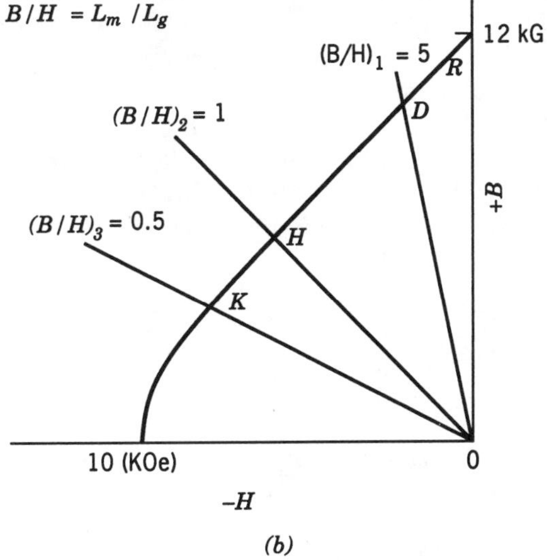

(b)

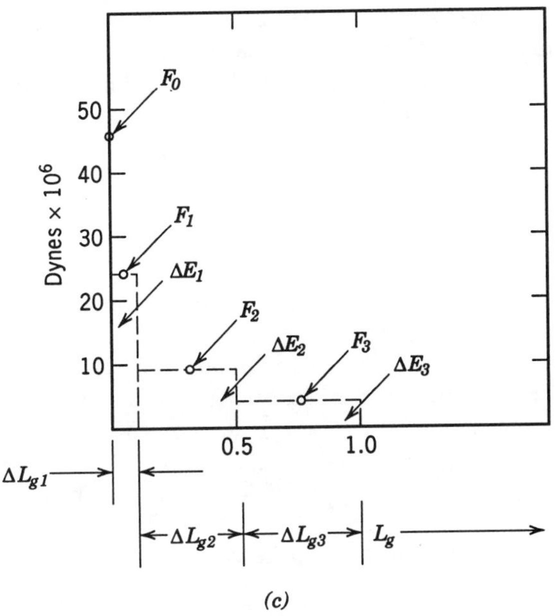

(c)

Figure 6.32 (*Continued*)

In the magnetic circuit we have the concept of flux lines in a given cross-section. If B represents a constant flux density and A is the area of the cross-section, then the total flux $= BA$. Also the work done in carrying a unit pole around a path along a particular flux line is $H \cos\theta \, dl$, and is the magnetomotive force of the circuit. At any point in the circuit $H = B/\mu$ or $H = \phi/\mu A$ since $B = \phi/A$. Multiplying both sides of the above expression by dl, an elementary length of magnetic circuit, we have $H\,dl = \phi\, dl/\mu A$, if H and dl are assumed to be in the same direction the term $\cos\theta$ may be dropped and integrating over the complete circuit we have

$$\int H\,dl = \phi \int dl/\mu A$$

or

$$\phi = \frac{\int H\,dl}{\int dl/\mu A} = \frac{\text{Magnetomotive force}}{\text{Reluctance}} \tag{6.23}$$

In the electrical circuit, moving electrical charges constitute current flow in a conductor. The electromotive force is the work done in moving a unit charge around the electrical circuit. The conductor resistance is given by $L\rho/A$, where ρ is the specific resistance of the conductor material and L and A are the length and area of the conductor. Ohms law relates current, voltage, and resistance

$$I(\text{Current}) = \frac{\text{Electromotive force}}{\text{Resistance}} \tag{6.24}$$

In the electrical circuit, current is confined in the conductor due to the great difference between the resistivity of the conductor and the resistivity of surrounding space. In the magnetic circuit the difference between the permeability of magnetic materials and the space surrounding them is relatively speaking, much less. There is no known magnetic insulator, and consequently we encounter magnetic leakage flux. Also, while ρ is essentially constant, permeability μ is a function of the flux in the circuit and cannot be easily evaluated.

We have seen how the permanent magnet must operate into a particular value of unit permeance if it is to establish maximum external field energy. The demagnetization curve may be compared to the volt-ampere characteristic of a battery having an internal resistance and delivering power to a load resistance (Figure 6.33a). In the battery, the open circuit voltage is analogous to the coercive force of the permanent magnet and the short circuit current I_{sc} is the equivalent of the residual induction B_r (Figure

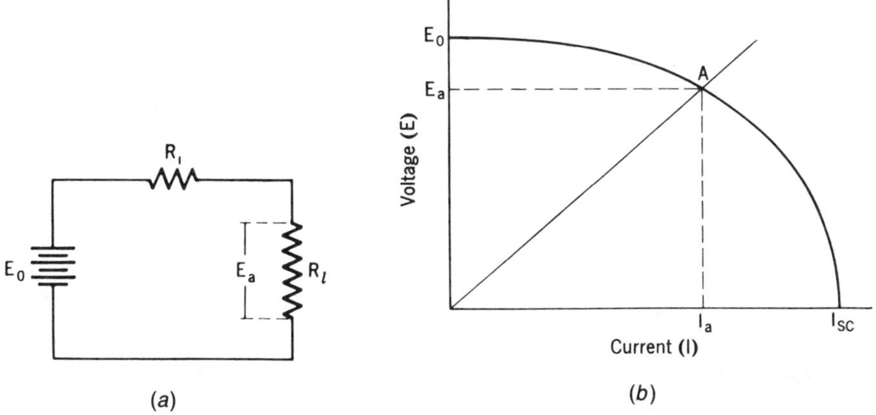

Figure 6.33 (a) Elementary electrical circuit analog of the permanent magnet. (b) Volt-ampere characteristic of the permanent magnet.

6.33b). The load line $R = E/I$, intersecting with the battery characteristic gives the voltage and current delivered to the load, and the product of E_a and I_a is the power delivered. Similarly, the permanent magnet load line

$$\frac{B}{H} = \frac{L_m}{A_m} \frac{A_g}{L_g} \tag{6.25}$$

where the slope of the line is the air-gap reluctance and the product of flux density B_m and magnetizing force H_m is proportional to the air-gap energy per unit volume of permanent magnet material. In the case of the battery, Theveron's theory shows us that maximum power is developed in the load resistor when it is equal to the internal resistance of the battery. Similarly, with the permanent magnet, maximum external energy is established when the reluctance of the gap is equal to the internal reluctance of the magnet. Consider the general case of the permanent magnet in a circuit containing an air gap in which the flux is to be maximized, combined with leakage paths. Figure 6.34 represents this circuit. We know that optimum energy will exist in the gap when the load is matched to the source

$$R_g = \frac{R_L R_m}{R_L + R_m} \tag{6.26}$$

or when the gap reluctance is equal to the parallel reluctance of magnet and leakage path. For type II magnets this simple representation of the magnet as a magnetomotive source with an internal reluctance is possible due to the constant magnetization irrespective of load conditions. The magnetic energy supplied may be expressed as

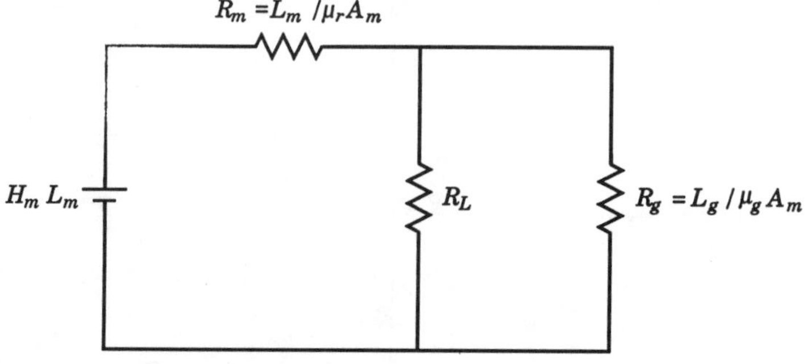

R_m = Magnet reluctance
R_L = Leakage reluctance
R_g = Useful air gap reluctance

Figure 6.34 Equivalent permanent magnet circuit.

$$\frac{H_m L_m^2}{4R_m}\left[1 - \frac{1}{(1+X)^2}\right] \tag{6.27}$$

where $X = R_l/R_m$. In this equation the fraction outside the bracket is an expression for the energy of a type II magnet operating at $B/H = 1$ with the idealized condition of $R_L = \infty$. If the magnet volume and weight are to be a minimum then R_L must be maximized and X made large.

In order to appreciate how easy it is to work a problem with type II magnets compared to earlier type I magnets, consider the circuit of Figure 6.35a. The various parallel leakage permeances are lumped together as P_L, which is in parallel with useful gap permeance P_g. This parallel combination P_t is in series with P_m, the magnet internal permeance and F, the magnetomotive force. For a type II magnet like $SmCo_5$ the demagnetization curve is linear and reversible. It may be expressed as

$$B_m = \mu_r H_m + B_r \tag{6.28}$$

with μ_r being used for magnet permeability. In addition

$$F = H_m L_m = \frac{L_m B_r}{\mu_r} \tag{6.29}$$

and $\phi = FP_t$, and since all permeances are in parallel one may find the flux of the gap from

$$\phi_g = \phi_t P_g / P_t \tag{6.30}$$

and flux density from

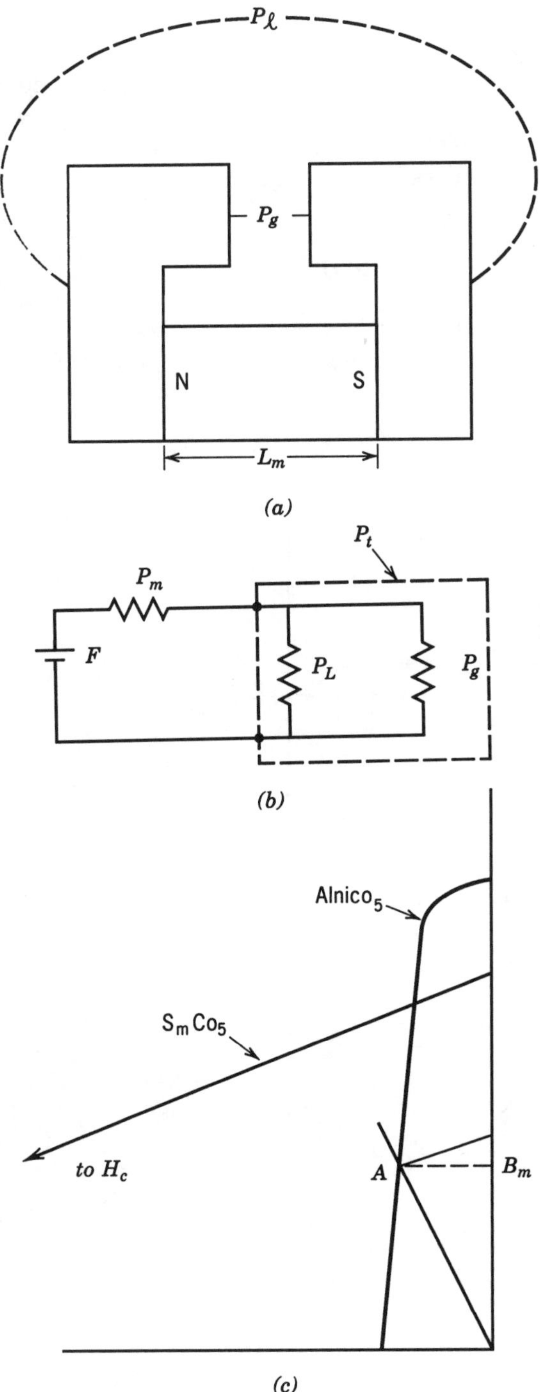

Figure 6.35 Comparison of solutions.

$$B_g = \phi_g / A_g \tag{6.31}$$

This is very straightforward since everything is linear. We do not really need the demagnetization curve of the material. In contrast, if we substituted a type I material such as Alnico 5 in the magnet configuration of Figure 6.35a we must obtain the operating load line from

$$\frac{B}{H} = \frac{L_m}{A_m} P_t \tag{6.32}$$

The interaction of the load line with the demagnetizing curve, point A (Figure 6.35c) allows the determination of B_m. The total flux ϕ_t is $B_m A_m$ and

$$\phi_g = \frac{\phi_t P_g}{P_t} \tag{6.33}$$

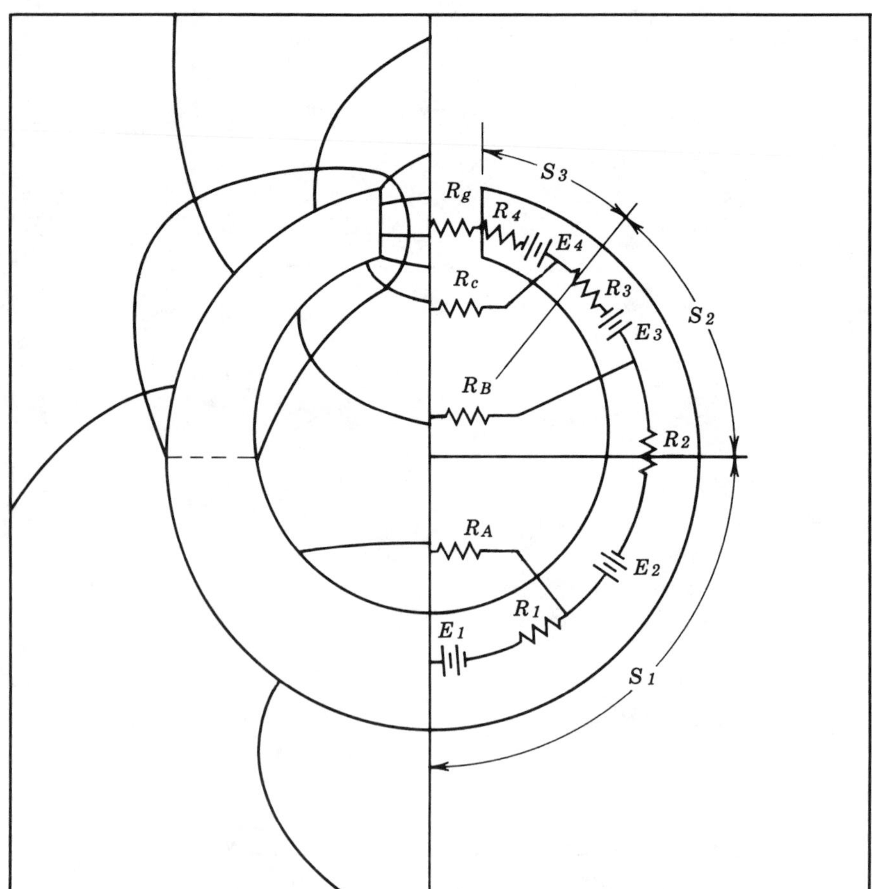

Figure 6.36 Electrical analogy of type I magnet.

Because the Alnico 5 curve is nonlinear and irreversible, we resort to the graphical analysis requiring additional steps. With type I properties it is often necessary to zone the magnet and consider several load lines. Figure 6.36 shows a C-shaped long limbed Alnico 5 configuration and equivalent electrical circuit. The flux plot is on the left-hand side and the equivalent circuit on the right. Here the approach is to divide the magnet into three zones and lump the circuit parameters. To represent the situation in a type I permanent magnet where the magnetomotive force is not constant, we use batteries that produce different potentials and different resistances to represent the changing internal permeance from zone to zone. The circuit variables allow exploration and fine tuning of magnet length and area to maximize gap energy. The optimum configuration is a tapered limb which tends to keep the entire magnet volume operating near $(BH)_{max}$.

6.5 USE OF HIGH PERMEABILITY MATERIALS IN PERMANENT MAGNET CIRCUITS

Flux conducting members in a permanent magnet circuit are used to (i) complete a return path for the flux, (ii) change the flux density in a circuit, or (iii) from multiple magnet poles. Figure 6.37 shows the circuits involved. In Figure 6.38 some widely used high permeability materials are shown.

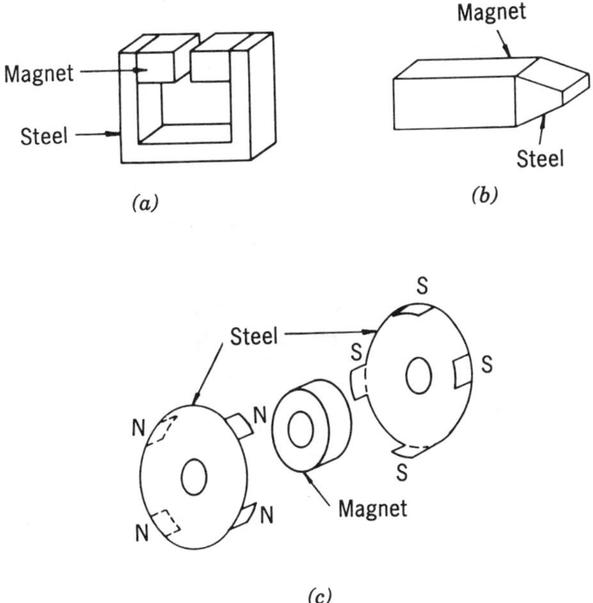

Figure 6.37 The use of high permeability steel in permanent magnet circuits.

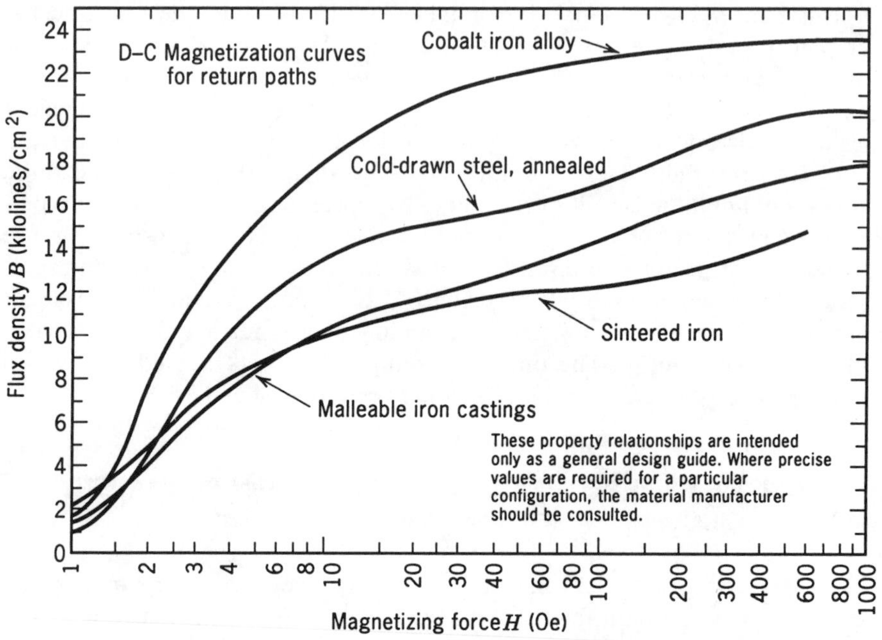

Figure 6.38 Magnetization curves for some high permeability materials.

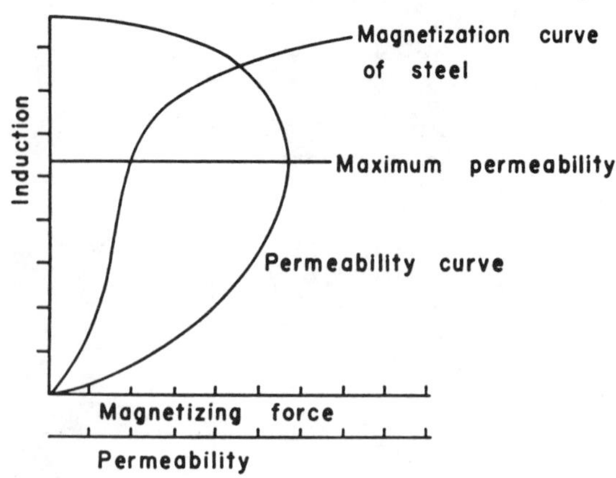

Figure 6.39 Relationships between permeability and induction.

Usually cost is an important factor in the choice and a high percentage of permanent magnet circuit elements are ordinary cold rolled steel. Cobalt-iron alloy is used in some high performance circuits. It is well to operate these materials near their maximum permeability. This will tend to minimize the cross-sectional area and at the same time keep the magnetic potential loss to a low level. If the cross-section is too small then the H drop in the material will tend to be high and in a long length of circuit an appreciable part of the magnet's potential would be dropped in the circuit member and be unavailable as potential to use across the air gap. The relationship between permeability and magnetic induction is shown in Figure 6.39.

Another fundamental principle in permanent magnet circuits is illustrated in Figure 6.40. Here the location of the magnet in the circuit is of importance. The magnet should be as close to the gap as possible, otherwise the magnets full potential will lead to excessive flux leakage between the circuit elements or pole pieces.

6.6 ECONOMIC CONSIDERATIONS IN DESIGN AND PROPERTY SELECTION

Permanent magnet prices are a function of many variables, such as weight, shape, mechanical tolerances, degree of freedom from chips and cracks. Prices also depend on the magnet plant and its efficiency in manufacturing a given material and shape. Often to evaluate one material with respect to another, it is convenient to generalize and reduce the prices to a $/lb basis. This approach can be misleading because of the relationships shown in Figure 6.41. In the magnet plant the relationship between magnet weight and cost can be broken down into two components, the relatively fixed cost of processing events plus the variable raw material cost. For micromagnets, the cost to form is the major part of the total. As weight is increased, the raw material cost may become the dominant part. It is now clear that very small magnets in terms of $/lb are very high while large magnets have a lower $/lb ratio. From another viewpoint, this fundamental helps us select magnet materials. If we are trying to purchase a micromagnet, we will tend

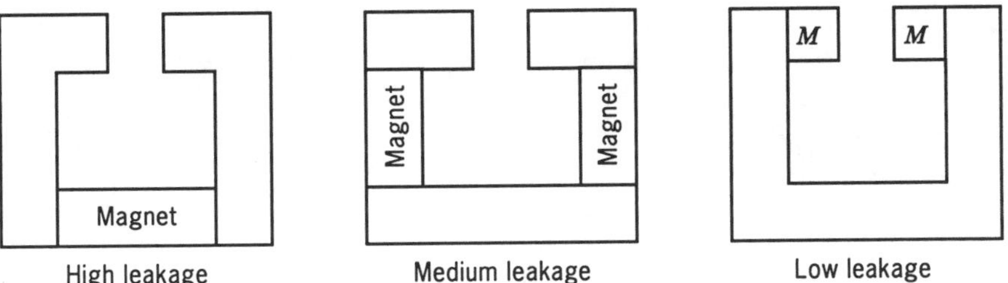

Figure 6.40 Circuit element location.

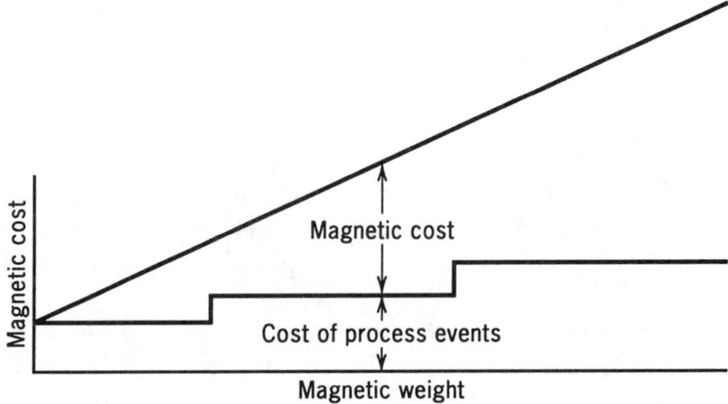

Figure 6.41 Fundamentals of magnet costing.

to get nearly the same component cost for all materials as determined by processing events. To slice and dice a material or to press a small part, the cost is quite independent of the material involved. In a sense, all micromagnets, regardless of raw material, cost the same. Consequently the best value in magnetic field energy will be to buy the magnet with the highest magnetic properties. As new higher energy materials have been developed, the first markets for these magnets are always in micro devices. A way to obtain lower cost magnets is to form them in multiple, for example, ductile magnets in wire form so that the user can cut to size. Cast Alnico magnets are often formed in multiple and broken off by the user. In rubber bonded magnets, the user may cut out small magnets from a continuously processed strip. These are examples of distributing the processing events over several small magnets in order to obtain a lower cost micromagnet. In planning the layout of a magnetic circuit it is usually better to use one larger magnet than two smaller magnets, if possible. The one larger magnet avoids paying for the fixed processing events twice. This, of course, is only true for quite small magnets.

Another economic principle which often leads one to a particular set of magnetic properties is shown in Figure 6.42. The issue is to consider how permanent magnet property energy density influences the cost of the total magnetic circuit in magnetoelectric machines over a size range. The example is for permanent magnet d.c. motors in sizes from a few watts to a few kilowatts, or from small consumer appliance motors to medium size industrial motors. In the display, we consider unit properties from about 1 MGOe in flexible rubber magnets to a 20 MGOe $SmCo_5$ magnet. In a 10 W motor the lowest cost magnet leads to the lowest cost magnetic circuit. In the 100 W automotive accessory motor, the best valued sintered ceramic magnet leads to the lowest cost magnetic circuit. In the larger machines of 1000 W or more, one finds that a high energy density magnet has great leverage on the cost of the return path components and in general on the total motor size.

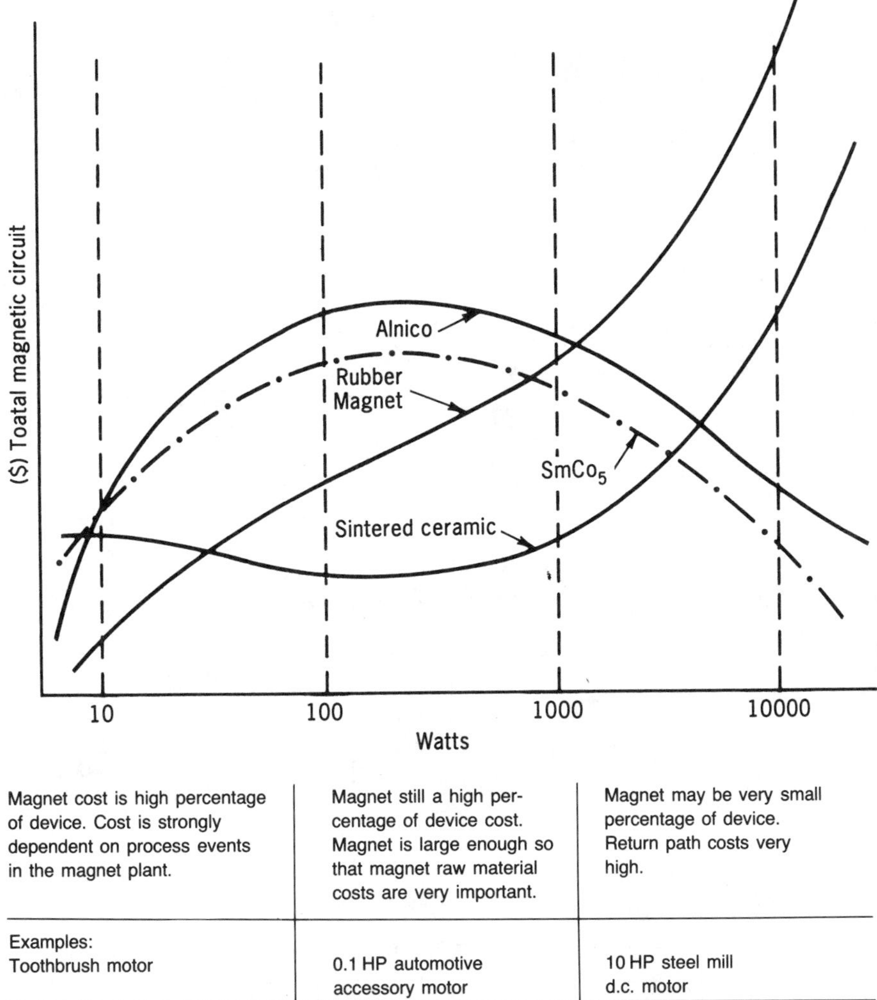

Magnet cost is high percentage of device. Cost is strongly dependent on process events in the magnet plant.	Magnet still a high percentage of device cost. Magnet is large enough so that magnet raw material costs are very important.	Magnet may be very small percentage of device. Return path costs very high.
Examples: Toothbrush motor	0.1 HP automotive accessory motor	10 HP steel mill d.c. motor

Figure 6.42 Cost of permanent magnet circuit related to material and size.

The conclusion is that one cannot afford the lowest cost magnets because of the adverse influence low properties would have on the machines total energy density and circuit component costs. This is an excellent example of the impact that magnetic properties can have on the whole device and it clearly shows the need for systems thinking and analysis. The permanent magnet is too sophisticated and functional to allow for pricing totally in terms of $/lb ratios.

The designer now has a tremendous range of unit properties to choose from. Figure 6.43 is an interesting comparison of magnet volume and weight for operating at $(BH)_{max}$. The relative weights and volumes are referenced

Relative Magnet Volume — Sm (Co, Fe, Cu, Zr)$_{7.4}$=1

31	31	17.2	8.6	7.0	5.3	5.2	3.9	3.6	1.6	1.0

Relative Magnet Weight — Sm (Co, Fe, Cu, Zr)$_{7.4}$=1

20	28	19	5.2	6.6	4.8	3.0	3.8	7.0	1.6	1.0

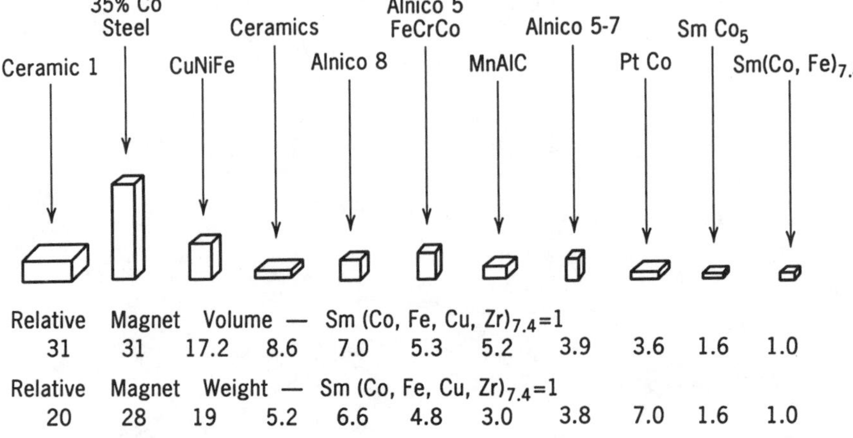

Figure 6.43 Economic comparisons of magnet materials. From K. J. Struat.

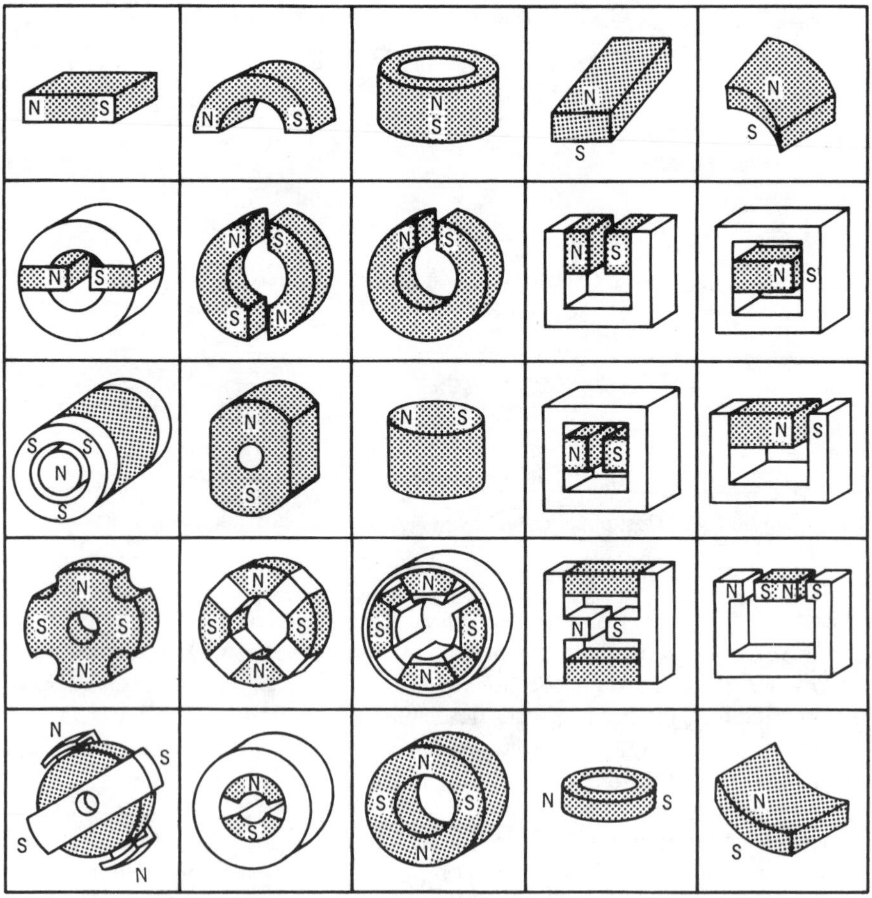

Figure 6.44 Magnetc circuit arrangements.

to Sm(Co, Fe, Cu, Zr)$_{7.4}$. The designer also has great freedom in a choice of magnetic circuits. Figure 6.44 shows several permanent magnet circuit arrangements. The property and circuit combinations are many. However, often one predominant end device parameter or characteristic will narrow the choices. Stability, magnetizability, or a required physical property may be very important in addition to economic considerations.

6.7 PERMANENT MAGNETS AND THE LAWS OF ELECTROMAGNETIC SCALING

In Figure 6.45 simplified models of a permanent magnet structure and an electromagnetic structure are shown [12]. In applying the scaling law, it is first interesting to consider the case of the fixed gap (scaling area only) and then the case of scaling all dimensions (full scaling). In the fixed gap case, the volume of the permanent magnet will be proportional to the second power of the scaling factor K, and inversely proportional to $(BH)_{max}$ for a given gap energy. For the electromagnet, which is part conducting material and past magnetic core, we find the core portion scales as the second power of K, while the conductor portion scales as the first power of K. This fixed gap scaling is the typical situation in scaling rotating electrical machines

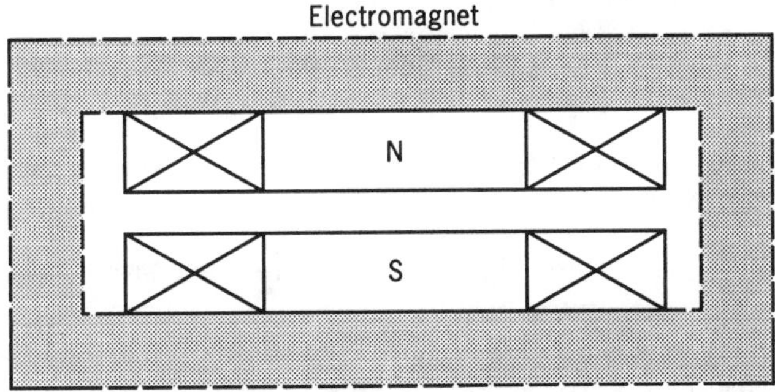

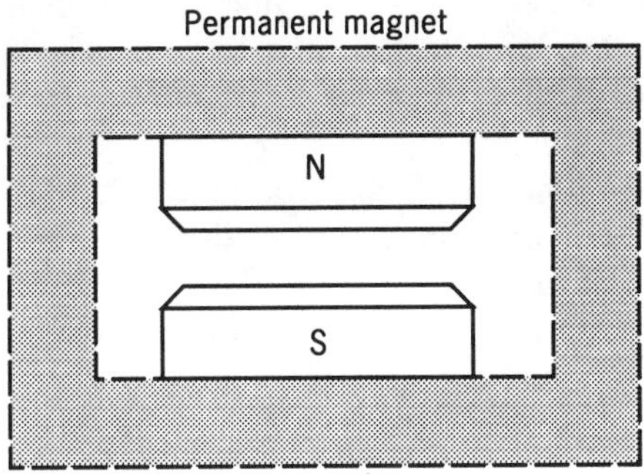

Figure 6.45 Simplified models of electromagnets and permanent magnets.

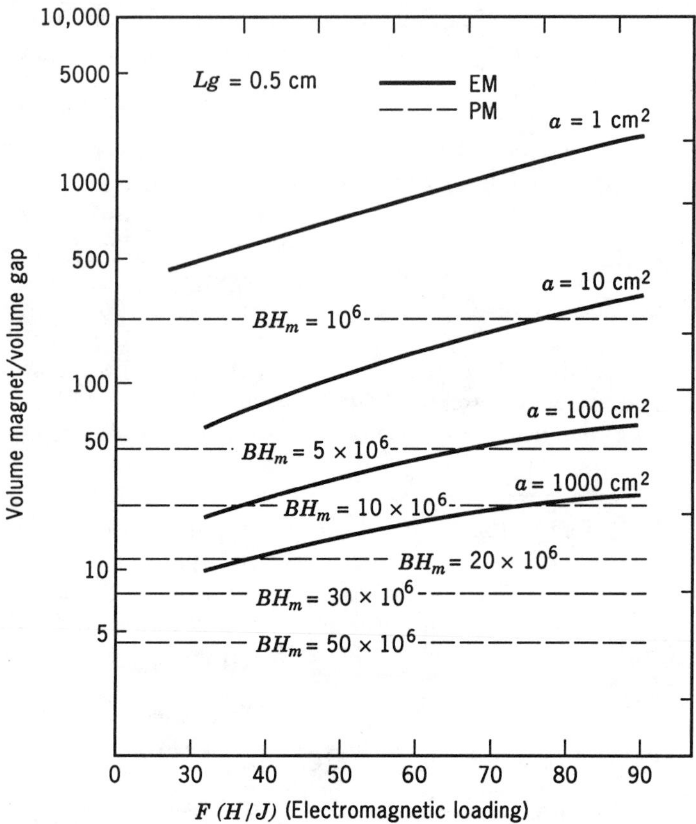

Figure 6.46 Volumetric efficiency of electromagnets and permanent magnets.

since the air gaps are rather independent of machine size. Under fixed gap scaling, it follows that the scaling law favors electromagnetism in large structures with high permeance gaps. Our traditional experiences tend to confirm this situation. We generally think of permanent magnets for small devices and electromagnetism in large structures. In the case of scaling all dimensions including the air gap, we find the permanent magnet volume now scales as K^3. Now the electromagnet requires more ampere-turns and the conductor cross-section must be increased as the second power of K, thus both conductor and core change as K^3. Consequently, in full scaling the permanent magnet is not at a disadvantage. This is particularly true in devices that have low permeance gaps such as microwave tubes and beam control structures. As we scale structures, we find current density is inversely proportional to scale factor and the permanent magnet gains with respect to electromagnets in small devices. The ability of the magnet to establish a field without ohmic loss is very significant. In many small devices the increasing current density in current carrying conductors leads to insurmountable cooling problems and field levels cannot be maintained. The permanent magnets ability to maintain a field is unchanged with scale. This feature makes high energy rare-earth magnets capable of solutions not possible or feasible with even a superconducting coil under some magnetic circuit conditions [13]. Figure 6.46 gives a useful framework comparing electromagnets to permanent magnets. The model involved here is one of partial scaling with a fixed gap of 0.5 cm and a gap flux density of 10 kG. The volume efficiency of the electromagnet and permanent magnet (V_m/V_g) are plotted for cross-section areas of 1, 10, 100, and 1000 cm^2. An additional factor F, the ratio of H/J is also displayed. For a fixed field the horizontal axis becomes effectively a range of current densities. In the model J the current density is varied from 100 to 300 A/cm^2. Permanent magnet properties from 1 to 50 MGOe are displayed. Consistent with our experiences, this display shows that it takes very high property magnets to volumetrically compete with a large electromagnet having a short gap length. In small structures, even a 1 MGOe magnet competes favorably with electromagnets. In summary, todays best magnets can compete favorably with electromagnetism. Indeed, high energy magnets represent an important interim step up from the energy density of electromagnetism toward the energy density of the superconductor. Even in large systems, when one considers volume efficiency, operating costs and support facilities, the permanent magnet may well be the best solution. This is particularly true in transportation systems since here energy density improvements translate directly into system efficiency.

REFERENCES

[1] D. C. MacDonald and B. Perch, Finite Element Magnetic Analysis of Certain Three Dimensional Devices with Two Dimensional Software, 4th Joint MMM-Intermag Conference, Vancouver, 1988.

[2] A. E. Paladino, N. J. Dionne, P. F. Weihrauch and E. C. Wettstein, Rare-earth cobalt permanent magnet technology, Goldschmidt Inf. 35 (1975) 63.

[3] H. A. Leupold and F. Rothward, Design of a tuneable magnetic circuit for K and Ka band microwave filters, Proc. 2nd International Workshop on Rare-Earth Cobalt Permanent Magnets and Their Applications, 1976, p. 187.

[4] A. R. Hand, General Electric Technical Information Report No. R50GL221.

[5] F. G. Spreadbury, Permanent Magnets (Pitman, London, 1949) p. 101.

[6] H. C. Rotors, Electromagnetic Devices (John Wiley, New York, 1941) p. 120.

[7] S. Evershed, J. Inst. Electr. Eng. 1920 780–837.

[8] F. E. Carter, J. Elect. World (November 1901).

[9] W. Cramp and N. I. Calderwood, J. Inst. Electr. Eng. 61 (1923) 1061.

[10] R. L. Balke, British Thomson Houston Company, Electron. Eng. 380 (October 1953).

[11] R. J. Parker, Understanding and predicting permanent magnet performance by electrical analog methods, paper read at AIEE Conference on Magnetism, Washington, 1957; J. Appl. Phys. 29 1958 409–410.

[12] R. J. Parker, Paper No. II-5, 6th International Workshop Rare-Earth Permanent Magnets, 1982.

[13] R. L. Gluckstern and R. F. Holsinger, Linear accelerator Conference, 1981.

7

APPLICATIONS OF PERMANENT MAGNETS

7.1 Introduction and Classification of Applications
7.2 Applications Based on Coulomb Force Law
 7.2.1 The Compass and Other Torque Producing Devices
 7.2.2 Mechanical Force Applications
 7.2.3 Repulsion Mode Devices
 7.2.4 Synchronous Torque Couplings
7.3 Applications Based on Faraday's Law
 7.3.1 Alternator/Generator/Tachometer/Magneto
 7.3.2 The Microphone
 7.3.3 Eddy Current Devices
7.4 Applications Based on Lorentz Force Law
 7.4.1 Loudspeakers
 7.4.2 Permanent Magnet Linear Force Devices Used in Computer Peripherals
 7.4.3 Direct Current Motors
 7.4.4 The Servo Motor
 7.4.5 Alternating Current Motor
 7.4.6 The Electronically Commutated Motor
 7.4.7 The Hysteresis Motor
 7.4.8 The Stepper Motor
 7.4.9 Electrical Measuring Instruments
7.5 Applications Based on Lorentz Forces on Free Electron Charges
7.6 Miscellaneous Applications Based on a Variety of Physical Principles

7.1 INTRODUCTION AND CLASSIFICATION OF APPLICATIONS

In this chapter the intent is to describe the major devices and systems that use permanent magnets. The focus will be on relating a magnet's properties to the device parameters. In some instances the sensitivity of the magnet's unit properties to the function of the device will be explored. In reviewing the history of permanent magnets, it is interesting to note how the advance in properties have influenced the magnetic circuits, the energy density, stability and cost of many energy conversion devices and systems. The evolution of several devices are described in this chapter.

The first uses of permanent magnets were in applications where a magnet was essential to the device function. The compass, the D'Arsonval meter movement and the magneto are examples. The weak fields established by rather excessive volumes of early permanent magnet steels were tolerated in these early uses. As properties have improved over the years, designers have continually explored replacing earlier magnets or electromagnets. Volumetric efficiency, cost and stability requirements had to be achieved. In Chapter 6, the volumetric comparison between electromagnetism and permanent magnets of various energy products was developed. The modern permanent magnet was shown to have great advantages and now is positioned to compete with larger electromagnetic structures. Traditionally, permanent magnets have been used only in small devices. The stability achievement of today's magnets is a compelling reason to use the permanent magnet in many devices. It can be said that the places where one can now use electromagnetism to advantage are rather limited. Of course, field control easily achieved with electromagnets is an inherent advantage. However, with the high coercivity magnets (type II) and the freedom from irreversible loss, we can use flux switching paths to turn the modern permanent on and off. Figure 7.1 shows the elements of such a structure applied to rotating machines. A permanent magnet radial subassembly is shown in the center of the figure (magnets are shaded). An axial 4 pole magnet and soft iron return path are on either side, assembled as shown. A soft iron pole of the center assembly receives magnet flux from four directions. If the pole areas of the radial and axial magnets are made equal, one can, by rotating the two axial magnets and return paths 90°, effectively turn this assembly on and off. This approach is useful in large motors where the forces of removing rotor from stator are large. Servicing a large rare-earth machine in case of bearing failure would be an example of where this control feature would be very desirable. An incidental advantage is the flux concentration feature due to the parallel paths.

In grouping applications into categories there are many choices. Applications could be grouped by major markets or industries, by type of energy conversion, or by physical principle involved. It is the author's opinion that grouping by physical principle would be the most meaningful. Knowing the relationship between the physical principle and magnet properties allows one to gain insight as to how improved magnet properties could be used in a particular device. The major grouping based on physical laws and principle are as follows:

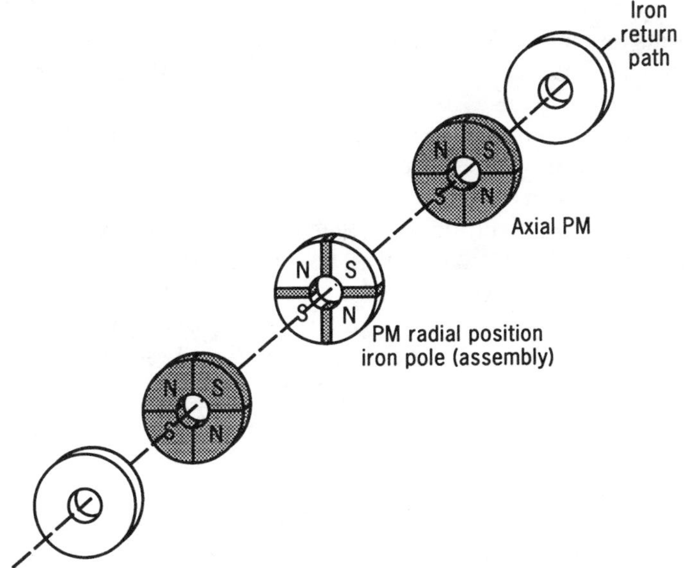

Figure 7.1 Turn on-turn off permanent magnet rotor.

1. applications based Coulomb force principles;
2. applications based on Faraday's law of induced voltage;
3. applications based on Lorentz force principle with solid conductors;
4. applications based on Lorentz force principle with free election charges.

7.2 APPLICATIONS BASED ON COULOMB FORCE LAW

The Coulomb force law is the principle involved in the conversion of magnetostatic field energy to mechanical work. The force occurring with holding and lifting magnets, magnetic couplings, magnetic bearings, and magnetic separators rely on Coulomb forces. These forces act between magnetically charged bodies. Depending on polarity, we can have either attractive forces or repulsive forces between permanent magnetic bodies. In the case of induced poles in high permeability magnetic bodies only attractive forces are possible. The forces and torques are proportional to the intrinsic magnetization B_i^2.

7.2.1 The Compass and Other Torque Producing Devices

The compass, which indicates the horizontal component of the earth's magnetic field, is generally considered to be the oldest applications of the permanent magnet. In the compass, the torque is expressed as $B_i H V_m \sin \theta$. This is a maximum when the magnet is at 90° to the field and zero when in alignment with the field. The compass needle should operate near B_r if maximum sensitivity is to be achieved. Due to mechanical problems in making a long slender needle, the early steel magnets and sintered Alnico 5 are the materials most often used.

Related to the compass, in principle, are polarity indicators and devices for locating ferromagnetic objects; the magnetometer and gaussmeter described in Chapter 8 are additional examples. In the case of the gaussmeter, lower B_r materials can be tolerated since the higher fields being measured compensate for the low B_r. The coercive force is very important to avoid reversal of the magnetization. Ferrite, silmanil, and rare-earth magnets are the materials of choice for measuring high fields.

7.2.2 Mechanical Force Application

The permanent magnet finds wide use and great diversity in attracting and holding ferromagnetic objects. The field of use ranges from simple novelties to important industrial applications such as the magnetic chuck used in machining operations. The words holding, lifting, and attracting are used rather widely and at times inaccurately. It is very important to make the distinction between the situation in which the magnet exerts a force on a material with which it makes contact and the situation where it exerts a force through a distance on a magnetic body. The design for these two purposes is quite different. If we design for maximum contact holding force the equation

$$F = KB^2A \tag{7.1}$$

is used and it is common practice to maximize B^2 by tapering the magnet poles or by adding iron pole pieces to drive the flux density to a high level. If we design for attracting a magnetic body over a distance, then we are concerned with maximizing the work, which is force times distance. In the latter case we need configurations that allow a deep field pattern to exist. The poles need to be far apart. Figure 7.2 shows four Alnico 5 magnetic circuit configurations. The weight of magnetic material is the same in each case. In the bar magnet the poles have the maximum spacing and hence the greatest force at large air gaps. Type C and D both use iron pole pieces to concentrate B at point of contact and hence these design configurations give high contact force but very small work or energy levels. The shunt gap in the type C assembly is an interesting way to load the magnet at a high level of B and keep it at that level when the armature or keeper is removed.

The total area under the force-air gap curve represents the total work done in pulling the magnet from the armature. This area is proportional to magnet volume and the potential energy change incurred due to the change

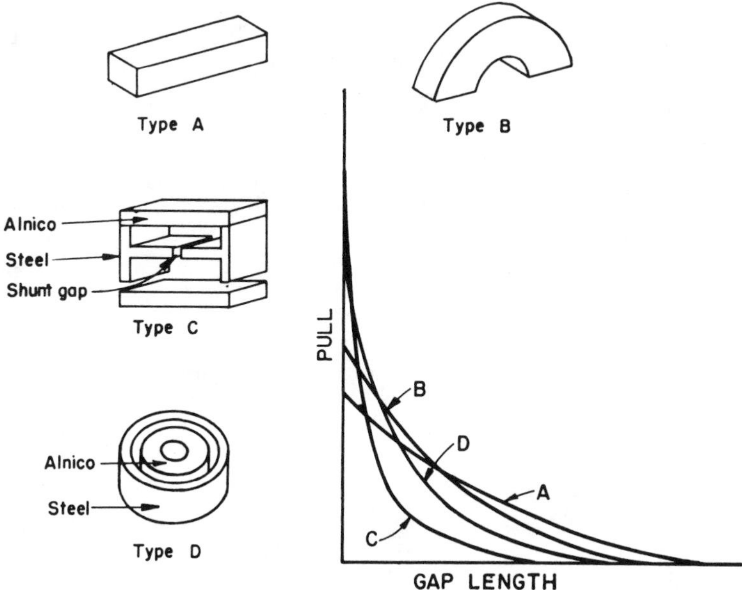

Figure 7.2 Comparison of various permanent magnets for mechanical work application.

in load permeance. The zero gap force value is proportional to B^2A where A is the total magnet or pole piece area at contact for both poles. There is a point on each force distance curve which represents the maximum work point. It is represented by the largest rectangle which can be drawn under the curve. The shape of the curve depends on the rate of change of magnetic permeance; or the rate at which leakage flux and or shunt flux is being transferred to the armature as a function of gap change. In Chapter 6 some detail was given on how to design a magnetomechanical system for maximum effeciency at a particular air gap.

Figure 7.3 shows how the laws of scaling can be used to compare the performance of a family of lifting magnets. Magnets B and C are scaled up versions of A. The factor K represents the change in linear dimension. Since the pull curve shapes are a function of geometry, we have exactly the same shape of pull curve for all members of this family of magnets. On curve A we see that the maximum work point is at M and so the efficiency of the magnet can be expressed as $(P)(G)/V_m$ (work per unit volume of magnet material). All magnets in this family have the same efficiency since P is proportional to K^2, the gap G is proportional to K, and the magnet volume V_m is proportional to K^3. However, the maximum work point will occur at different values of G. Since force is proportional to K^2 and the gap is proportional to K, we can define the ratio $\sqrt{P}/G$ as being a constant work factor for any given geometry.

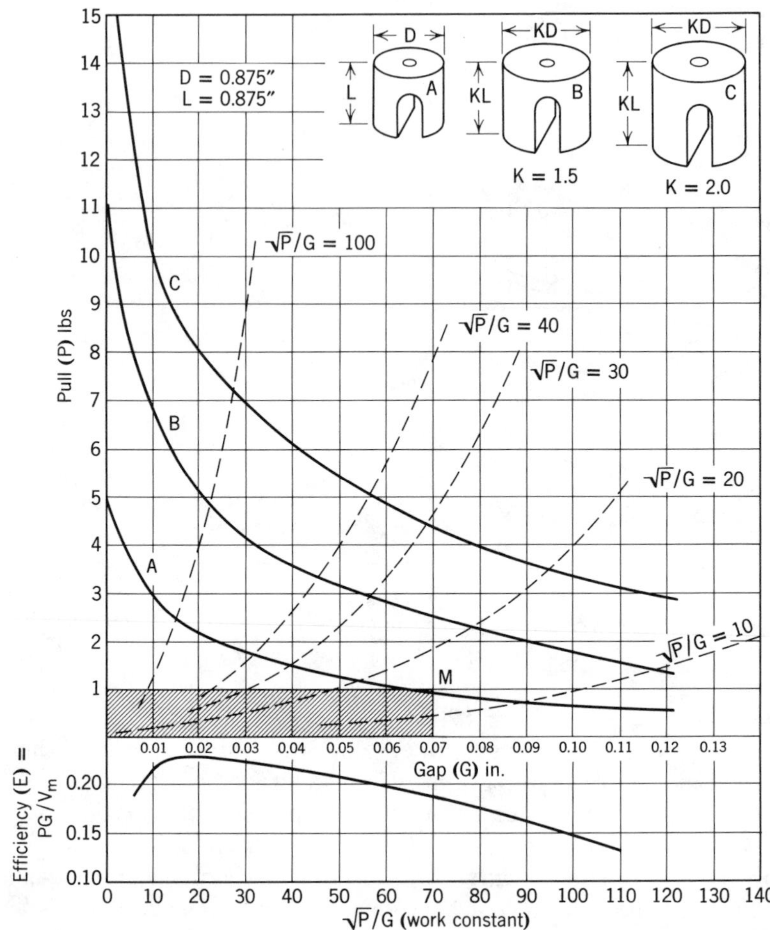

Figure 7.3 Performance and mechanical work relationships for a family of permanent magnets.

In mechanical force applications, ferrite magnets tend to dominate because of their low cost per unit of energy. Unless there is need for miniaturization, a low cost solution is achievable with oriented ferrite, bonded ferrite or flexible ferrite. In Figure 7.4 three pole piece arrangements are shown for a common volume of oriented ferrite. The contact pull and the maximum work point can be varied widely by choice of the magnetic circuit.

Permanent magnets are in wide use in industrial plants to sort and convey steel objects. A typical magnetic conveyance system is shown in Figure 7.5.

Another widely used magnetic system in industry is the magnetic separator. Permanent magnets are used to separate foreign material of magnetic nature from sand, coal, and oil in industrial processes. In iron ore

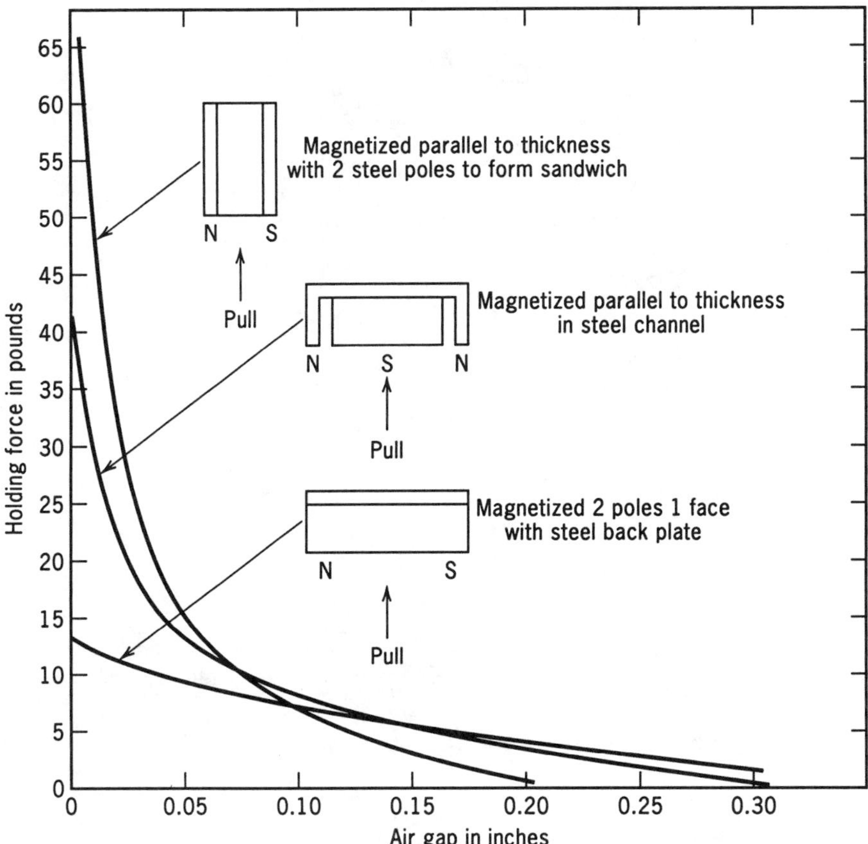

Figure 7.4 Ferrite magnet with variations in pole pieces to alter force-distance curves.

processing, magnets are used to separate iron oxide from the ore body. Magnets are usually configured in the form of a rotating drum or else arranged as shown in Figure 7.6 where the drum carries the material by a stationary magnet system. The concentrate stays with the drum while the nonmagnetic material falls off. A good separator design achieves the correct compromise between holding force at the drum surface and attracting the particle at a distance from the drum. A relationship that approximates the force on a small particle is given by

$$F = KB(\Delta B / \Delta X) \tag{7.2}$$

where F is the pull on a small volume of a ferromagnetic material, K is a variable coefficient depending upon shape and volume of the small ferromagnetic particle and on magnetic geometry, B is flux density and $\Delta B / \Delta X$ is the slope of the flux density curve at the point in question.

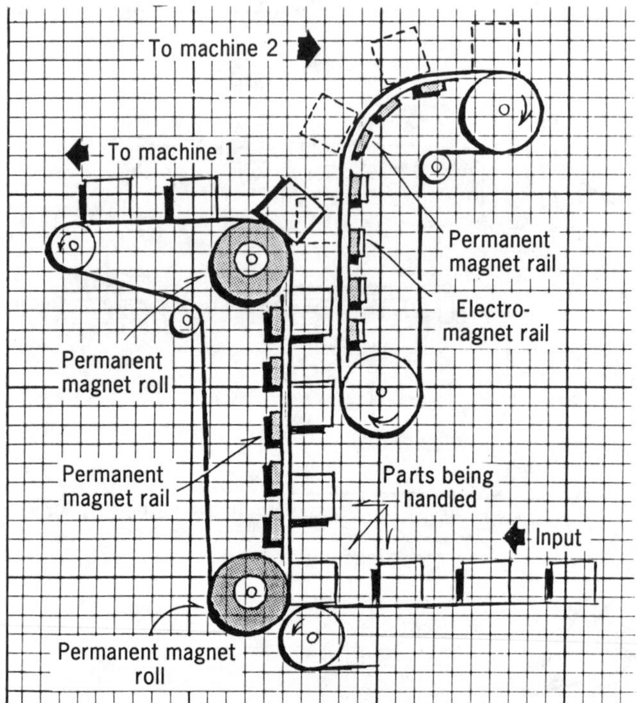

Figure 7.5 Magnetic conveyor.

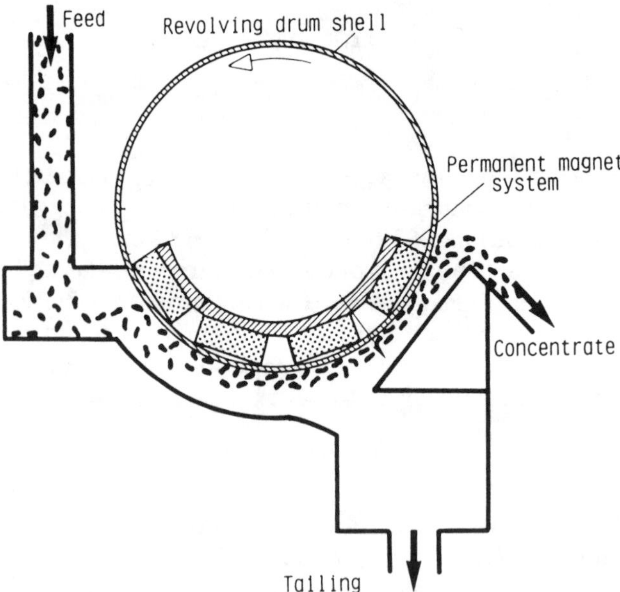

Figure 7.6 Wet drum separator with permanent magnets.

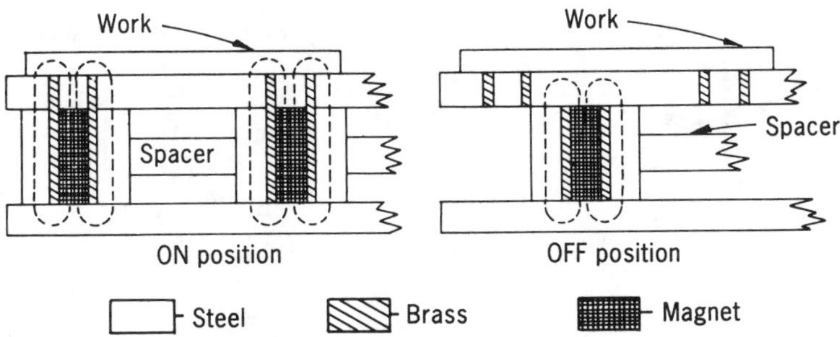

Figure 7.7 Arrangements of permanent magnets in holding chucks.

Another important mechanical-work application is the magnetic holding chuck. A typical magnetic circuit arrangement is shown in Figure 7.7. A flux switching technique is used to reduce the flux through the work piece. The entire bank of magnets is moved to a point where the return path is through the top plate of the chuck rather than through the work piece. One major advantage of a permanent magnet chuck over an electromagnetic chuck is its ability to maintain the machining set up in case of a shop power disruption.

Magnets to actuate reed switches find wide use in a great variety of control circuits. The reed switch consists of two ferromagnetic flat metal reeds hermetically sealed in an inert gas atmosphere. The reeds are supported as cantilevers, overlapping at the center of the glass cylinder, where they are separated by a small air gap. This arrangement forms a normally open single pole, single throw (SPST) switch. This switch can be closed by moving a permanent magnet close to the element. Poles are induced in the vanes and the switch is closed. Many combinations of magnet position and movement are possible. In Figure 7.8 some widely used arrangements are shown. Parallel and perpendicular magnet motion are indicated in Figure 7.8a and b. The switch can be actuated as a result of the motion of a shunt as shown in Figure 7.8c. The latching type control function is displayed in Figure 7.8d. In this case the latching magnet is chosen and placed relative to the switch so that the magnetic flux induced is insufficient to close the contacts, but is adequate to hold the contacts closed until its field is cancelled by the control magnet field of reversed polarity.

7.2.3 Repulsion Mode Devices

Magnets approaching each other with unlike poles repel each other with the same force with which they attract each other if the pole strengths remain constant. It is only since the development of magnets having $H_{ci} \geq B_r$ that pole strengths are stable in repulsion. Repulsion forces can be achieved between permanent magnets, electromagnets or between a permanent magnet and an electromagnet. Magnets can induce like poles in steel plates which will repel each other. A novel use is for separating steel plates to be

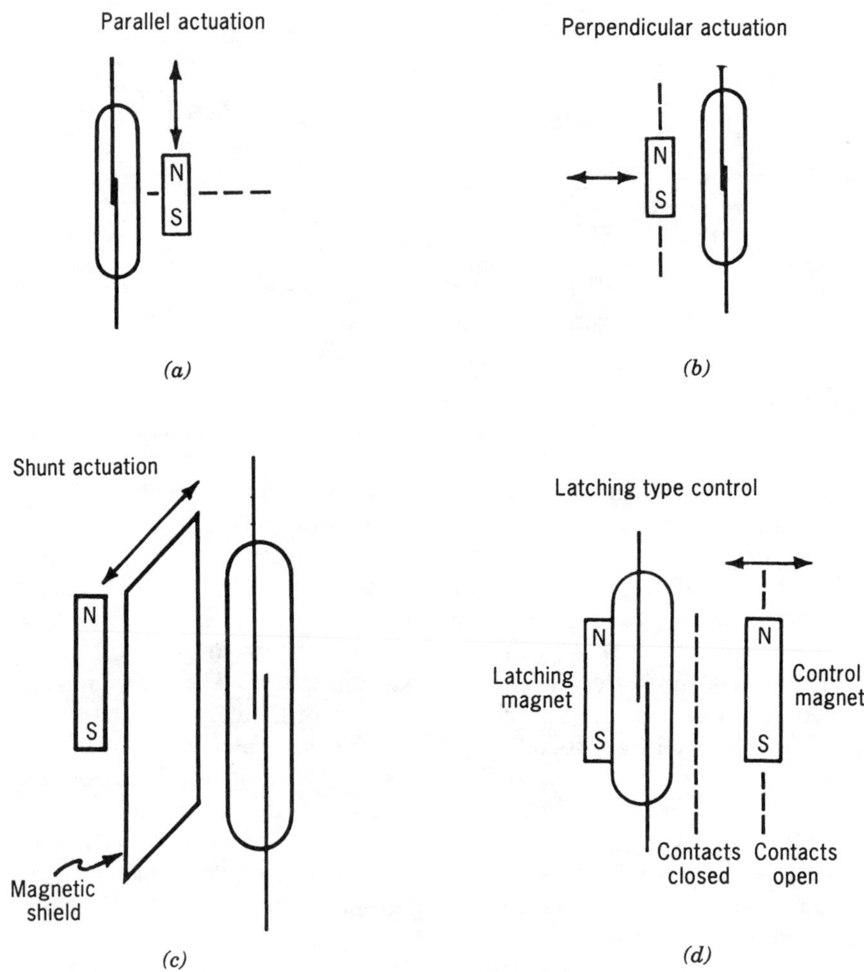

Figure 7.8 Permanent magnets for reed switches.

transferred to a machining operation as shown in Figure 7.9. The major use of magnets to produce repulsion forces is in magnetic bearings. The arrangement of magnets is shown in Figure 7.10 for a household watt-hour meter. The design objective here is to support the eddy current disc magnetically, but without attempting to levitate the disc. A graphite guide pin is used to cancel out radial forces and give stability. This arrangement gives a low friction bearing that is rather constant due to a minimum of wear over very long periods of time. Ferrite magnets and bonded Alnico magnets have been used in meter bearings. Some development work has been done on the use of rare-earth magnets, but at this time they are not in commercial use. Whatever magnet material is used must be very uniform or else a

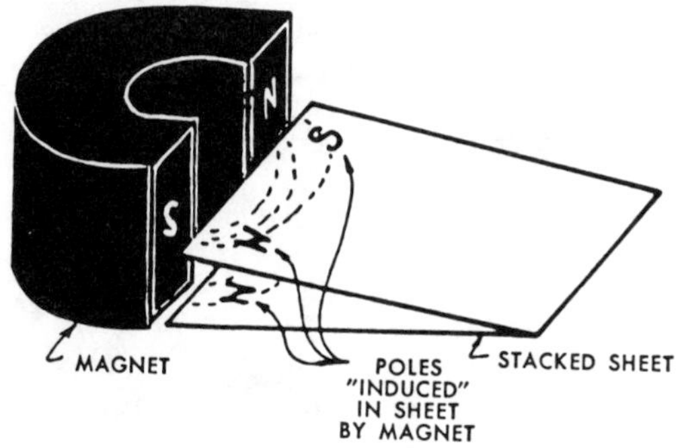

Figure 7.9 Sheet steel repulsion device.

hysteresis torque will be encountered and meter accuracy would be impaired.

Modern high coercive force magnets are also the object of extensive study in bearings for flywheels used in space satellites. There are several advantages to using a magnetic bearing in space such as freedom from wear and lubrication. Additionally, very low drag torques lead to low spin power requirements. Very high speeds are possible, which improve the momentum/mass relationship.

One simple form of bearing uses two concentric ring magnets with radial magnetization in a repulsion mode. Figure 7.11 shows the arrangement. If

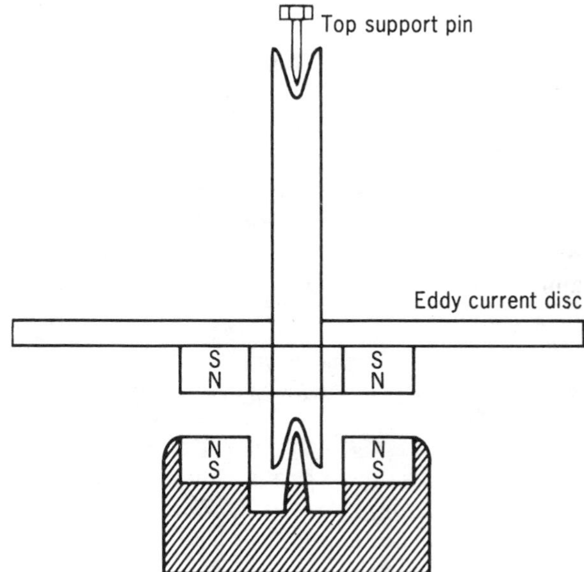

Figure 7.10 Watt-hour bearing.

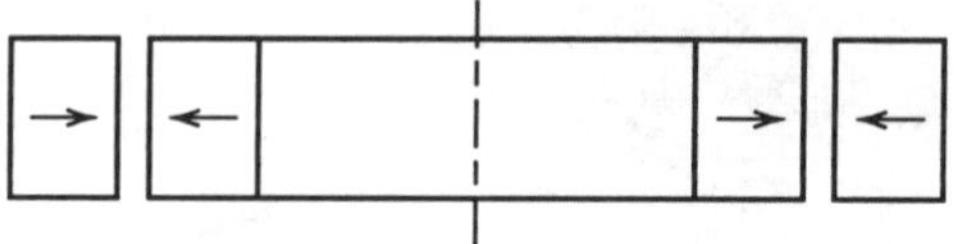

Figure 7.11 Radial magnetic bearing cross-section.

the rings are perfectly centered, there are no radial forces. If one magnet is displaced radially, forces act to center the system. This arrangement has radial stability. However, in the axial direction the two rings will not be in the same plane and hence the system is unstable axially. A fundamental electromagnet principle known as Earnshaw's theorem [1], teaches that any bearing system must have one active control loop to stabilize the system. Real space bearing systems have five degrees of freedom, and a combination of passive permanent magnets and active electromagnets are used. A system may use magnets in either the attractive or repulsive mode. Rare-earth cobalt and rare-earth iron magnets are very well suited to bearing needs. Yonnet [2] shows that the bearing stiffness depends drastically on magnetization B_i and that magnets are submitted to fields composed of their own self-demagnetizing field and to the field created by the other magnets involved.

7.2.4 Synchronous Torque Couplings

Permanent magnets find extensive use in transmitting motion across a partition or diaphragm. Often the task is to have a magnet follow in a fluid or in a gas or vacuum. There are two basic configurations of drives used. The axial type and the concentric or radial type as pictured in Figure 7.12. The axial drive develops large axial forces which must be absorbed by the bearing system. These forces are at a max-when the torque transmitted is at a minimum value. The concentric or radial drive in its simplest form involves isotropic magnets with poles on the outer diameter of one member and on the inner diameter of the other. For larger, more efficient couplings, individual anisotropic magnets with return paths of iron are used. Figure 7.13 shows a large radial type coupling using $SmCo_5$ magnets. This unit is used in a chemical pump.

7.3 APPLICATIONS BASED ON FARADAY'S LAW

Permanent magnets find wide use in the conversion of mechanical energy. Faraday's law is the physical principle involved in this kind of energy transformation. The change in magnetic flux within a conductor turn is linked to the induced or generated voltage. The voltage is proportional to the flux density B. In the special case of eddy current torque the force becomes proportional to B^2, since both an induced voltage and a Lorentz force are involved.

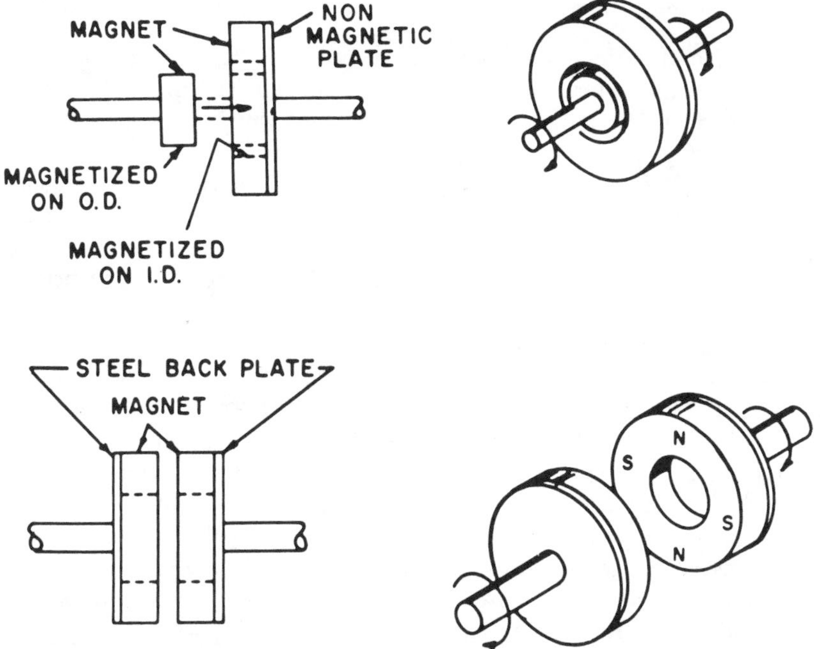

Figure 7.12 Basic drive configurations.

Figure 7.13 Large synchronous coupling.

7.3.1 Alternator/Generator/Tachometer/Magneto

For some time, permanent magnets have been used extensively in a variety of small electrical machines to convert mechanical energy to electrical energy. The earliest form was the d.c. dynamo, which could be used either as a motor or generator. One special form of d.c. generator is known as the tachometer. The voltage output is proportional to speed and hence the voltage is used in servo control systems with a tachometer being driven by a servo motor. In the tachometer, the principle issue involving magnet selection is to use a magnet with a low reversible temperature coefficient, so that the tachometer calibration will hold over a wide temperature range.

In a.c. alternators, the magnets are in the rotor and a sine wave output voltage is achieved. One of the major uses of permanent magnet alternators has been to supply excitation power for the field coil of large alternators. With the properties of $SmCo_5$ and NdFeB, there is appreciable interest in development of larger permanent magnet alternators. There is a trend toward generating power close to its use in the home, in aircraft, and in specialized water and air driven alternators. Often the alternators power density is very important and high energy permanent magnets are preferred over wound excitation. Figure 7.14 shows the cross-section of a wound alternator. The force can be expressed as $F = B_g J V_g$ where B_g is the gap density, J is the current density and V_g is the gap volume. The power W is $B_g J V_g K R$ where K is a constant, R is the radius of the rotor and $K\sqrt{R}$ is the peripheral velocity. A fundamental way to increase the machines power density is to increase its speed. In a wound machine, the strength of the copper conductors limit the speed. By going to a rotating magnet, with high strength material supporting the relatively low strength magnet material, very high speed machines are feasible. In aircraft alternators, 40,000–60,000 rev./min machines have been developed. An interesting use for small, efficient rare-earth magnet motor/alternators is shown in Figure 7.15. Off-peak power is used to put energy into the flywheel and then it is used as

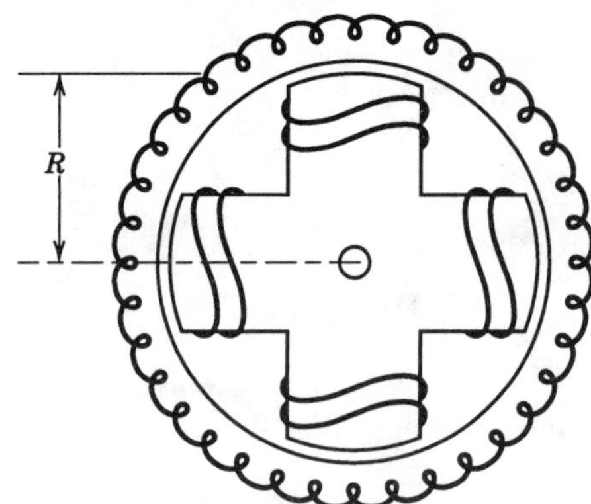

Figure 7.14 Alternator cross-section.

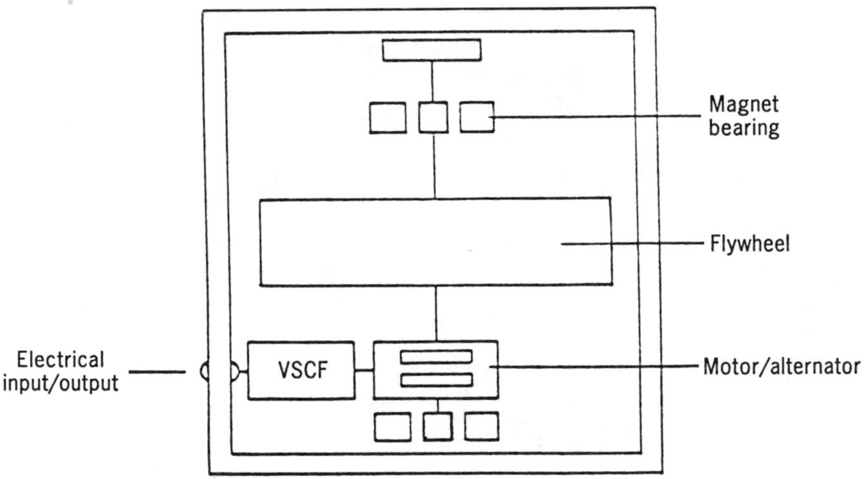

Figure 7.15 Flywheel energy storage.

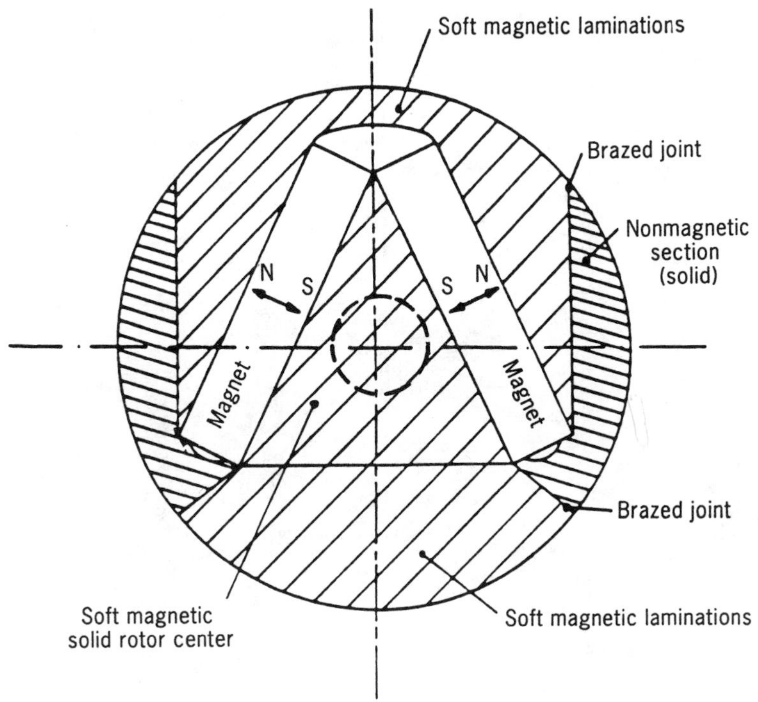

Figure 7.16 Aircraft high speed rotor.

required in the household. The alternator output is fed into a variable speed constant frequency (VSCF) power conditioning network to allow useful constant frequency power for the household. This apparatus also uses magnets in a bearing system which contributes to system efficiency. The flywheel is in a vacuum chamber.

In order to improve the power density in aircraft systems, $SmCO_5$ and NdFeB magnets are being placed in high speed rotors. Figure 7.16 shows a two pole structure where parallel paths are used to obtain an improvement in gap density.

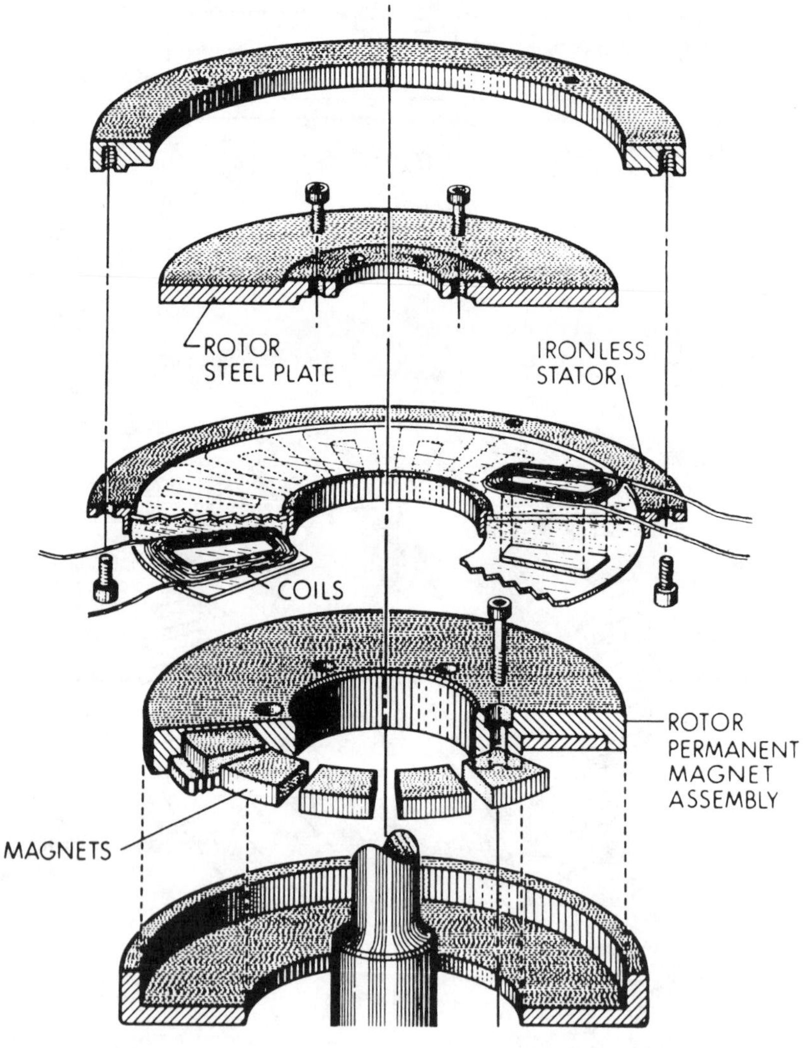

Figure 7.17 Ironless stator alternator. From A. R. Miller, M. I. T. Lincoln Lab.

In small alternators, rare-earth magnets have led to the design trend of the ironless stator as shown in Figure 7.17. The very high levels of magnetomotive force per unit length allows large air gaps in which the stator coils are located. The power density of this 500 W machine is approximately 150 W/lb and the efficiency is 94%.

The magneto is a specialized generator that produces high voltage pulses and is used with small internal combustion engines that do not have a battery to supply ignition needs. The magneto was one of the earliest devices to use permanent magnets and tracing the evolution of the magneto and how it has been influenced by magnet properties is very interesting. The magneto has gone through three basic magnetic circuit changes as magnet properties have improved (Figure 7.18). Without question, the major change occurred when the volumetric effeciency of magnets improved so that the magnet could be moved to the rotor position. This replaced the wound armature, collector rings, and brushes. Figure 7.19 shows the elements of a modern magneto ignition system. In order to show the influence of the cycle of operation on the magnet refer to Figure 7.20. With the rotor in the maximum permeance condition, the breaker is closed and point M represents the load line of the magnet. Point P represents the minimum permeance condition (rotor removed from stator). As the rotor turns 90°, minimum permeance occurs and the magnet operates at point P, current flows in primary winding in a direction which tends to maintain the stator flux. As the rotor continues to rotate, the permeance increases, however the flux due to the primary current opposes the magnet polarity, further demagnetizing the magnet to point D. The primary current continues to increase as the rotor position approaches the maximum permeance condition again and at the current maximum the breaker opens. The energy stored in the primary circuit is $1/2\ Li^2$ where L is primary inductance and i is the current maximum at the time of breaker opening. This energy is transferred to the secondary and the plug is fired. The useful energy is proportional to area $MKER$ and point D should be chosen to be optimum for recoil conditions as discussed in Chapter 6.

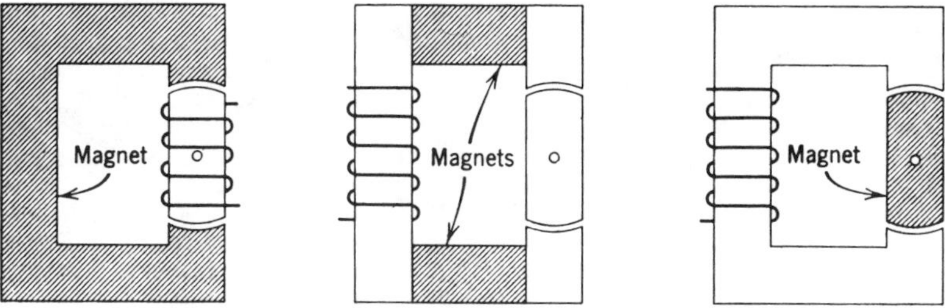

Figure 7.18 The evolution of the modern magneto.

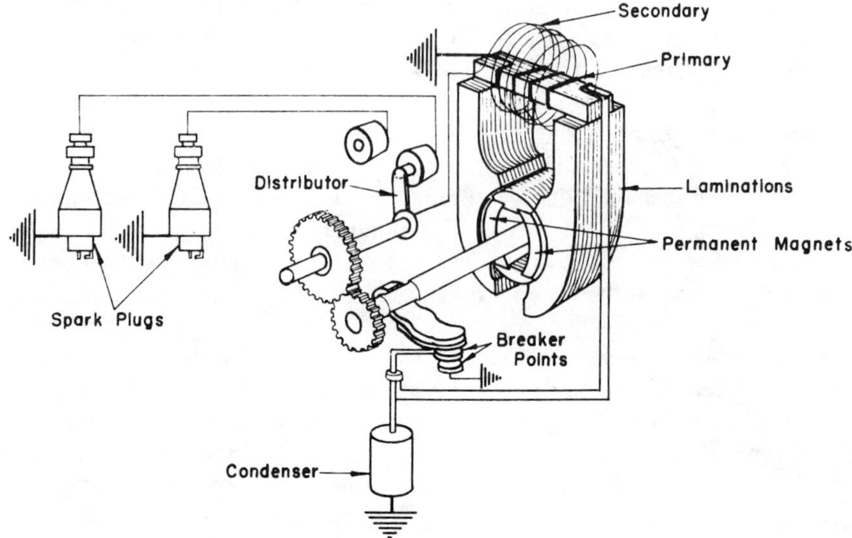

Figure 7.19 Elements of the magneto ignition system.

7.3.2 The Microphone

Magnets find use in both the velocity and dynamic or moving coil type microphones as shown in Figure 7.21. In the velocity type, a thin conducting metal ribbon moves in a magnetic field and generates a voltage proportional to the sound pressure. The moving coil unit is essentially an inverted

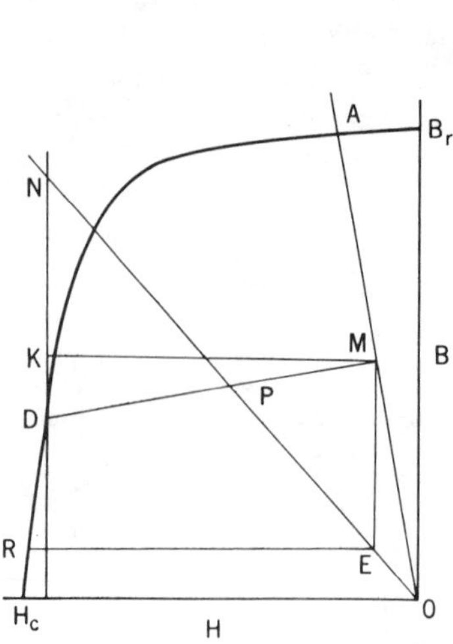

Figure 7.20 The operation of the magneto in terms of permanent magnet properties.

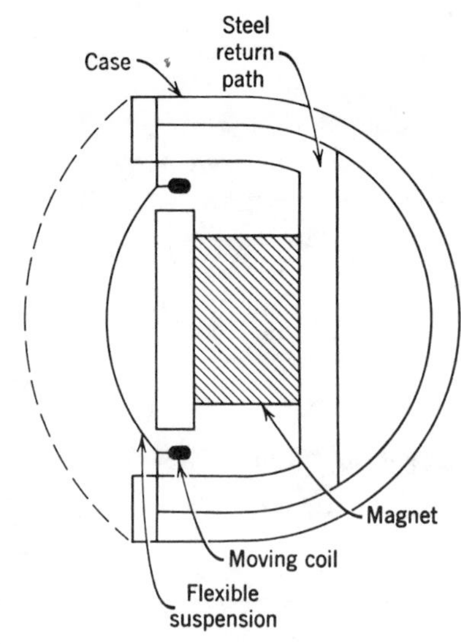

Figure 7.21 Microphone.

loudspeaker. In many intercommunication systems the loudspeaker also serves as a microphone. In both types of microphones, design emphasis is on achieving high gap densities with a minimum magnet volume. Consequently, materials with high $(BH)_{max}$ values are most often the choice.

7.3.3 Eddy Current Devices

Eddy current devices are a special case of the Faraday law. Induced voltages in a conducting member lead to circulating currents in the conductor and to a Lorentz force interaction with the magnet [3]. Torque transmission, retarders and damping devices all use eddy currents. The automotive speedometer is a unique torque device. The speedometer is pictured in Figure 7.22. The magnet and return path are driven from the drive train. An aluminum cup is restrained from rotation and the developed torque works

Figure 7.22 Automobile speedometer.

against a spring calibrated to indicate speed. Eddy current torque is characterized by a linear torque-slip relationship.

Figure 7.23(a) shows in an elementary way how eddy currents are set up in a metallic conducting plate moving with respect to a magnet pole; the instantaneous electromagnetic poles induced due to currents flowing as shown, have polarities with respect to the fixed magnet pole, such that a force opposing the movement of the metallic plate is set up. A basic torque expression applicable to rotating devices is as follows

$$T = \phi^2 a^2 X\omega / 100 A\theta\rho \tag{7.3}$$

T = torque (dyne cm)
X = disc thickness (cm)
a = length torque arm (cm)
ω = angular velocity (rad/s)
A = area magnet pole (cm^2)
θ = variable coefficient depending on eddy current paths
ρ = resistivity of disc (μohm cm)
ϕ = total flux at magnet pole (lines)

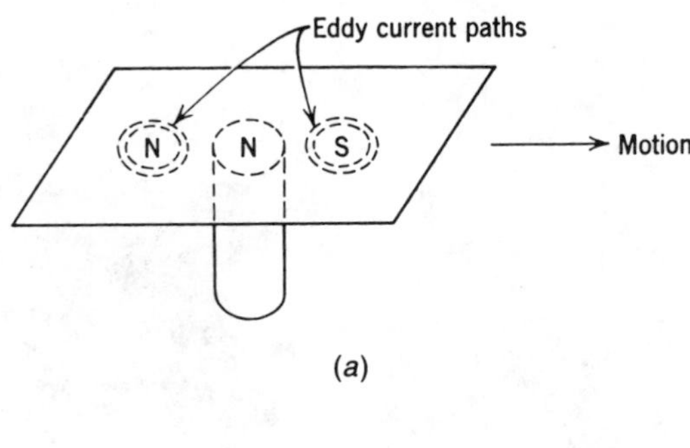

(a)

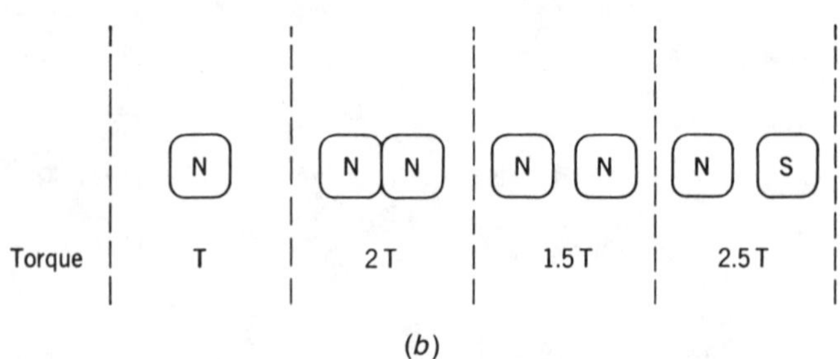

(b)

Figure 7.23 (a) Eddy current formation. (b) Effect of location and polarity of flux cutting eddy current disc.

The coefficient θ varies greatly dependent on the disc or cup relationship to the magnet pole; in typical revolving disc meter and relay applications, θ is of the order of 10–15 and maximum torque is developed with the magnet pole at approximately 0.8 of the disc radius. As the edge of the disc is approached, the eddy currents are in effect open circuited and the retarding force diminishes rapidly. The shape and arrangement of the poles greatly influence the torque. A circular pole or one with well rounded corners is most desirable since the eddy current path will then have minimum length and resistance. Figure 7.23b shows possible pole arrangement. If T is the torque produced with an isolated pole moving past a conducting plate then values of torque ranging from 1.5 to 2.5 T may be obtained depending on the location and polarity of the second pole. The case of flux piercing the disc with unlike polarity allows the most efficient eddy current paths. Since the eddy currents diminish as the edge of the disc is approached it is obvious that concentrating the field is highly desirable. It must be recognized that although it is desirable to concentrate the flux into small areas piercing the disc, the high reluctance of the gap under such conditions will have a counteracting adverse influence and sufficient area of gap to allow efficient operation of the permanent magnet will merit consideration in a practical design. From the torque expression, we find that torque is proportional to gap flux squared, ϕ^2 and conductor thickness X. It is of interest to explore how the gap dimensions and conductor thickness should be chosen to obtain maximum results. Referring to Figure 7.24 let G equal total gap at one end of the magnet, let M equal the air part of the gap, then $G = M + X$. Since ϕ diminishes linearly as the gap length increases over a small range

$$T \propto \phi^2 X \quad \text{and} \quad T \propto X/G^2$$

then

$$T \propto (G - M)/G^2 \propto (1/G) - (M/G^2)$$

Differentiating T with respect to G

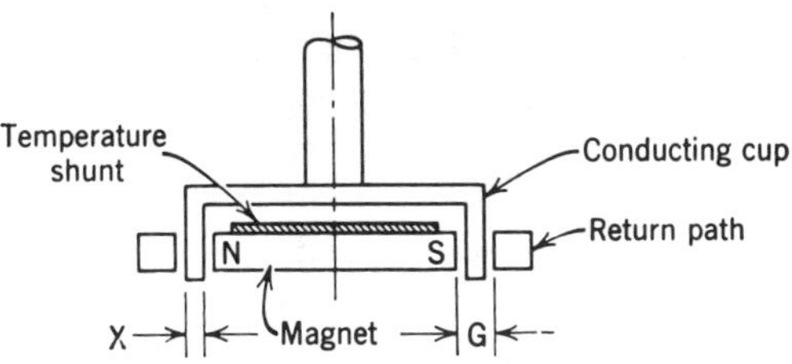

Figure 7.24 Relationships of permanent magnet circuit in eddy current device.

$$dT/dG = (1/G^2) + (2M/G^3) = 0$$

or $G = 2M$ for maximum torque since

$$G = M + X\,, \qquad 2M - M - X = 0$$
$$M - X = 0\,, \qquad M = X$$

for maximum torque

7.4 APPLICATIONS BASED ON LORENTZ FORCE LAW

The Lorentz force law involves the interaction between current and magnetic flux, the basic principle involved in converting electrical energy to mechanical energy. Figure 7.25 shows three variations of current and flux interactions to produce a torque, a transverse force, and an axial force.

By far the largest use of permanent magnets is to convert electrical energy into a linear force or a torque in the case of rotary motion. Motors are growing rapidly in numbers produced and in types that can use magnets to advantage. In addition to the traditional motor function to develop torque and power, the permanent magnet motor has become a critical component in the motion control industry. There is in progress a major industrial revolution involving time. In Figure 7.26 an electric motor family tree is shown. There are a great number of specialized types, each of which has functional characteristics and economic factors which uniquely make it a choice for a particular application. It is estimated that permanent magnets are in only 20% of the motors currently made. As magnets improve in properties and value and as motor parameters become more demanding, permanent magnets may well be used in nearly all types of motors.

7.4.1 Loudspeakers

The loudspeaker was the first major application of permanent magnets. With the introduction of Alnico 5 in the early 1940s, the permanent magnet became competative with wound field electromagnetic speakers both in performance and cost. Today, a half century later we find several materials in wide use. Ferrite magnets are quite dominant. However, for small speakers and applications where leakage fields are troublesome, Alnico is still used. In microspeakers and earphones, rare-earth cobalt and rare-earth iron magnets have wide acceptance. In many advanced wide range audio systems, the high energy density magnets are being used. The dynamic moving coil speaker has remained essentially unchanged for 50 years. Its principle advantage is the large motion and high efficiency in reproducing low frequency tones in contrast to other reproducing systems using diaphragms where motion is quite restricted.

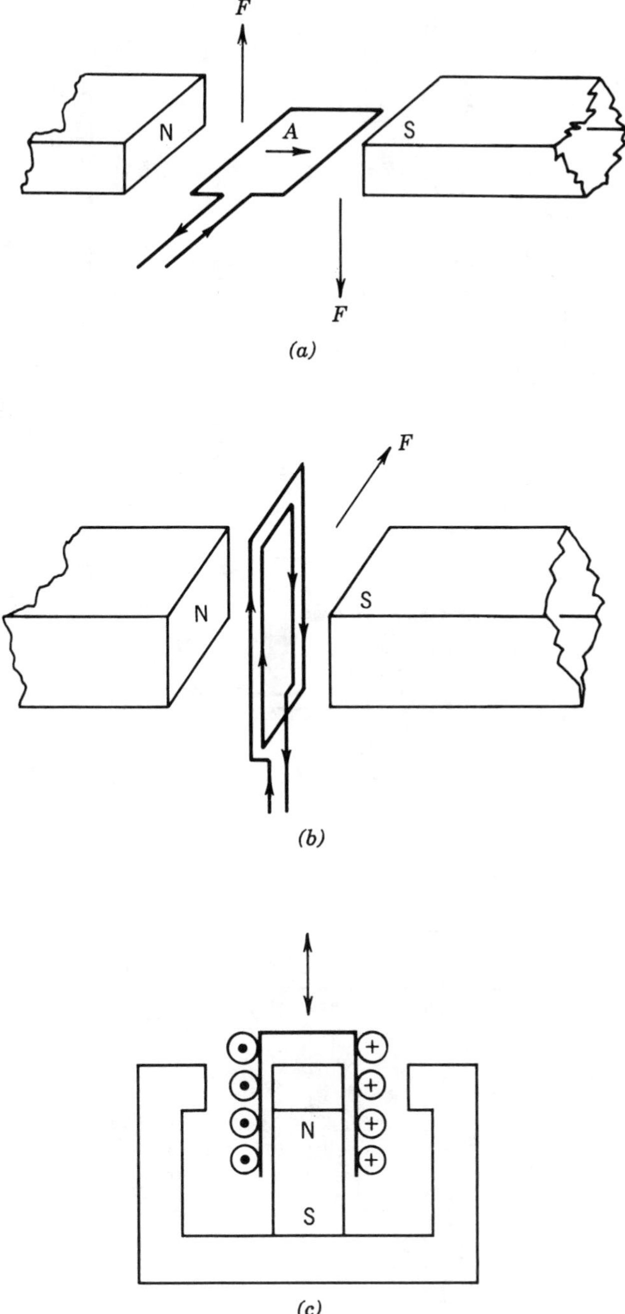

Figure 7.25 Flux and current interactions. (a) Component arrangement to produce torque; (b) arrangement to produce linear transverse force; (c) voice coil arrangement for linear axial force.

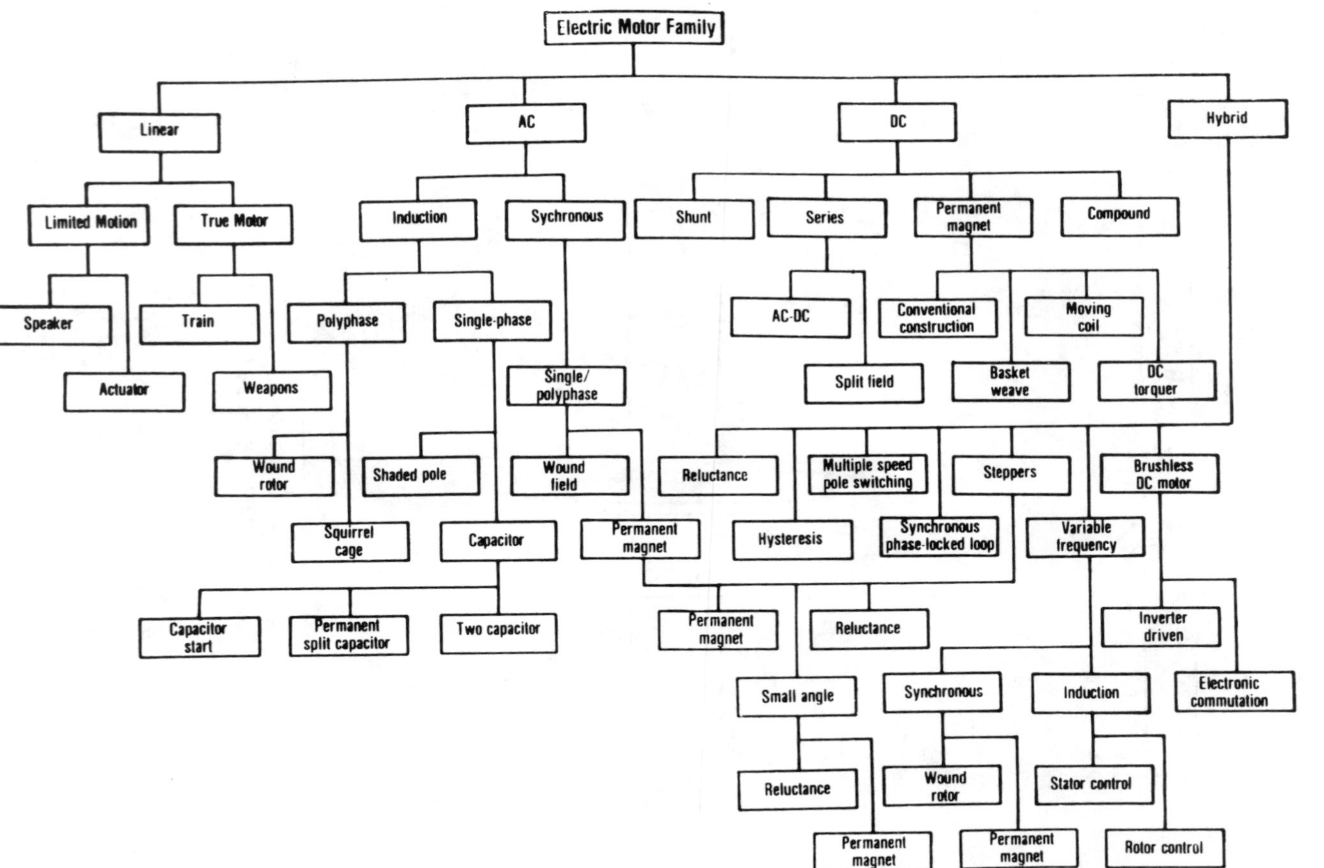

Figure 7.26 Motor family tree.

With NdFeB magnets the energy density is high enough to consider inverting the speaker structure. A moving magnet reacting with a heat sinked stationary coil would be a superior thermal design. There is development work in progress and such a structure might be much easier to manufacture due to looser tolerances between magnet and coil.

The following considerations are developed to link important speaker parameters to magnet unit properties. We must consider how electrical energy is delivered to the voice coil. The speaker is basically a device which converts electrical energy to mechanical work in the form of acoustic power. The load is the restraining influence of the moving coil plus the radiation resistance which the cone establishs by distributing the air. If no magnetic field existed, all of the voice coil watts would be dissipated as heat and no motion and hence no energy conversion would result. If a field exists across the air gap, the voice coil will vibrate and a counter e.m.f. will be generated. The watts lost in heat will be less and acoustic output will result. It can be shown that power output of the speaker is $P = i^2 R_m$, where R_m is the acoustical radiation resistance and i is the voice coil current. The power output is also proportional to $R_m/(R + R_m)$, R being the resistance of the voice coil, or we may write $P = C/(1 + (1/X))$ where C is a constant and X is the ratio acoustical resistance to coil resistance, R_m/R. It is thus desirable to make R_m/R large and X becomes a measure of speaker efficiency. R may be expressed in terms of the constants of the voice coil $R = \rho l/a$, where ρ = specific resistance, l = winding length, and a = wire cross-section. R_m is proportional to driving force squared $N(B_g li)^2$ or $N(B_g l)^2$ for unit current where N is a constant determined by cone stiffness, suspension stiffness and frequency; therefore

$$X = R_m/R = NB_g^2 l^2 a/l\rho = \frac{N}{\rho} B_g^2 la = \frac{N}{\rho} B_g^2 V_c \qquad (7.4)$$

where V_c is volume of conducting material in the air gap. From this analysis we see that the output of a speaker is a function of flux density squared and consequently in the speaker application great efforts are exerted to use the magnet efficiently and by design arrangements make the gap flux density B_g as high as possible. Since efficiency is proportional to B_g^2 it is also inversely proportional to $(BH)_{max}$. The basic design relationships in Chapter 6 gave $B_g^2 = V_m/(BH)_{max}$ as the relationship between magnet volume, properties, and B_g^2. The selection of a magnet will also depend on the B_r of the material. A small wide range speaker must have a voice coil that weighs less than a gram and for good tonal balance a B_g of at least 5000 G. These boundary conditions mean that certain magnet materials can only be used with specific magnetic circuit arrangements. Figure 7.27 compares two permanent magnet speaker magnetic circuits, in (a) the ring magnet design is shown and in (b) the cylinder magnet is shown. Cylinder magnets are inside the voice coil diameter and tend to be very efficient. Approximately 65% of

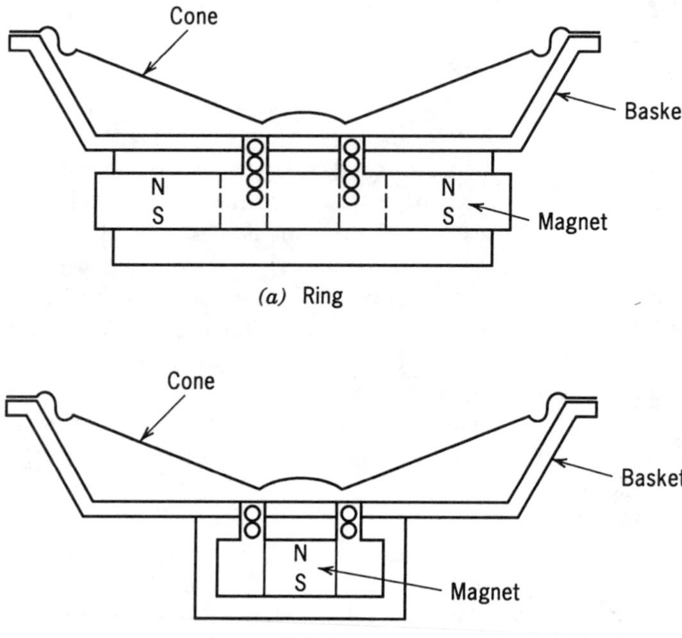

(a) Ring

(b) Cylinder

Figure 7.27. Speaker circuit comparison.

the total magnet flux will be measured in the gap area. To locate the magnet inside the voice coil requires a relatively high magnet B_r since magnet area is limited. It follows that if we need B_g levels of 5 kG for a good sounding speaker and we can obtain only 65% efficiency, we must start with B_d levels of 7 kG which means B_r levels of perhaps 8–10 kG. Note that the cylinder magnet has a disc shaped soft steel pole cap which alows the flux lines to turn 90° and move into the radial air gap. Depending on the ratio of diameter to thickness of the pole cap, the flux density in the circuit can be adjusted by about ±20% without encountering efficiency problems. Alnico 5 permanent magnets with B_r levels of 12 kG have been the predominant material used with the cylinder shape. This combination gives high circuit efficiency with very low leakage flux.

The ring magnet design has the magnet material outside the voice coil and the magnetic circuit efficiency is much lower, only 35–50% of the total flux is useful. However, the ring design allows the magnet area to be very large and hence lower B_r materials can be used. The total flux is collected over the large circular pole pieces and directed into the air gap. It is this approach which has allowed ferrite magnets with very low B_r levels to become widely used. Ferrite magnet speakers have low circuit efficiency but still give acceptable solutions because of the low cost of the ferrite magnet. Today most large speakers use ferrite magnets. The low cost ferrite magnet compensates for two major disadvantages; (i) the soft iron parts for the ferrite ring design are large and expensive; (ii) the high leakage is difficult to tolerate in some application environments such as in small television sets and in automotive dashboard locations. For the ring design, a circuit modification can be made as shown in Figure 7.28. Now the leakage is cancelled by the second ring magnet. This is an effective functional solution but the economics are questionable. very often the best solution when leakage must be reduced is to go to the cylinder design and use a more expensive magnet material.

The speaker designer must in his design consider power handling and sensitivity requirements to arrive at the gap energy level that meets equipment demands. The widely accepted figure of merit in a speaker is the gap energy expressed as

$$E_g = B_g^2 V_g / 8\pi \quad \text{(erg)} \tag{7.5}$$

The figure is a direct measure of a speakers quality and efficiency. A speakers sound pressure output is directly proportional to B_g. Often sound pressure differences are expressed in terms of a change in B_g, a change in $(BH)_{max}$ or an adjustment to magnet volume. To compare sound pressure changes resulting from changed B_g due to volume or energy product adjustments, denoted by subscripts 1 and 2, the following relationships are useful

$$\text{Decibel (db)} = 20 \log_{10} (B_{g1}/B_{g2}) \tag{7.6}$$

$$= 20 \log_{10} [(BH)_{max-1}/(BH)_{max-2})]^{1/2} \tag{7.7}$$

$$= 20 \log_{10} (V_{m-1}/V_{m-2})^{1/2} \tag{7.8}$$

One decibel is the smallest change in sound pressure detectable to the human ear. In addition to being of critical importance in generating sound pressure, high levels of gap density B_g have other desirable influences. First

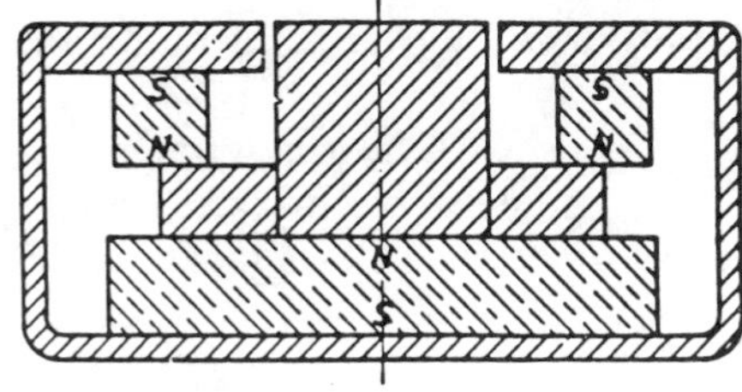

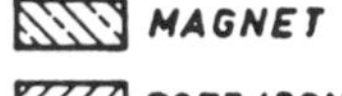

Figure 7.28 Low leakage speaker circuit.

the speaker's transient response is a measure of a speaker's ability to follow rapidly changing passages or inputs. A speaker with low gap density will be sluggish and will not faithfully reproduce the brilliance and sharpness of note of an orchestra. Another measure of quality of a speaker is its dampening factor. If the speaker cone continues to move after the electrical signal, representing a note, decreases then the speaker will be lacking in definition, and hangover and distortion will be present.

7.4.2 Permanent Magnet Linear Force Devices in Computer Peripherals

The trend in todays computer peripherals is toward very small printers and storage devices. There is great competition to increase storage density, decrease the access time and increase the data transfer rate. Permanent magnets having high volumetric efficiency have made major contributions in reducing device size, response, and reducing thermal losses.

In high performance disc drivers as shown in Figure 7.29 the read-write head positioning device is usually a permanent magnet voice coil actuator. The permanent magnet takes up a minimum of space, allows rapid acceleration and when used in a closed-loop servo system can position on a data track with great precision.

The fundamental motion equations are:

$$F = Ma = B_g li \tag{7.9}$$

where F is force, M, head and carriage mass, a, acceleration, B_g is air gap flux density, l, coil conductor length and i, coil current. In addition,

$$s = 1/2\ at^2 \quad \text{or} \quad F = 2Ms/t^2$$

where s is the distance or head travel and t is the time allotted for positioning. In general the goal is to achieve lowest access time at the least power dissipation. Forward acceleration is used for approximately 1/2 the required head displacement, then acceleration is reversed until head is in desired position in the allotted time t. For a given coil current, a high energy permanent magnet will give higher B_g levels, hence higher acceleration and improved access time.

Also of importance is the inductance of the coil which determines the electrical time constant of the system. It is desirable to have a low inductance which leads to short current rise time. This allows rapid current reversals and rapid positioning on a data track.

Higher B_g values can influence coil geometry to minimize inductance.

Another important performance factor is the power dissipated in the coil. It is desirable to minimize the heating so that air cooling requirements can be reduced. A higher B_g will allow a current and wattage reduction. The maximum current density is limited by the coil's thermal environment.

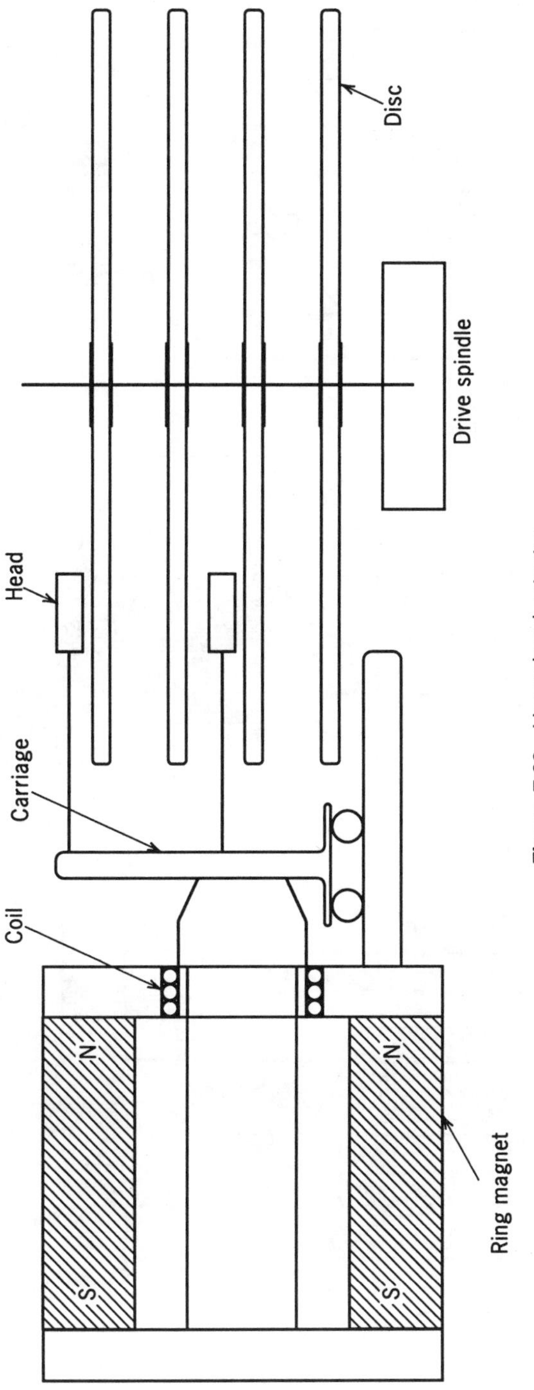

Figure 7.29 Linear head actuator.

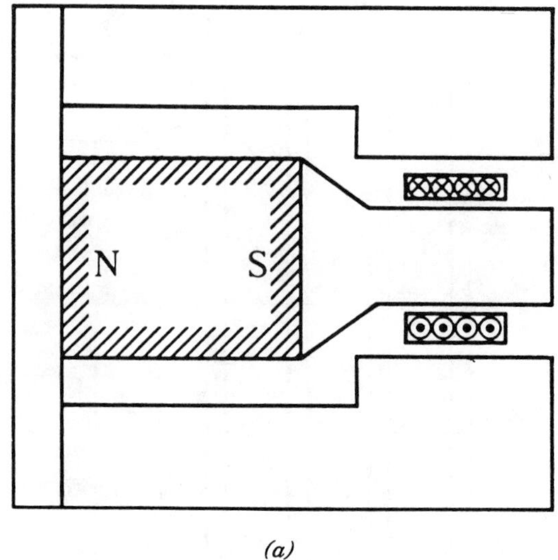

(a)

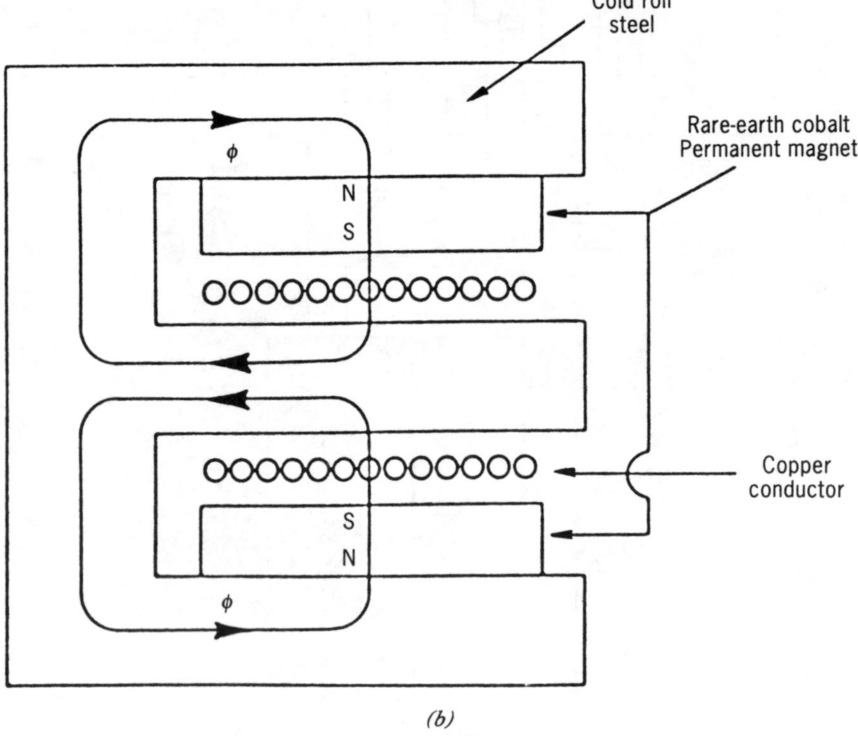

(b)

Figure 7.30 Disk actuator designs.

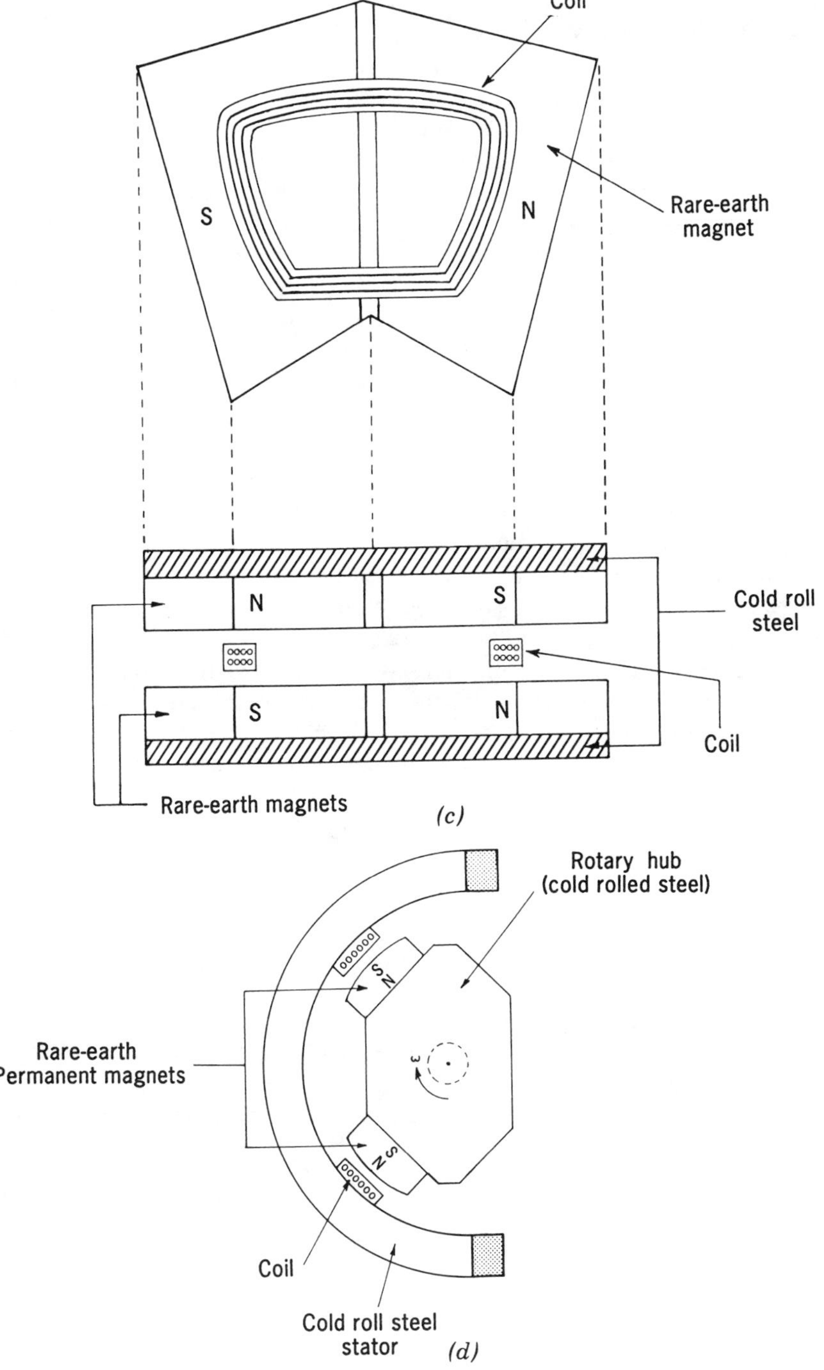

Figure 7.30 (*Continued*)

As the magnet properties have improved and the system demands have increased, the linear actuator has evolved. In Figure 7.30a the structure used Alnico 5 magnet material. A very large structure was necessary. The "E" core flat coil design shown in Figure 7.30b came into wide use with the availability of rare-earth magnets. With magnets next to the air gap the circuit efficiency was greatly improved. A further improvement in package size resulted with the flat meter like structure shown in Figure 7.30c. This rocking motion device uses rare-earth cobalt or rare-earth iron magnets.

Figure 7.30d is an inverted design, limited motion rocking actuator which may represent future design direction as still higher energy density magnets become available. The magnet moves and due to the high magnetic moment per unit mass, the dynamic motion capabilities are excellent. The coil is stationary and its thermal environment now allows very high current densities. The age-old tradition of massive stationary magnetic structures with thermally limited moving coils may be over. The high energy density magnets will give the actuator designer some new degrees of freedom.

The Lorentz force principle is utilized in line printers (Figure 7.31). This force exerted on the print hammer is due to the interaction of coil current and a permanent magnet field ($F = Bli$). Since the current reverses in the two legs of the coil, the flux must be of opposite polarity for the two portions of the coil for an additive force pattern to be developed. In the Dataproducts printer which formerly used Alnico 8 magnets. NdFeB magnets have had major impact on line printer designs. The force has been increased and also drive current levels have been decreased to reduce power supply costs. The reduced heating also eliminated forced cooling of the printer. This application of magnets is an excellent example of a high performance, relatively high cost magnet being cost-effective in a sophisticated system.

7.4.3 Direct Current Motors

The earliest motor, which is still widely used, is the conventional d.c. motor, offering speed control and greater output per unit size than a.c. motors. In Figure 7.32 the motor variables are defined and a graphic display relating speed, torque, and current is shown.

Using the basic relationships between flux, current, and voltage, the motor parameters can be expressed as follows

$$\text{Stall torque } (T_s) = K_1 I_s \phi_a Z \tag{7.10}$$

$$\text{True no load speed } (S_{tnl}) = \frac{K_2 V - I_n R_a}{Z\phi_a} \tag{7.11}$$

$$\text{Armature current } (I_a) = \frac{7.46 \times 10^{-4} S_{tnl}(T + T_{in})}{E} \tag{7.12}$$

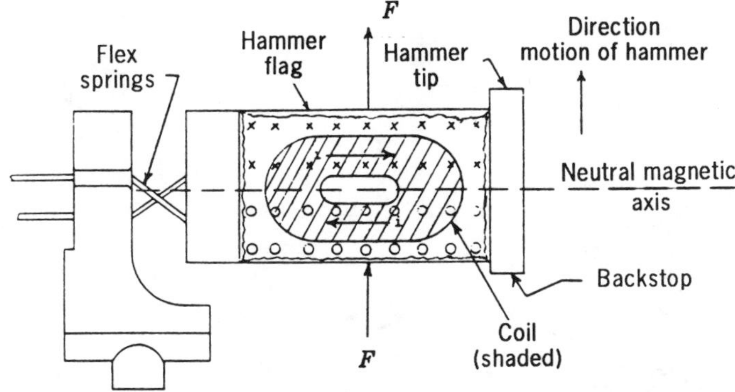

(a)

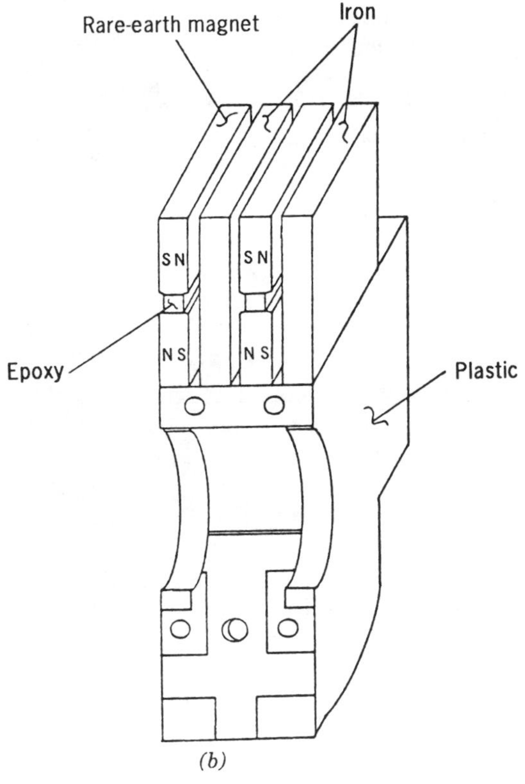

(b)

Figure 7.31 Line printer. (a) Coil and hammer detail; (b) magnet configuration.

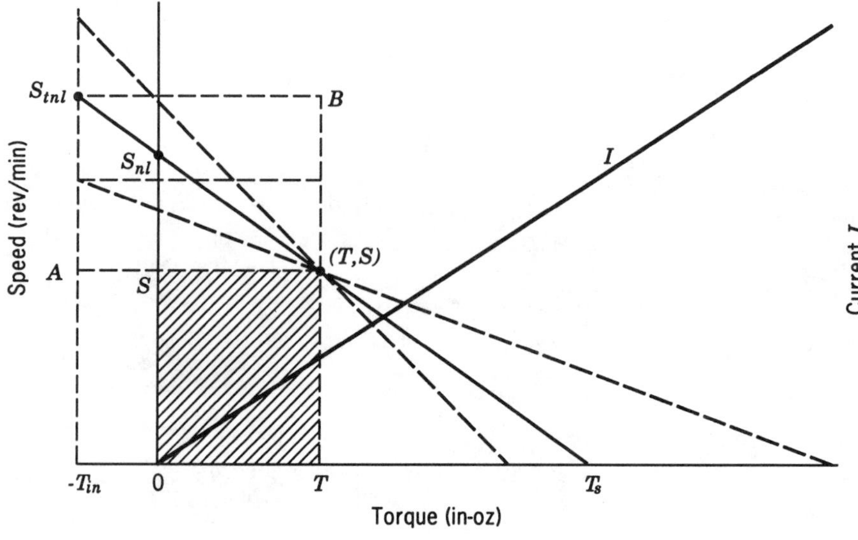

V	= Applied volts	M	= Slope of speed-torque curve
I_n	= No load armature current	D	= Diameter of armature wire
R_a	= Armature resistance	E	= Efficiency
T_{in}	= Internal torque	S	= Speed
S_{tnl}	= True no load speed	T	= Torque
S_{ne}	= Observed no load speed	I_s	= Stall current
K	= Conversion constants	I_a	= Armature current
Z	= Armature conductors	ϕ_a	= Armature flux

Figure 7.32 Graphical analysis of d.c. motors parameters.

The speed torque curve may be projected with negligible error until it intersects a negative value of torque T_{in}, which may be defined as the internal torque developed to supply internal losses. The intersection of the speed torque curve with T_{in} is defined as true no load speed, S_{tnl}. Motor efficiency E is output ST divided by input $S_{tnl}(T - T_{in})$ or

$$E = \frac{100ST}{S_{tnl}(T + T_{in})} \tag{7.13}$$

The internal torque T_{in} can be estimated by reference to previous data on a similar motor. It is also apparent that for a given efficiency level, one can work back to the ratio of the rectangular areas in Figure 7.32 and estimate T_{in}.

It is of interest to explore the variables that influence the slope M of the speed torque curve. If $M = T_s/S_{tnl}$, then

$$M = K_3 \frac{I_s Z^2 \phi_a^2}{V - I_n R_a} = K_4 \frac{Z^2 \phi_a^2}{R_a} \tag{7.14}$$

Since M is proportional to ϕ_a^2, we see that increasing armature flux is a very significant way to increase output and efficiency.

To relate permanent magnet unit properties to the permanent motor load line refer to Figures 7.33 and 7.34. A good approximation for P_c or B/H is simply the L_m/L_g ratio. Assessment of K_1 and K_2 from Figure 7.34 will give somewhat greater accuracy in load line determination. The total armature flux ϕ_a will be $B_d A_m$. In order to determine the length of the magnet L_m, we need to determine the maximum armature reaction expressed as F_m, which will occur at start or stall conditions

$$F_m = \frac{0.495 Z I_s d}{p(L_m + L_g)K_2} \tag{7.15}$$

where F_m is the external demagnetization field due to armature reaction, Z is the total number of conductors, I_s is the stall current, d, the demagnetization factor, p, the number of parallel paths, L_m the magnetic length, L_g the air gap and K_2, the reluctance factor.

To determine d we need to consider half of the ratio of the armature slots within the magnet arc α to the total armature slots. For armature slots greater than 10, d is closely approximated by $d = \alpha/720$. Using the F_m value

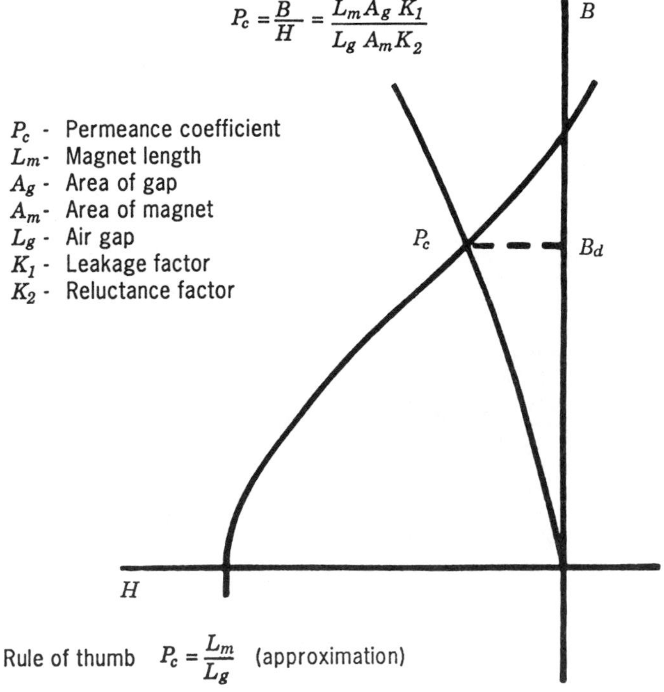

Figure 7.33 Motor design relationships.

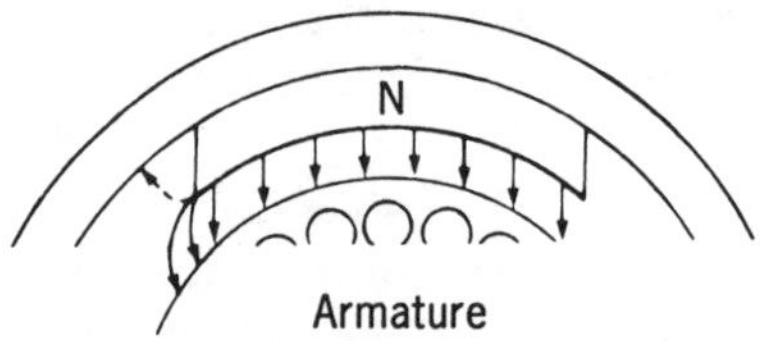

Leakage Factor
K_1 — Leakage is the Ratio of The Total Flux to Air Gap Flux

Typical Motor Values

K_1 — 1.1 to 2.0

Reluctance Factor
K_2 — Effective Increase in Gap Length Due to Curvature of Flux (Fringing or Spreading)

Typical Motor Values

K_2 — 1.1 to 1.5

Figure 7.34 Motor design factors.

we can go to the intrinsic demagnetization curve (Figure 7.35). F_m, the total armature reaction must be balanced by $L_m H_a$ where H_a is the largest potential per unit length the magnet can encounter without suffering severe irreversible demagnetization. Knowing F_m and H_a from the intrinsic curve, we can solve for magnet length

$$L_m = \frac{F_m}{H_a} \tag{7.16}$$

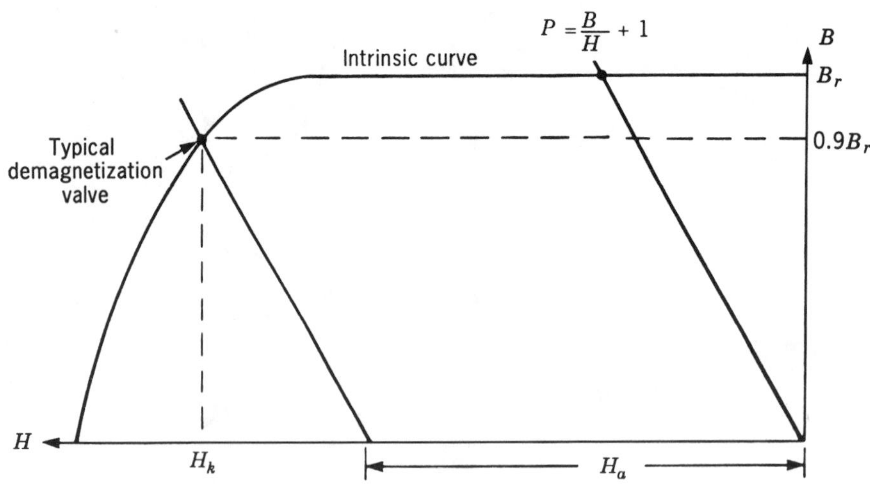

H_a- Change due to armature reaction

Figure 7.35 Magnet—armature reaction relationship.

To be accurate, we need to estimate the operating temperature range of the motor in service and then use the unit properties of the magnet at the temperatures involved. Figure 7.36 shows the property temperature displays for ceramic 8. In the motor, the magnet is exposed to dynamic loading, the magnet area is inversely proportional to B_r, and the magnet length inversely proportional to H_k. Thus the volume will be inversely proportional to the dotted rectangular area shown in Figure 7.35. H_k is by definition the value of $-H$ which decreases the magnetization by 10%. It is a convenient way to approximate the largest rectangle that one can draw under the intrinsic curve. In motors intrinsic energy product becomes a useful figure of merit. Figure 7.37 illustrates how three magnets having the same $(BH)_{\max}$ can have widely different capability in the d.c. motor due to intrinsic property differences.

The use of H_k values is somewhat misleading when comparing materials having a wide range of characteristics. H_k values are a good measure of area under the intrinsic curve as long as the intrinsic curve is essentially horizontal. For materials having appreciable intrinsic curve slope, a better index of capability is the product B_rH_x. The H_x component is determined by using

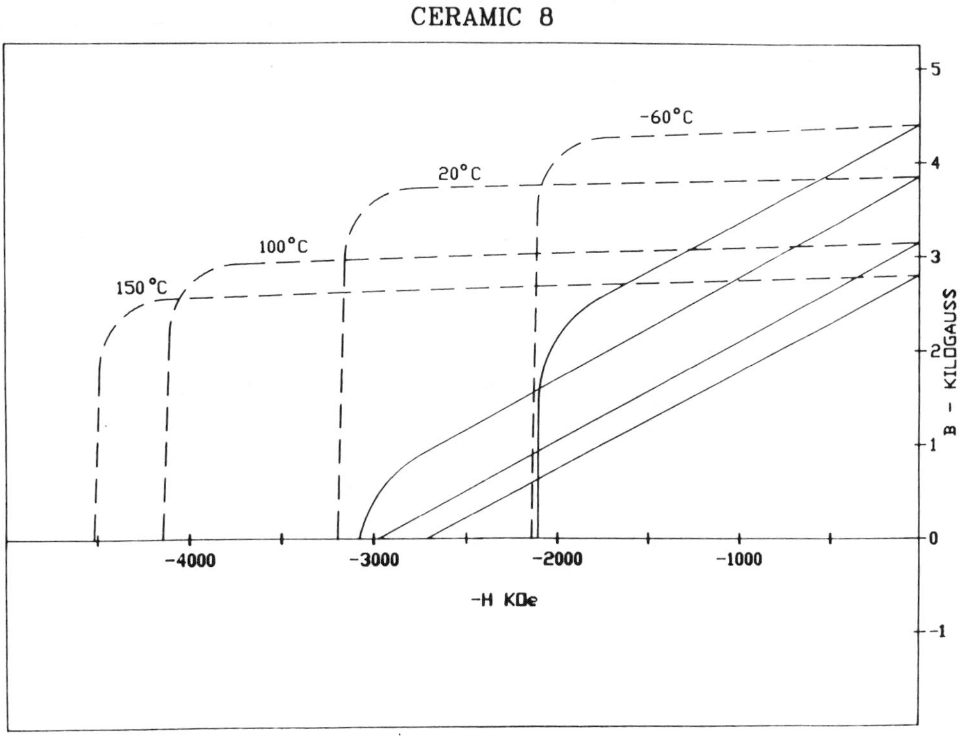

Figure 7.36 Demagnetizaion curves at temperature.

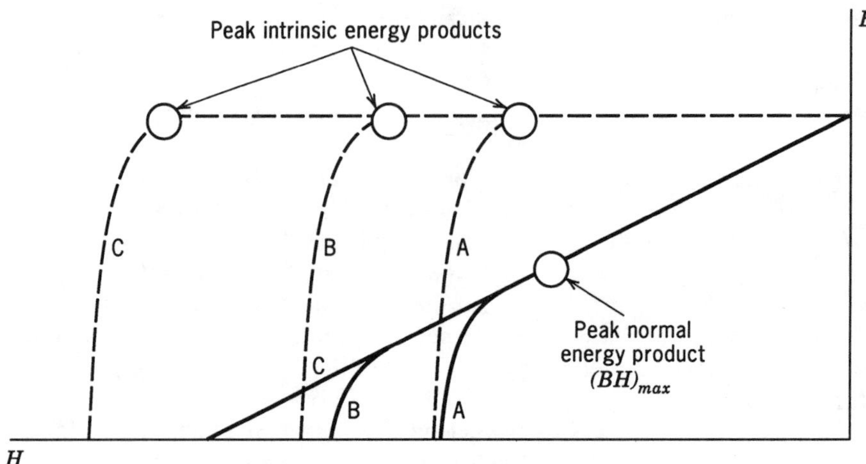

Figure 7.37 Energy product variation.

the slope of the intrinsic curve to locate the H value representing 10% demagnetization. A line representing the intrinsic slope (Figure 7.38) is drawn through the B axis at a point which is $0.9B_r$. The point where this recoil line intercepts the intrinsic demagnetization curve is H_x. From H_x a horizontal line is drawn intersecting the B axis. A B_r multiplying factor, m, is now computed by dividing the horizontal line intercept by the level of B_r. This m factor can be used to rapidly locate H_x values for a given material.

Referring again to Figure 7.38 we can express the load line

$$(B/H) + 1 = \frac{(m)B_r}{H_x - H_a} \tag{7.17}$$

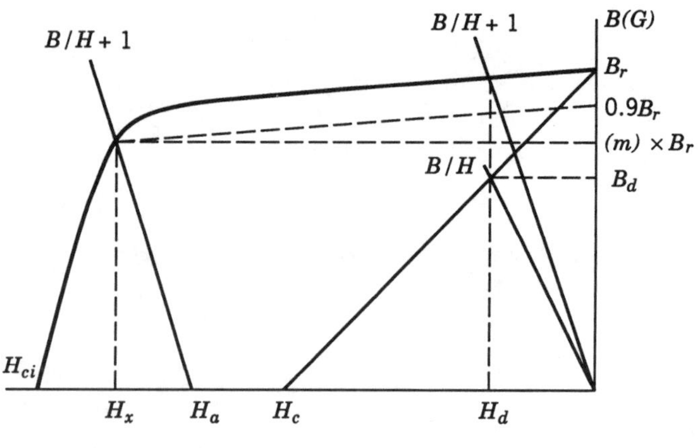

Figure 7.38 Intrinsic and normal demagnetization curves.

Magnet area will be inversely proportional to B_r and magnet length inversely proportional to H_x. Magnet volume will be inversely proportional to the B_rH_x product. In the magnet industry, B_rH_x values are rather widely accepted as a figure of merit for magnets used in the d.c. motor operating environment. It works with materials having different shaped intrinsic curves resulting from various degrees of material orientation. Table 7.1 is a summary of magnetic property values for several widely used magnet materials.

In the small d.c. motor the armature wave winding pattern is used in a very high percentage of motors. The wave winding has only two armature paths. There are fewer conductors than in a lap winding and only one pair of brushes is required. Two pole machines are most popular. However, in machines having high levels of armature reaction the use of multipole designs can reduce magnet length and back iron thickness. As a general statement one can halve the magnet length and back iron thickness for a doubling of the poles. This tends to halve the machine volume. These size advantages must be weighed against the cost of additional components and assembly time in the overall design decision-making process.

The permanent magnet d.c. motor has nearly replaced wound machines due to the many advantages of the magnet machine:

higher stall torque, linear speed torque curve;
reduction in size and weight;
lower current, no stator losses;
simpler construction; better reliability;
higher efficiency.

Table 7.1 Comparison of properties for use in d.c. motors.

MATERIAL (SOURCE) \ PROPERTY	$B_r H_x$ MGO	B_r GAUSS	H_x OERSTEDS	H_k OERSTEDS	H_{ci} OERSTEDS	$B_d H_d$ MGO	DENSITY LBS./CU.IN.
BONDED FERRITE (3M B—1062)	10.4	2760	3775	3775	5300	1.9	0.13
CERAMIC 8 (MMPA)	11.2	3850	2925	2860	3050	3.5	0.177
CERAMIC 7 (MMPA)	12.5	3400	3700	3540	4000	2.75	0.177
BONDED Sm_2Co_{17} (3M 2001)	35.9	5700	6300	5100	10000	7	.19
SINTERED Sm_2Co_{17} (MMPA)	110	10500	10500	8300	11000	26	0.303
SINTERED NdFeB (MMPA)	114.5	12300	9310	9000	10000	35	0.268
MELT SPUN NdFeB (GMC MQ—1)	74	6100	12250	4438	15000	8	0.217
MELT SPUN NdFeB (GMC MQ—2)	112	8000	14000	4200	16000	14	0.271

7.4.4 The Servo Motor

In motion control systems the motor is often used with a closed feedback control system. A motor used in such a system is a servo motor. The basic elements of the motor control system are shown in Figure 7.39. For motion control systems some important new parameters in motor design must be considered. One is the motor constant $K_m = T_{pk}/\sqrt{w}$ where T_{pk} = peak motor torque and w is the input power. Another important figure of merit in the servo motor is peak acceleration $\alpha = T_{pk}/J_m$ where J_m = rotor inertia. High energy density rare-earth magnets allow dramatic increases in T_{pk}/J_m value which leads to a decrease in time required for positioning in the control system. As the rotor diameter decreases, T_{pk}/J_m values improve, since rotor inertia varies as the fourth power of rotor diameter and the torque varies as the second power of rotor diameter. Servo motor armature construction may be quite different from a conventional armature. Figure 7.40 shows an axial printed circuit rotor, a moving coil type and a surface wound rotor. Armature mass and inductance can be varied widely to meet demanding response parameters.

7.4.5 Alternating Current Motors

In order to understand how the a.c. motor functions, we need to consider how a rotating stator field is established. In Figure 7.41 a two-phase stator

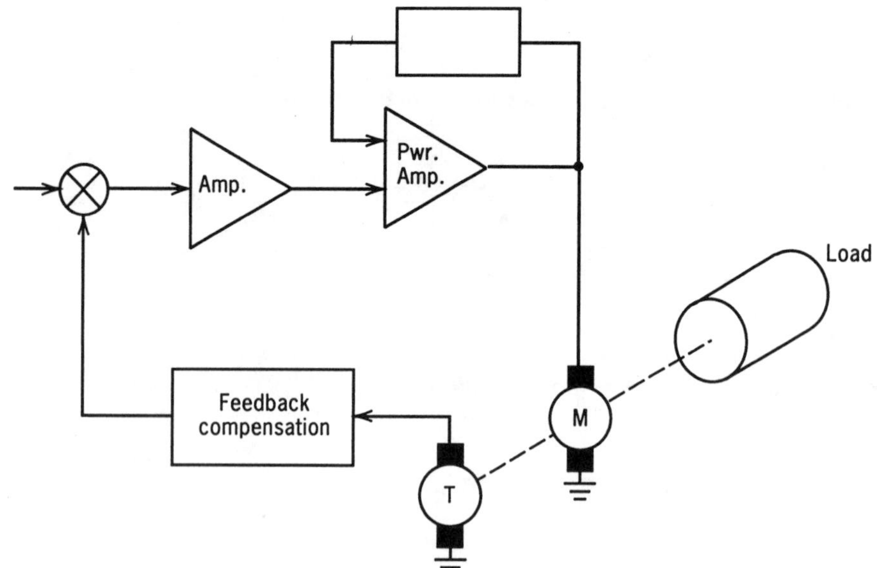

Figure 7.39 Block diagram of speed servo with tachometer feedback.

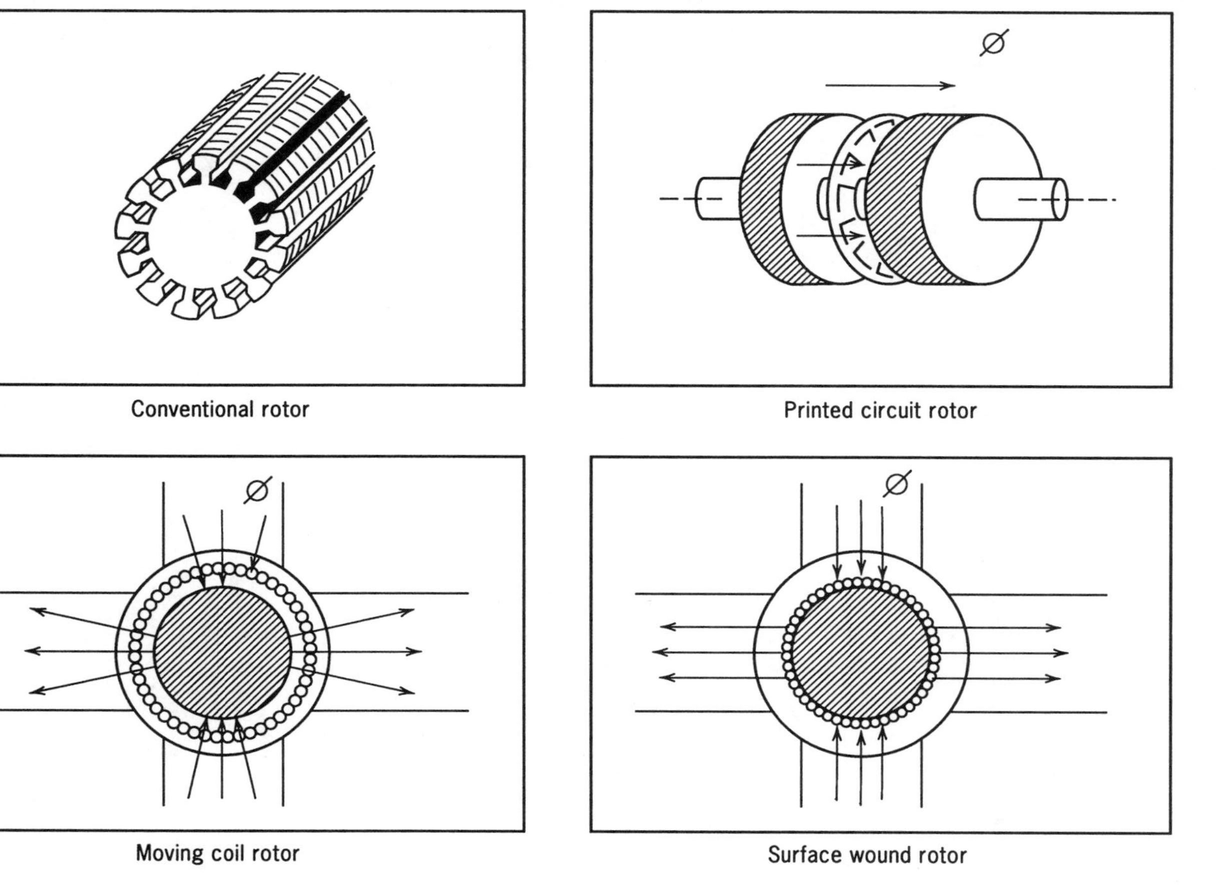

Figure 7.40 Rotor variations.

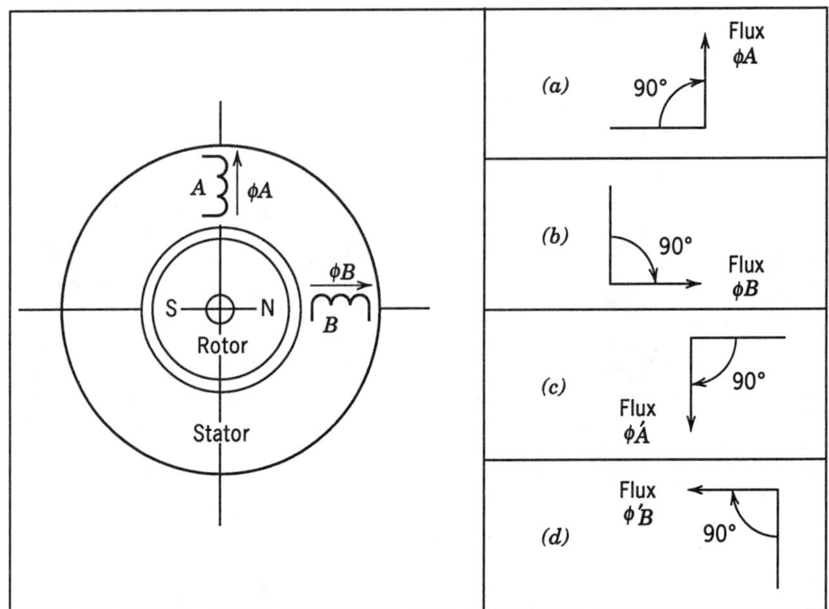

Figure 7.41 Concept of rotating stator field with respect to time.

field is set up by coils that are 90° apart. If we have a two pole permanent magnet rotor, we can, by switching on and off and reversing the two coils A and B, effectively produce a rotating magnetic field vector that will react with the permanent magnet rotor to produce rotation and torque. We have in principle a synchronous motor. The same effect can be achieved by supplying sine waves to coils A and B displaced by 90° as shown in Figure 7.42.

In the d.c. motor, the role of the commutator is to switch conductors in order to maintain a 90° angle between the armature flux vector and the permanent magnet field vector. The torque angle is fixed. By contrast in a.c. machines, the torque angle is not fixed. The speed is fixed by the frequency of the power supply. For example

$$N = \frac{60f}{P/2} \tag{7.18}$$

where N = speed (rev./min), f = frequency(Hz) and P = poles. The torque angle is small for light loads and increases with loading. The efficiency therefore increases with load. Consequently the d.c. motor is more efficient and has greater output than the equivalent synchronous motor. If the a.c. motor is of the induction type or hysteresis type, the d.c. motor advantage is more pronounced. If in the synchronous a.c. motor we replace the permanent magnet or wound rotor with a soft iron laminated cylinder surrounded

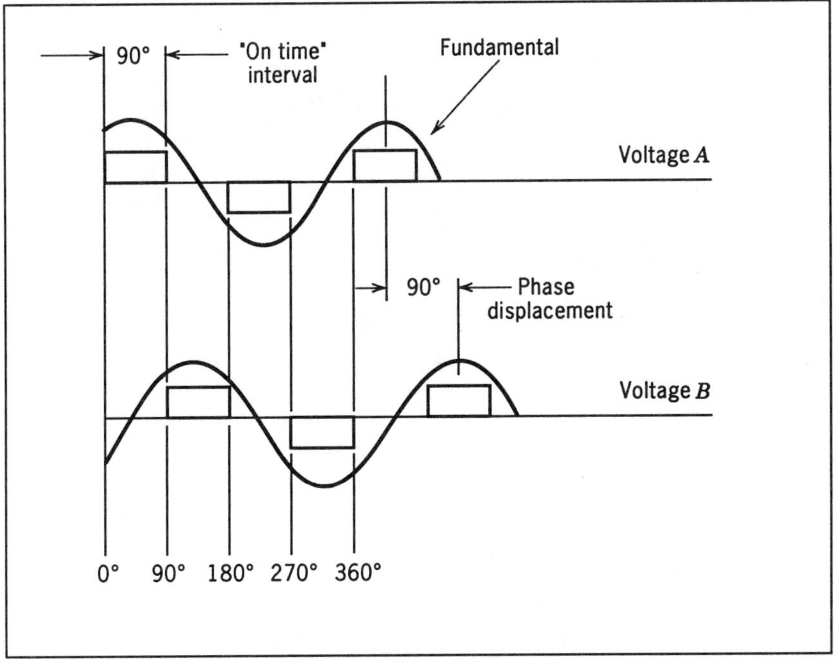

Figure 7.42 Field due to sine waves.

by a conductor arrangement known as a squirrel cage, we have a motor that runs by inducing voltages in the cage and hence current flows in the conducting bars and forms instantaneous poles that interact with the revolving stator field to produce a torque. This induction motor does not run at a fixed speed, but slips behind the synchronous revolving stator field by an amount depending on the motor loading. Its torque-speed characteristic is shown in Figure 7.43.

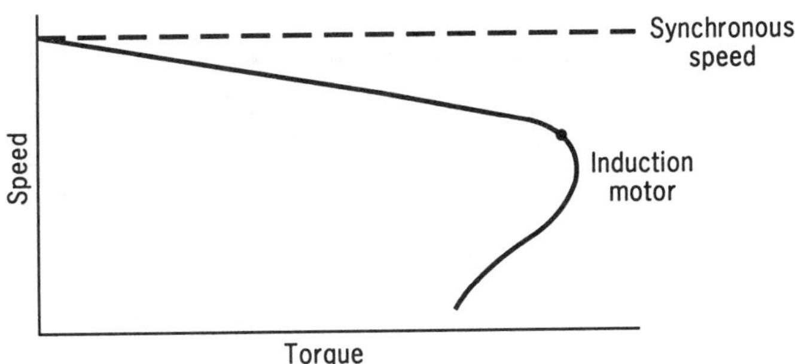

Figure 7.43 Induction motor-speed-torque curve.

In the induction motor the revolving stator field can be simulated during starting by adding a starting winding to a single phase machine (Figure 7.44). The capacitor current leads the current in the main winding and effectively allows a two-phase start mode. Large induction motors are three-phase and do not require a starting winding. The induction motor is by far the most popular motor in the world. It is simple in construction and low in cost. It has drawbacks in that its efficiency is quite low in smaller sizes and its speed regulation and power factor are poor. By placing a permanent magnet in the rotor we have a synchronous motor with constant speed and improved efficiency and power factor. Such a motor is termed a line start synchronous motor. Changing line start induction motors to synchronous motors represents singularly the greatest potential use for permanent magnets. NdFeB magnets may well offer cost/performance relationships necessary to serve this important market.

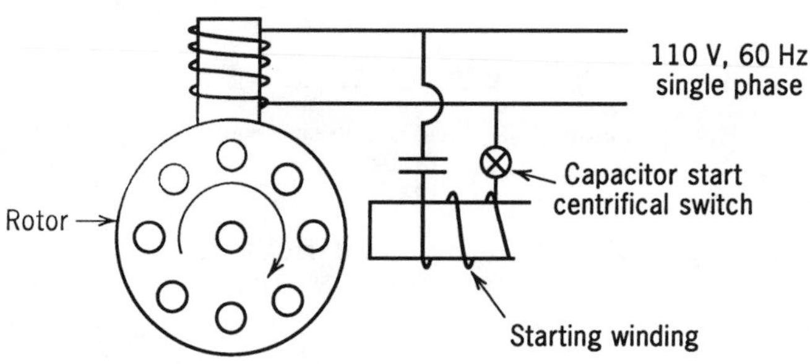

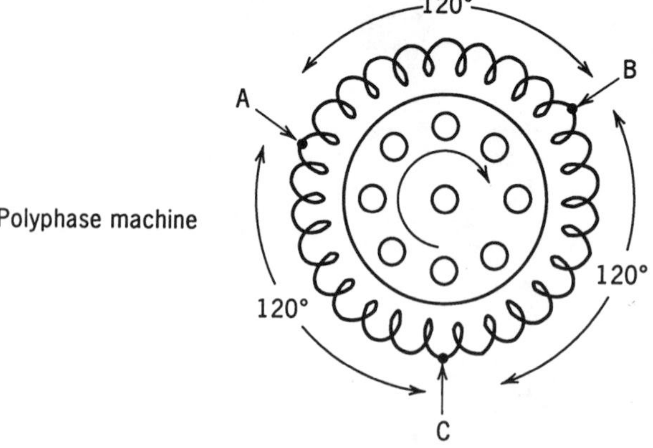

Figure 7.44 Starting the induction motor.

7.4.6 The Electronically Commutated Motor

For an a.c. synchronous motor with the same magnetic characteristics as a d.c. motor, there is a situation where the induced voltage is in phase with the applied voltage due to a specific load. The torque angle between rotor field and stator field will be 90°. Referring to Figure 7.45 this will happen at 15 inch ounces of torque and at half the no load speed. If a greater torque is applied the permanent magnet rotor will fall out of step and the torque would go to zero. If one can find a way to control motor speed and torque angle, then the a.c. motor becomes the same as the d.c. motor. The brushless d.c. motor or ECM development has solved this problem and a.c. and d.c. motor technology has merged together.

In Figure 7.46, the basic elements of the brushless or ECM are displayed. Some kind of position sensor is necessary to change the stator current at precisely the right time. These sensors may be optical, Hall effect or electromagnetic type.

With the trend to synchronous line start, constant speed requirements, and increasing need for variable speed ECM motors, there is need for a new and universal kind of moving magnet machine. This motor has a rotor with magnets and a squirrel cage (Figure 7.47). It has a conventional stator. This machine will line start and run at synchronous speed with much improved efficiency and power factor compared to an induction motor. The same machine can be the basis of the ECM with full control of speed, torque,

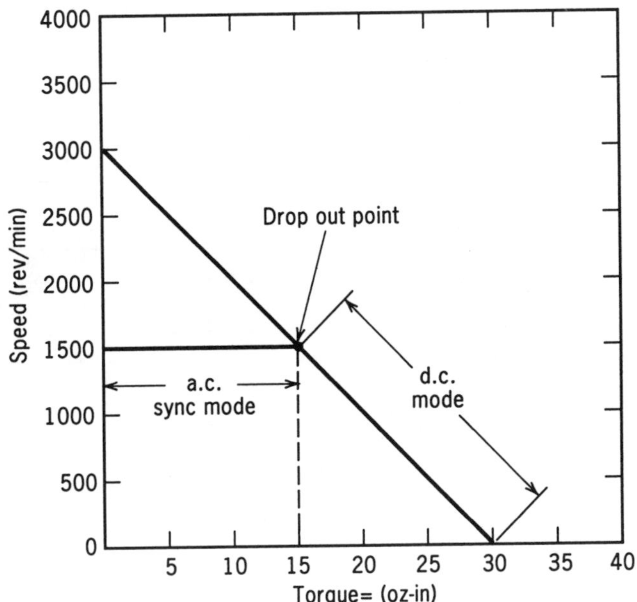

Figure 7.45 a.c. and d.c. modes.

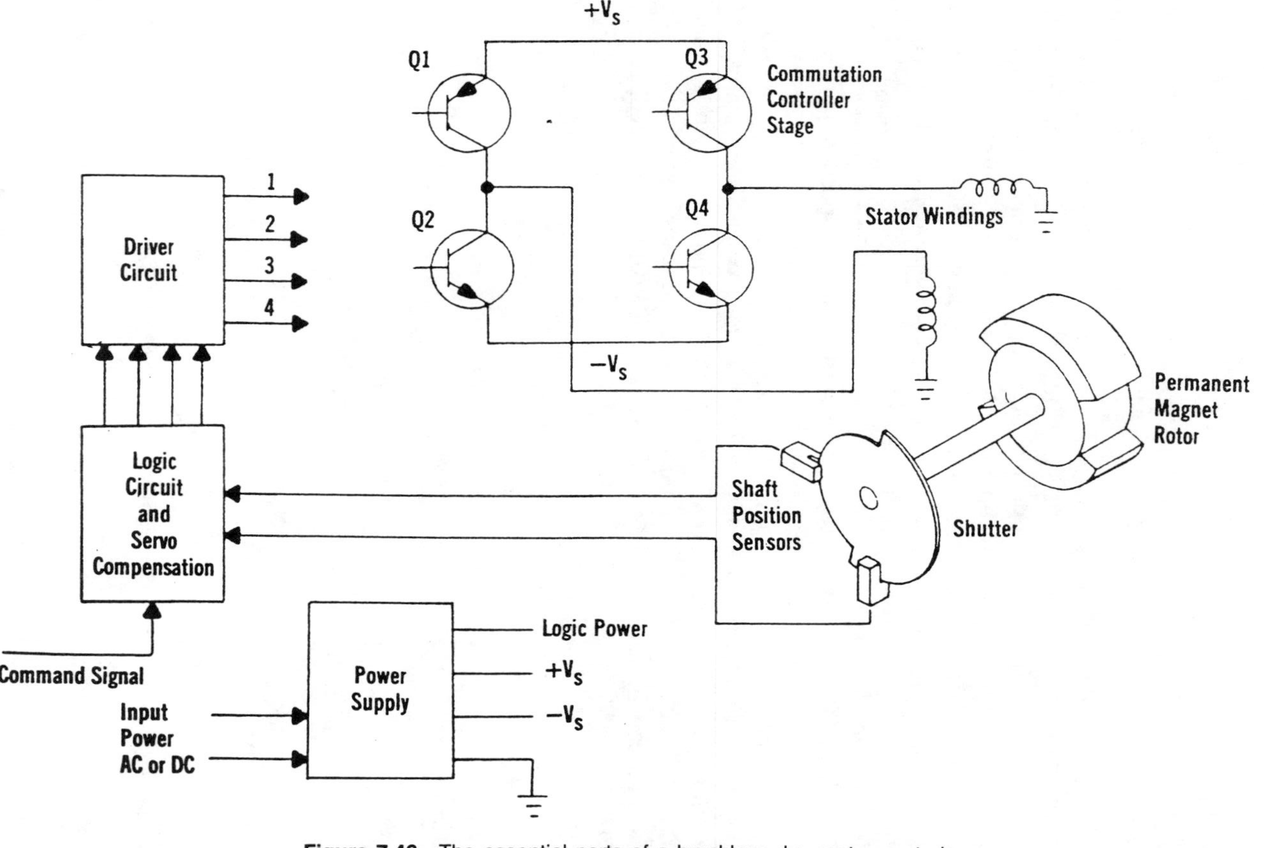

Figure 7.46 The essential parts of a brushless d.c. motor control.

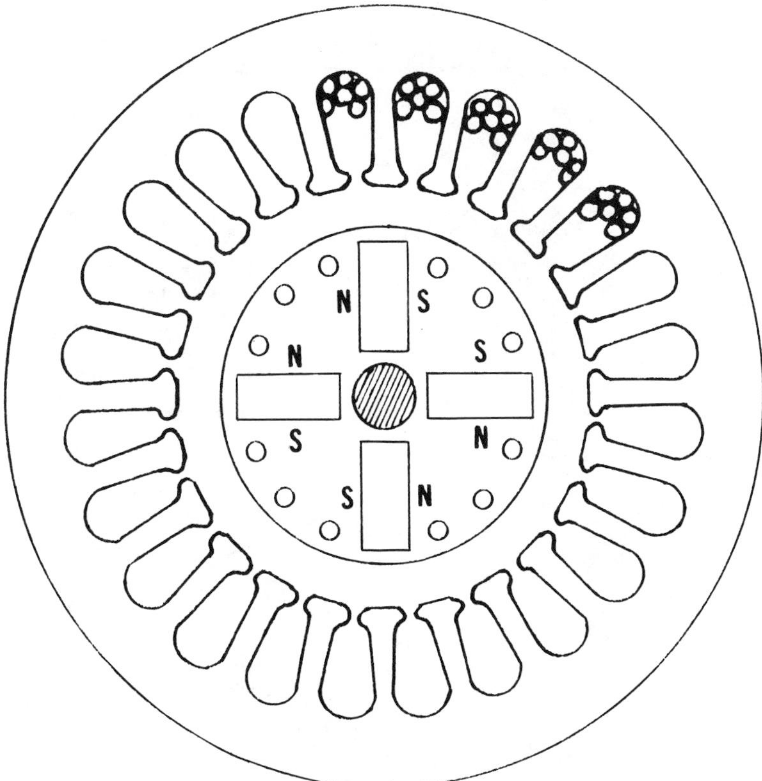

Figure 7.47 Permanent magnet universal motor.

torque angle, and efficiency. This same machine could run in stepper mode with the right controller. The construction of the rotor is illustrated in Figure 7.48a and b. In (a) the placement of a lower B_r magnet is shown. By staggering the magnets the effective area is increased. In (b) a lower coercive force material can be used due to the parallel paths.

Rare-earth magnet properties allow very small diameter rotors and hence very high response. Placing the magnet on the rotor is excellent thermally. The magnet has no ohmic loss and hence there are no losses to radiate from the interior. The stator windings are of course stationary and can be thermally coupled to the outer case. Many such designs actually use a heat sink around the stator.

The ECM is rapidly replacing other motor types due to the following features:

1. ECM can operate at any speed
2. ECM motors have good speed regulation
3. An ECM has linear speed torque curve
4. Brush and commutor limitations are removed

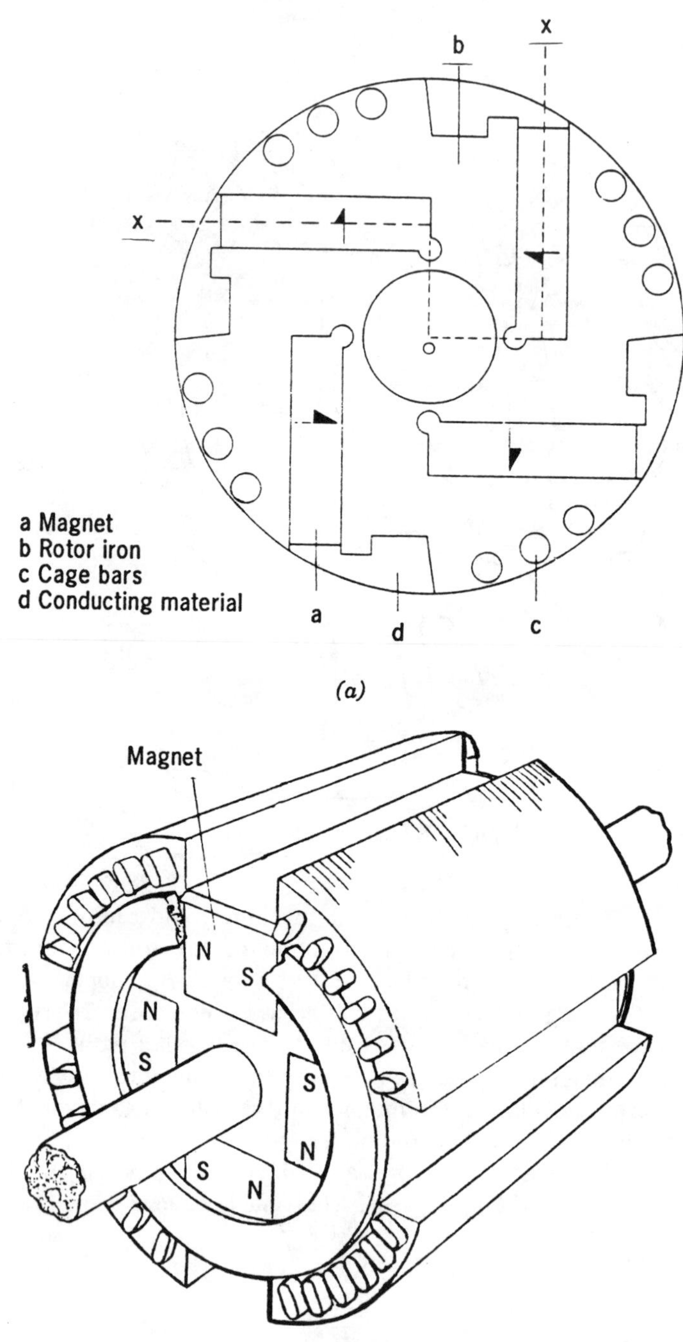

Figure 7.48 Moving magnet motor constructions. (a) Schematic diagram of geometry of typical rotor configuration; (b) alternate construction.

The principle disadvantage of the ECM is that magnet demagnetization can be more severe than in the commutated machine. This is due to the armature field acting over a greater region of the magnet. On the other hand, the drive electronics offer possibilities in limiting peak start or stall current levels.

This new universal kind of motor will be a very important part of electromagnetic energy conversion and control in the years ahead.

7.4.7 The Hysteresis Motor

The hysteresis motor is a synchronous motor which is capable of exhibiting high starting torque and can synchronize high inertial loads without auxiliary starting equipment.

The hysteresis motor develops torque by virture of the fact that the flux density lags the magnetizing current by an angle due to the nonlinear relationship between B and H in magnetic materials. A revolving magnetic field is established and induces magnetization in the rotor material. The flux in the ring will lag the magnetizating force by an angle depending on the nature of the rotor material. For soft materials, exhibiting negligible hysteresis loss, the angle would be zero and no hysteresis torque would be developed.

The maximum torque T which a hysteresis motor can develop may be expressed as follows

$$T = KfVE_h \tag{7.19}$$

where K is a constant depending on units, f is the frequency, V is volume of rotor and E_h is energy loss per cycle on a unit volume basis (proportional to hysteresis loop area).

Hysteresis motors use relatively low coercive force magnet material. The stator ampere-turns are required to drive the rotor material through the complete hysteresis loop.

Timing motors, clock motors, and constant torque drives are examples of the application of the hysteresis motor. Figure 7.49 shows principle component arrangement of a multiphase hysteresis motor.

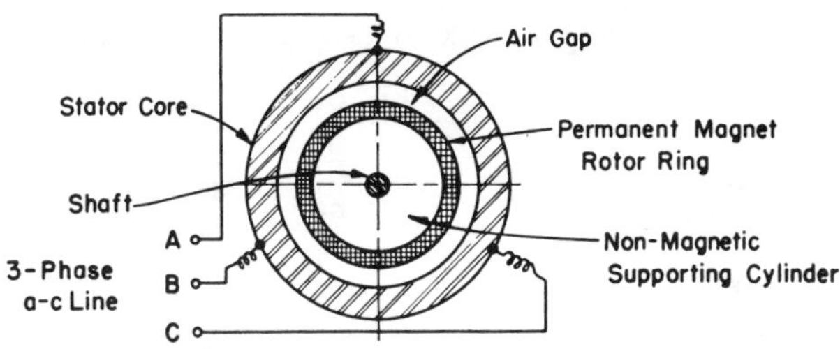

Figure 7.49 A schematic representation of a polyphase hysteresis motor.

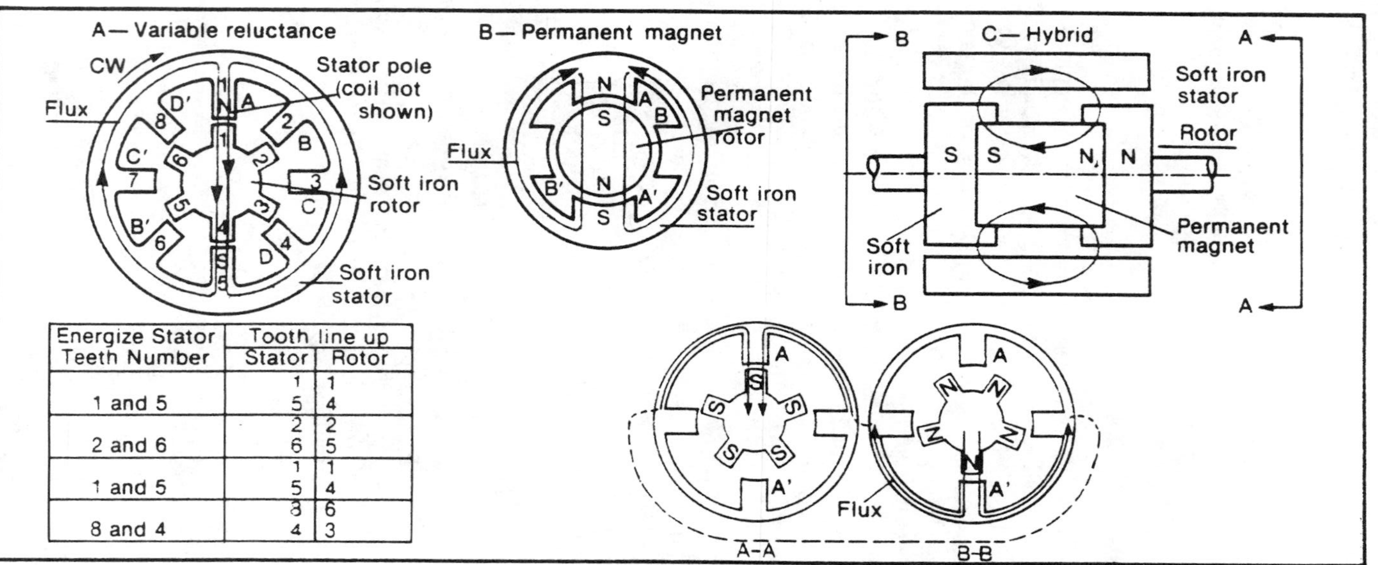

Energize Stator Teeth Number	Tooth line up	
	Stator	Rotor
1 and 5	1 5	1 4
2 and 6	2 6	2 5
1 and 5	1 5	1 4
8 and 4	3 4	6 3

Figure 7.50 Stepping motors.

7.4.8 The Stepper Motor

With the growth of digital electronics came the need for digital controlled actuators. Today stepper motors find wide use in all kinds of control systems, computer peripherals, robotics, and numerically controlled machine tools. The stepper motor is a discrete positioning device. The stepper may be of the variable reluctance type, which does not use magnets. Such a device uses a salient pole rotor which moves when stator poles are energized. This motor uses torque set up by the changing reluctance. The other type of stepping motor uses the field of the permanent magnet rotor to interact with the stator poles. Figure 7.50 shows reluctance and permanent magnet steppers and a combined or hybrid design. These arrangements offer a wide choice or torque and inertia. In simplest form we may express the stepper motor torque as

$$T_m \simeq \phi_m (NI)_s \tag{7.20}$$

where T_m is the stepper torque, ϕ_m is the magnet flux and $(NI)_s$ is stator ampere-turns. High energy magnets with their very high magnetic moment per unit volume can have great impact on a stepper motor's performance. In Figure 7.51 a state of the art rare-earth stepper is shown. The heart of the

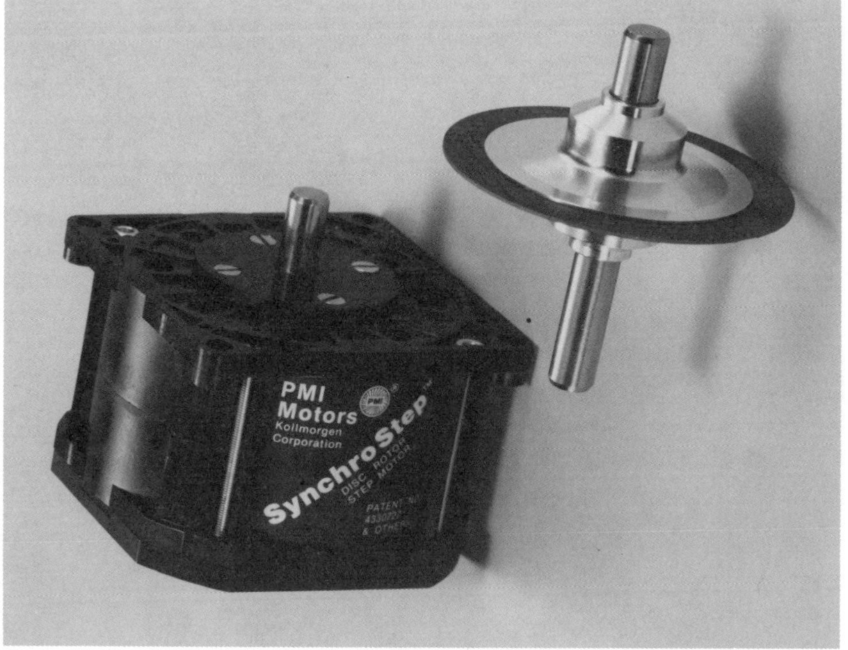

Figure 7.51 High performance stepper. Courtesy Kollmorgen Corp.

motor is an oriented rotor which has 50 pole pairs. This motor has low inertia, low mechanical time constant, and high torque and power rate.

7.4.9 Electrical Measuring Instruments

Permanent magnets play a vital role in electrical measuring instruments. They find use in both moving coil and moving magnet types. Variations of the magnetic structures for moving coil instruments are displayed in Figure 7.52. These magnetic circuits allow a uniform radial field in the air gap. The scale linearity, damping, and sensitivity depend on the magnitude and uniformity of the field. A major change occurred with the placing of the magnet inside the coil as shown in (d), the inverted core structure. Alnico 5, Alnico 8 and Lodex type magnets are used in this arrangement with Alnico 5 being widely used in arrangements (a), (b), and (c). Obviously the stability of the magnet is of great importance. Alnico 5 is used extensively since it can be stabilized to give fields which remain constant to one part in 10^2 or better over long periods.

The basic arrangement of a moving magnet type instrument is shown in Figure 7.53. This type of instrument does not use springs for counter torque, but uses control magnets to restore the moving magnet rotor. Figure 7.54 shows the restoring action with and without coil current. Since the mag-

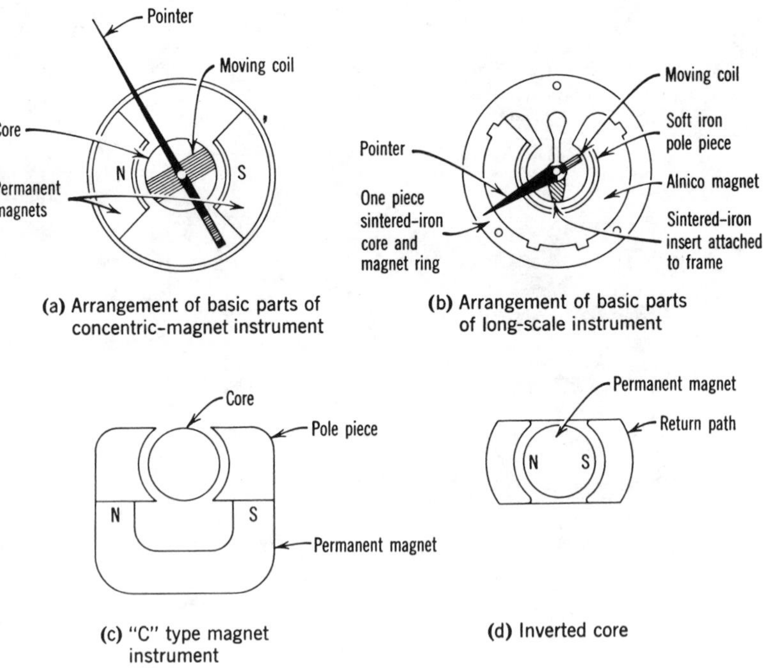

(a) Arrangement of basic parts of concentric-magnet instrument

(b) Arrangement of basic parts of long-scale instrument

(c) "C" type magnet instrument

(d) Inverted core

Figure 7.52 Permanent magnet circuits in moving coil instruments.

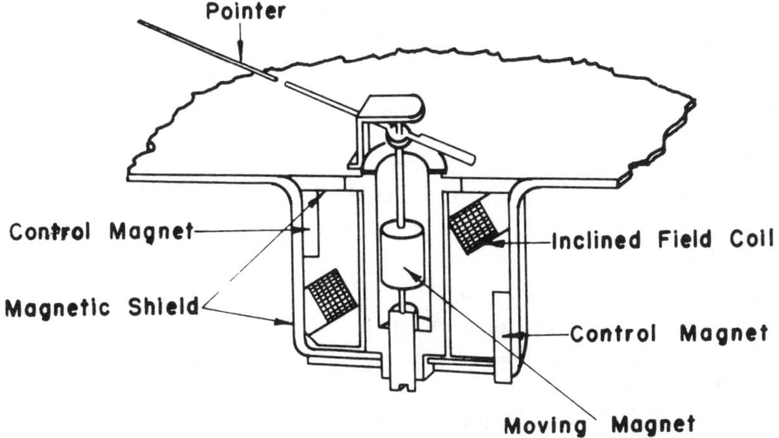

Figure 7.53 Permanent magnet moving magnet instrument.

nitude and direction of the resultant flux produced by moving magnet and control magnet are fixed, the direction of resultant flux with which the moving magnet aligns itself is dependent only on current magnitude in the coil. It should be noted that deflection angle in this system is dependent on the direction of resultant flux and not its magnitude. Moving magnet

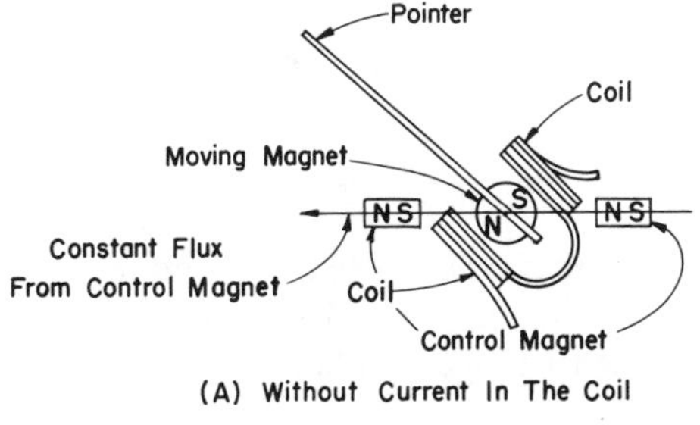

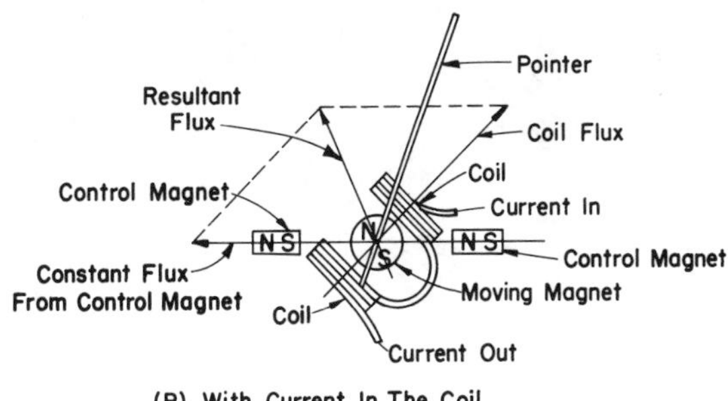

Figure 7.54 Schematic vector diagrams of moving magnet instruments.

instruments in the past have been limited by lower sensitivity but now there is renewed interest involving rare-earth properties. The moving magnet instrument is simple and of rugged construction with potentially lower manufacturing cost.

7.5 APPLICATIONS BASED ON LORENTZ FORCES ON FREE ELECTRON CHARGES

Permanent magnets are useful in deflecting and focusing a beam of electrons in a vacuum. Cathode ray tubes, magnetron oscillators, traveling wave tubes and klystron amplifiers are examples of widely used beam tubes using permanent magnets.

The force on a free electron in a magnetic field is expressed as

$$F = \frac{eVB}{10} \tag{7.21}$$

where F is the force (dyne), e is the charge in EMU units, and V is the velocity (cm/s). This equation gives the force on the particle moving at velocity V perpendicular to the field axis B. If the particle has a mass m it can be shown that the particle acceleration is eVB/m. From physical laws of motion, the particle will follow a circular path of radius r expressed as

$$r = \frac{mV}{eB} \tag{7.22}$$

These relationships are basic to the operation of the magnetron tube, which is a high frequency oscillator. The basic electrons of the magnetron are shown in Figure 7.55. Electrons leave the filament and move radially toward the positively charged plate system. The electrons are deflected and if the correct relationship beteween voltage and field exists, the electrons will not

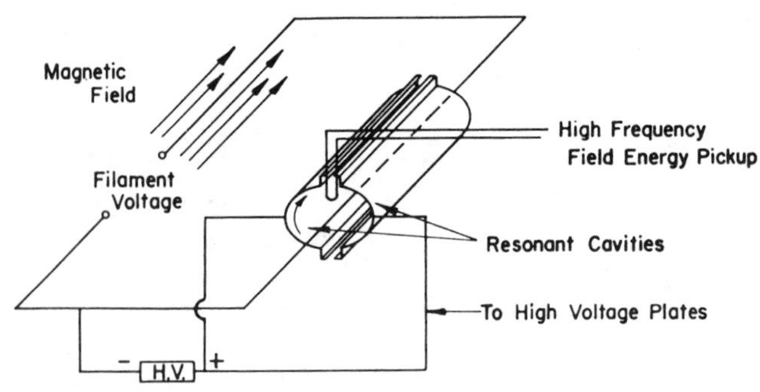

Figure 7.55 Basic elements of magnetron tube.

reach the plate but rotate within the plate system. An oscillating space charge is produced and high frequency energy is coupled out of the cavity to an antenna. An electron moving at high velocity in a field perpendicular to the velocity experiences centripetal acceleration and will move in a circular path. In the magnetron, angular velocity of the electron depends only on flux density and the ratio of electron charge to mass. Consequently, the permanent magnet field must be very stable and accurately calibrated. Figure 7.56 shows the outline of a magnetron using ceramic magnets in a microwave oven. Figure 7.57 shows the trend to smaller efficient field structures for magnetrons made possible with high energy magnets. Early Alnico 5 structures were limited to about 3 kG due to very high limb leakage. Only about 10% of the total flux was useful. Alnico 9 magnets improved the efficiency due to shorter magnet lengths. The use of $SmCo_5$ to control leakage by clading Alnico 9 further improved efficiency. Combinations of permendur poles and $SmCo_5$ gave entirely new levels of gap density and the figure of merit B_g^2/W improved by a factor of over 20. This is an excellent example of compounding the influence of a high energy magnet in a low permeance load. The energy density has improved by a factor of 4 but due to the reduction in leakage, the combined improvement is about a factor of 20.

In the focusing function in a beam tube, the field produced is axially symmetrical and influences the beam as follows. Any electron deviating

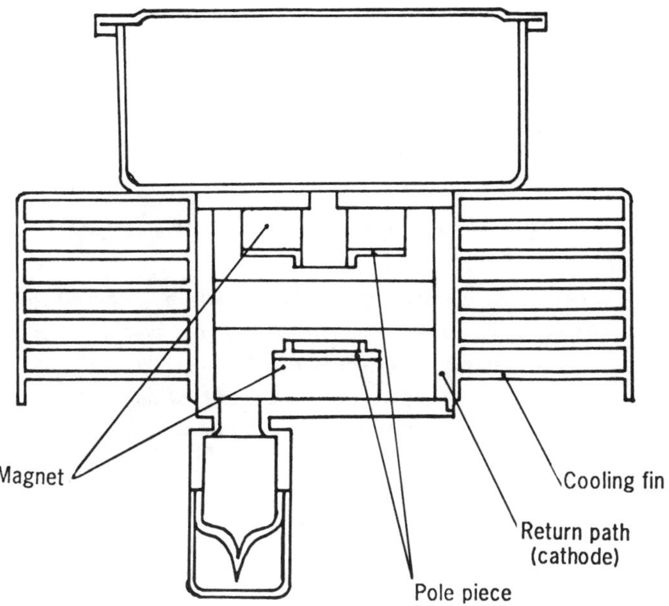

Figure 7.56 Cross-section of a magnetron.

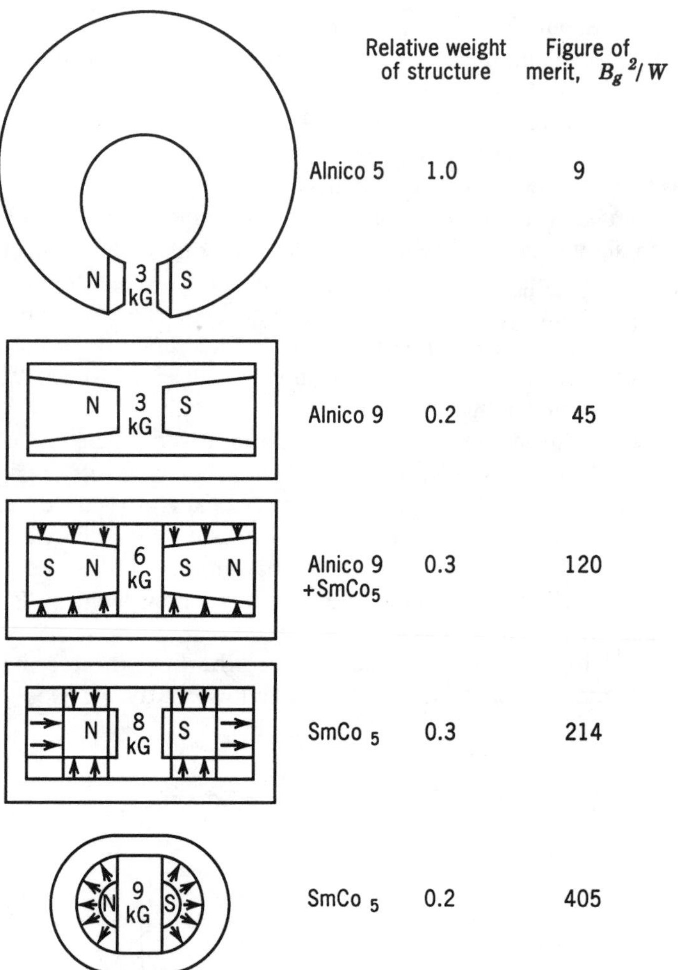

Figure 7.57 Voltage tunable magnetron field suppy evolution.

from the beam axis has a velocity component perpendicular to the field and consequently experiences a force which tends to make the electron spiral about the axis. The straying electron is forced back toward the axis and if the correct relationship between field strength and velocity exists, the electron will return to axis the instant it reaches the anode, target or collector. A general expression relating the factors that influence focusing with a magnetic lens is

$$\frac{1}{f} = \frac{e}{8mE} \int_A^B H_z^2 \, dZ \tag{7.23}$$

where f is the focal length, e is the electron charge, m, is the electron mass,

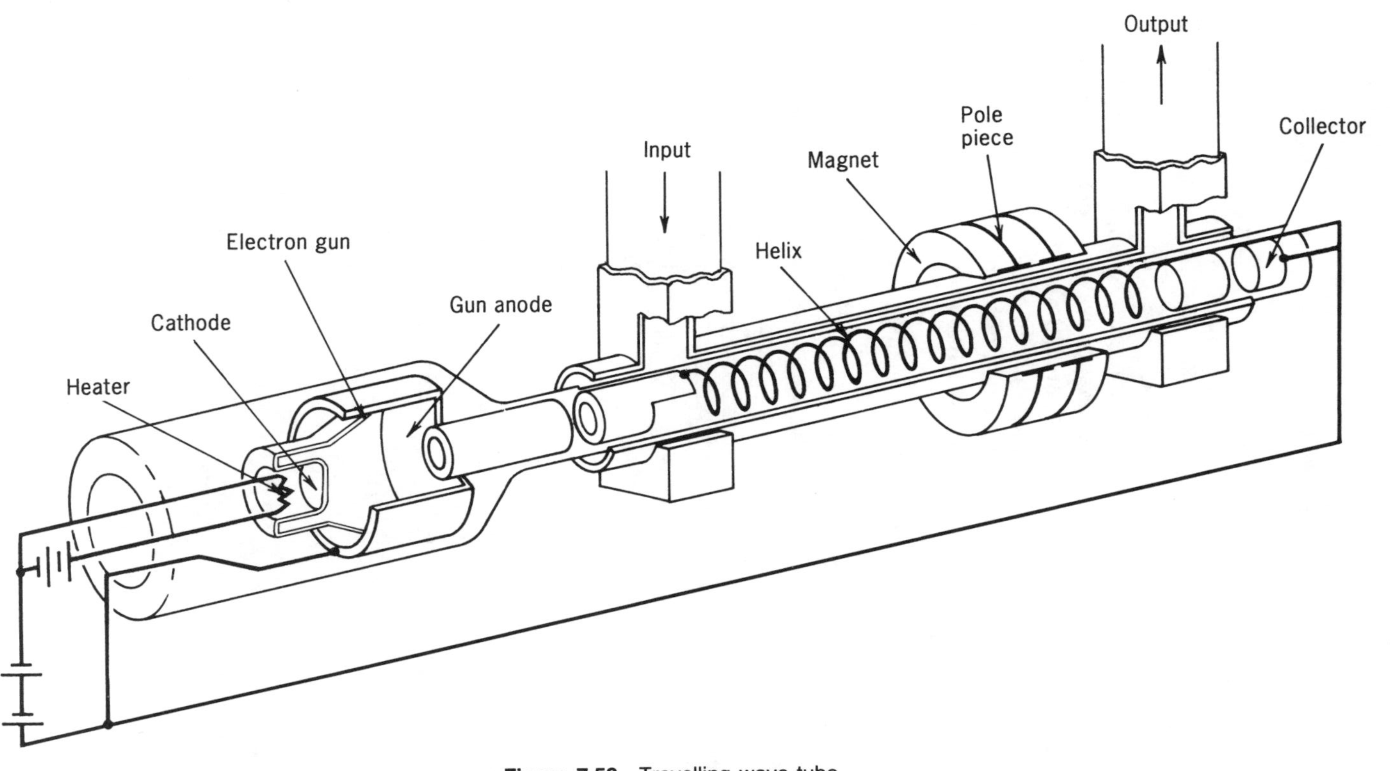

Figure 7.58 Travelling wave tube.

E is the accelerating voltage, and H_z is the Z component of field strength where z is the axis of symmetry.

The traveling wave tube is a wide band high frequency amplifier used as a pulse amplifier, frequency multiplier, or modulator. The basic components of this beam tube are shown in Figure 7.58. Between an electron gun and a collector there is a helical conductor. The high frequency wave to be amplified is placed on the helical path. The velocity of this wave is set by the d.c. potential of the helix with respect to the cathode. An axial magnetic field focuses the electron beam and confines it to the inner diameter of the helix. The velocity of the high frequency wave in the axial direction is reduced by the ratio of the pitch to circumference ratio of the helix. The velocity of the wave is set by voltage adjustments to be just slightly less than that of the electron beam. The interaction of beam energy and wave energy is such as to transfer d.c. beam energy to the *r-f* electric field and hence amplification is effected.

The first permanent magnet arrangements produced a uniform field and were in the form of a tubular magnet (Figure 7.59a). Such a permanent magnet system becomes very heavy for the extremely long region over which a uniform field is needed. A development which has drastically

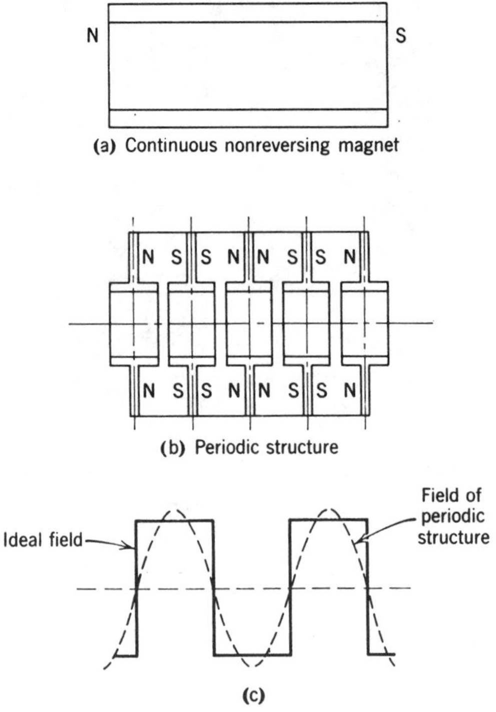

Figure 7.59 Beam focusing.

reduced the weight of the permanent magnet in this application is the use of periodic focusing. The basic principle of periodic focusing is that an electron beam can be equally well focused irrespective of the polarity sense of the field vector with reference to the electron flow. Figure 7.59b shows a cross-section of a periodic focusing structure, and Figure 7.59c shows the variation of field intensity along the tube axis. Periodic focusing offers a tremendous weight advantage in that the weight of the periodic system is proportional to the length of the field while the continuous uniform field magnet weight is approximately proportional to the third power of the field length so that the theoretical weight advantage becomes equal to the square of the field length. In Figure 7.60 a photograph of a typical focusing stack for a traveling wave tube is shown.

The traveling wave tube is a very demanding application for the permanent magnet. High power, high frequency tubes require very uniform properties and high coercive force at high temperature. A very low reversible temperature coefficient is also desirable. The traveling wave tube is definitely magnet limited. One cannot have better tubes without better magnets. In the development of $SmCo_5$ from 1965 to 1975, the United States Air Force support was to a large extent aimed at demonstrating what a high energy product, high coercive force magnet could do for a traveling wave tube. This device was the vehicle used to show how important the new properties could be. Periodic focusing required magnet lengths inversely

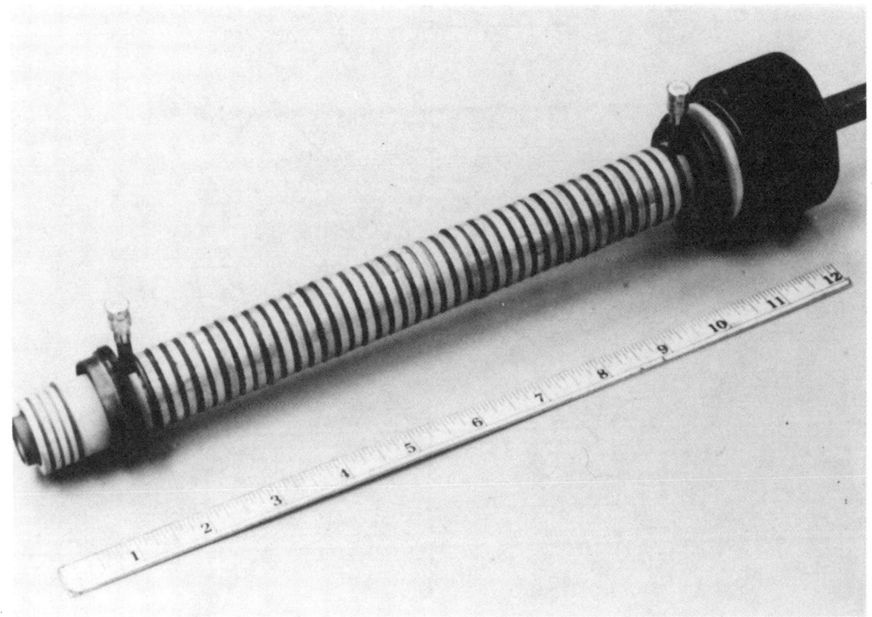

Figure 7.60 Traveling wave tube focusing structure.

proportional to frequency and peak fields are limited by the magnets' H_c level.

There has developed a new area of interest for permanent magnets, made possible by rare-earth properties. This is the use of magnets in charged particle beam focusing and bending applications [4] in synchronous radiation experiments. The quadrupole shown in Figure 7.61 is for beam focusing and separation use. The field in the bore is approximately 10 kG. This level of field would be impossible to obtain with water cooled conductors or even superconductivity at this scale. This device is an excellent example of the impact that permanent magnets can have due to their ability to provide energy density independent of scale. The quadrupole shown was constructed of 16 segments with 4 orientations associated with the segments.

The flux distribution of a rare-earth wiggler assembly is illustrated in Figure 7.62. This device generates the same spectrum of light from infrared to X-rays as a bending magnet, but the radiation is more intense with the wiggler. The charged beam is forced into a sinusoidal oscillation which allows the generation of more synchronous light energy.

There are several other applications of magnets based on Lorentz forces on free electrons. The mass spectrometer and ion pump are two that are

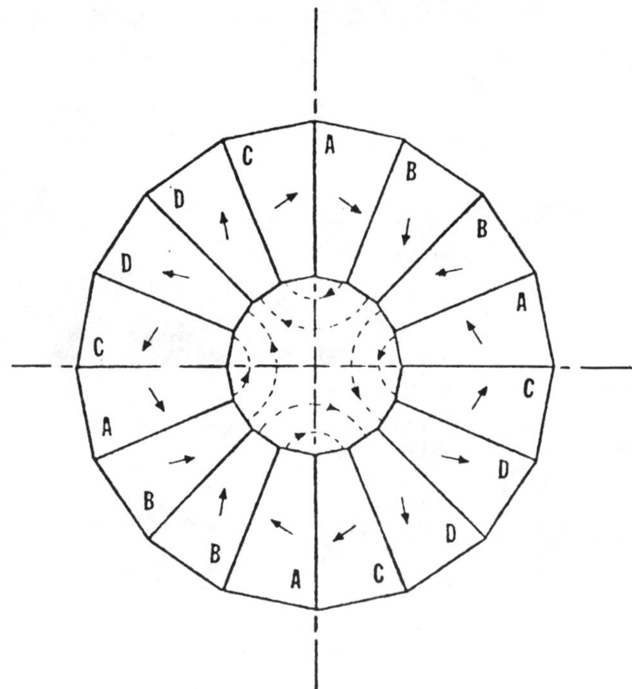

Figure 7.61 Schematic of 16 piece segmented REC quadrapole with 4 easy axis orientations.

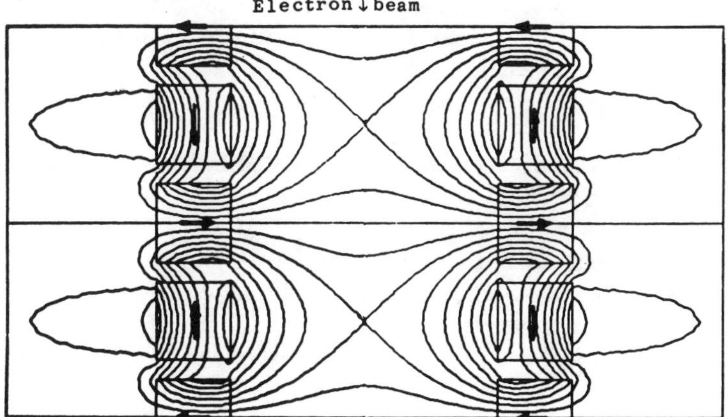

Figure 7.62 Flux in a pure rare earth cobalt wiggler.

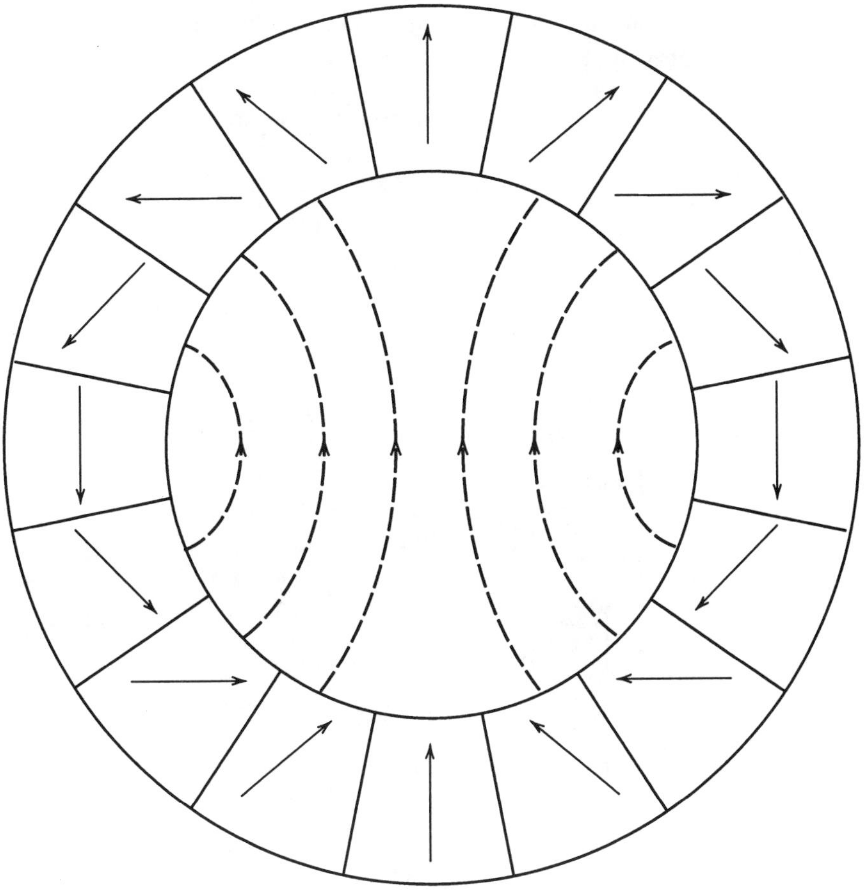

Figure 7.63 Possible MRI structure.

found in the laboratory. Permanent magnets are often used for arc extinguishing in electrical contractors. The field effectively lengthens the arc, which is helpful in the extinguishing process.

7.6 MISCELLANEOUS APPLICATIONS BASED ON A VARIETY OF PHYSICAL PRINCIPLES

Permanent magnets are often used to alter the characteristics of electrical and magnetic materials. The field can bias an electromagnetic core to change its apparent properties. In the Hall effect, a field changes the electrical properties of a crystal. There are also magneto-optical effects that use magnets.

A major potential use for permanent magnets is magnetic resonance imaging (MRI). The entire human body is placed in a uniform magnetic field and the body can be mapped for irregularities in body chemistry. Presently this equipment mostly uses large superconducting solenoids. However, permanent magnets offer an interesting compromise solution at a much lower value of flux density. A permanent magnet structure as shown in Figure 7.63 offers much lower cost and relative freedom from leakage

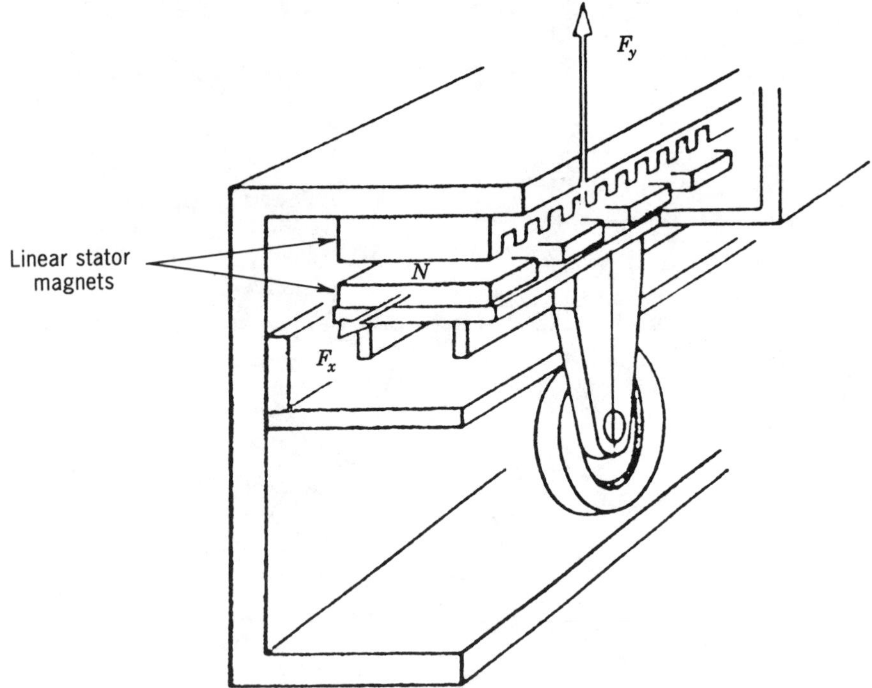

Figure 7.64 Levitated transportion system.

compared to a superconducting solenoid. NdFeB magnets are currently being considered for MRI.

One of the most unique and potentially the largest application of permanent magnets is the levitated train. Figure 7.64 shows a low speed people mover which uses the permanent magnet to perform two functions. Magnets hold the vehicle up against the bottom of a steel track and the

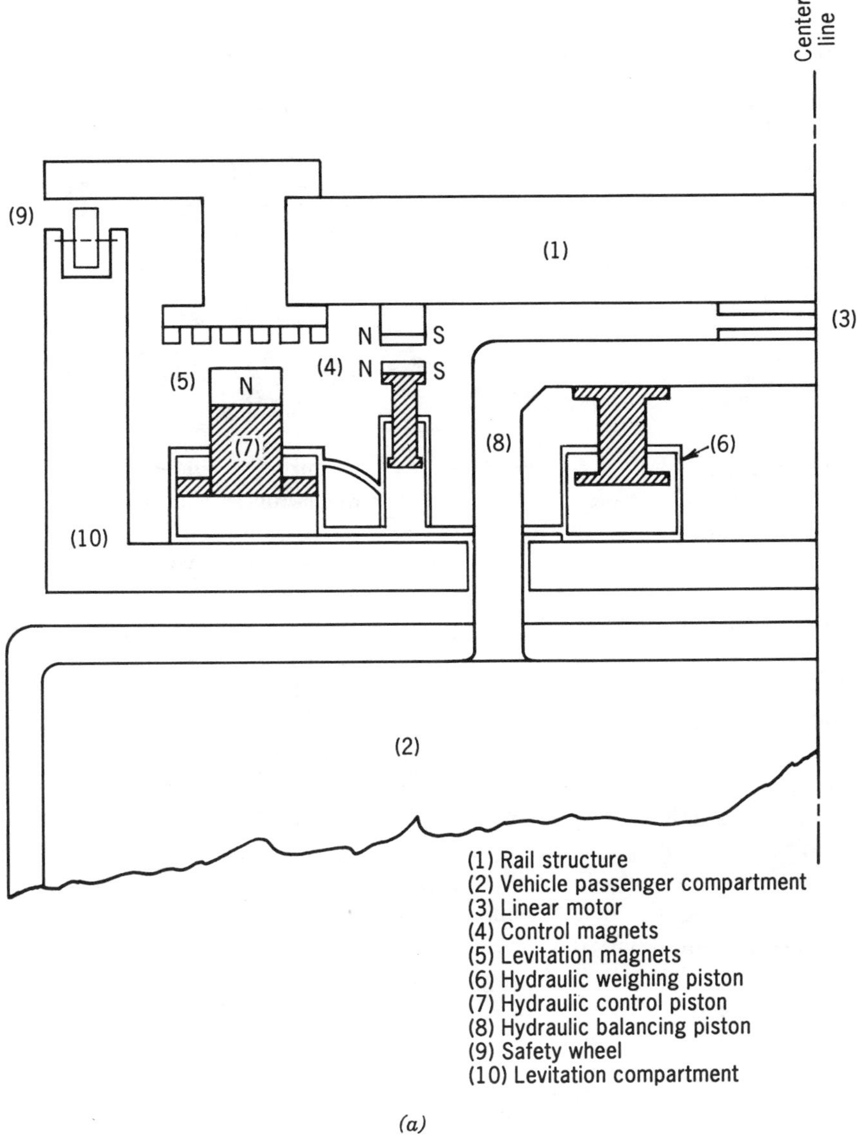

Figure 7.65 (a) Magnetic-hydraulic levitation system; (b) gravity magnet balance system.

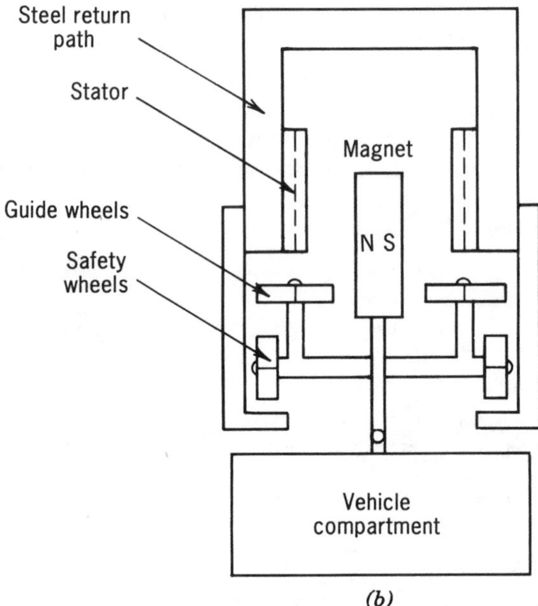

Figure 7.65 (*Continued*)

magnet also serves as the synchronous element in a linear motor. This system developed in West Germany, is known as the M-Bahm system. The vehicle is both propelled and braked by a linear motor. The linear motor stator is embedded along the track. Stator sections are activated or powered only as the vehicle passes over them. Depending on stator frequency and the direction of the moving electromagnetic wave, the vehicle can be moved forward at various speeds or braked to a stop. In this system, small guide wheels keep the vehicle aligned and spaced with respect to the track. These wheels are essentially unloaded. This system has several major advantages. The light weight vehicles lead to light track construction and hence, minimum investments. The system can be fully automated and thus, operating costs are down. The system is also very efficient, compared to normal rail transportation.

Two interesting variations of the levitated drive are shown in Figure 7.65a and b. Magnatrain Inc., Covina, California has developed a permanent magnet-hydraulic levitation system which has a unique hydraulic weighing and balancing feature as shown in Figure 7.65a. Figure 7.65b shows the basic elements of a Magnatrain, Inc. levitation system which uses a permanent magnet to work with gravity to balance the vehicle weight. Guide wheels with verticle axis give the system lateral stability without the need for servo electromagnetic balancing.

No description of permanent magnet applications would be complete without reference to a very different kind of use of magnetic particles and properties in the recording and reproduction of information. In audio, video and digital information recording, permanent magnet properties are of critical importance. In terms of dollars this is by far the major use of magnets. However, it is such a large and uniquely different use that recording and reproduction of information is not considered a part of the permanent magnet industry for this reason. Recording and reproduction is an industry in itself. The principles and figures of merit for a recording media are totally different than for bulk magnet material. Therefore, the reader is referred to the treatment of magnetic recording principles by Watson [5].

REFERENCES

[1] S. Earnshaw, On nature of molecular forces which regulate the constitution of the luminiferous ether, Trans. Cambridge Philos. Soc. 7 (1839) 97–112.

[2] J. P. Yonnet, Analytical calculations of magnetic bearings, Proceedings of the 5th Rare-Earth Cobalt Magnet Workshop, Roanoke, VA June 1981, pp. 199–216.

[3] R. J. Parker and R. J. Studders, Permanent Magnets and their Application (John Wiley, New York, 1962) p. 246.

[4] R. F. Holsinger, Rare-earth cobalt magnet assemblies for charged particle beams, Paper No. II-3 at the 6th International Workshop on Rare-Earth Cobalt Magnets and their Applications, Baden, Austria, 1982.

[5] J. K. Watson, Applications of Magnetism (John Wiley, New York, 1980) Chapter 7.

8

MEASUREMENTS

8.1 Introduction
8.2 Apparatus and Techniques for Generating Magnetic Fields
  8.2.1 Steady State Fields
  8.2.2 Pulsed Fields
8.3 Measuring Magnetic Fields and Magnetic Potential
  8.3.1 Traditional Force and Torque Measurements
  8.3.2 Electromagnetic Induction Methods
  8.3.3 Measuring Magnetic Potential
  8.3.4 The Hall Effect in Magnetic Measurements
  8.3.5 Nuclear Magnetic Resonance
8.4 Instrumentation for Magnetic Field Measurement
  8.4.1 Electromechanical Intergrating Devices
  8.4.2 Electronic Intergration
  8.4.3 Hall Effect Gaussmeters
  8.4.4 Moving Magnet Gaussmeter
  8.4.5 Helmholtz Pair Detection System
  8.4.6 Vibrating Sample Magnetometer
8.5 Open Circuit Measurement Techniques
8.6 Instrumentation Systems for Closed Circuit and Hysteresis Loop Measurements
  8.6.1 Equipment and Techniques
  8.6.2 Determination of Demagnetization Curve
8.7 Search Coil Arrangements and Characteristics
8.8 Special Measuring Techniques for Permanent Magnet Circuit Analysis and Design Optimization
8.9 Calibration, Standards, Precision, and Accuracy
8.10 Measurement Practice in Production Quality Control

8.1 INTRODUCTION

> "I often say that when you can measure what you are speaking about and express it in numbers you know something about it; but when you cannot measure it, when you cannot express it in numbers, your knowledge is of a meager and unsatisfactory kind; it may be the beginning of knowledge but you have scarcely, in your thoughts, advanced to the stage of SCIENCE, whatever the matter may be."
>
> —Lord Kelvin, 1883

Lord Kelvin's famous quotation was never more meaningful then when applied to magnetism. From research to production and end use, one finds measurements and the defined unit properties deduced from them to be the basis of evaluation and communication.

In preceding chapters, we have seen that the relationship between magnetizing force H, magnetization B_i, and induction B, is of great significance in evaluation, design and analysis of permanent magnets and the magnetic circuits and devices in which they are found. In the following pages, the instruments, equipment, and techniques for determination of defined unit properties are described. The measurement of fields in air space and the numerous uses of measurements in design optimization and production testing are explored. In addition, magnetic standards, accuracy, the role of the International Electrotechnical Commission, and the traceability of magnetic quantities to international standards is described.

8.2 APPARATUS AND TECHNIQUES FOR GENERATING MAGNETIC FIELDS

8.2.1 Steady State Fields

The field requirements for magnetizing and measuring permanent magnet properties are usually established with current carrying conductor systems. A simple and widely used conductor system is the air core solenoid. The solenoid may be air cooled, water cooled, cryogenic or superconducting, depending on field requirements. For many rare-earth magnets, fields of 30 kOe are required to magnetize and to allow the demagnetization curve to be drawn. To fully characterize (complete major hysteresis loop) fields in excess of 100 kOe are required.

Air cooled solenoids are limited to a few kOe. By water cooling, the field can be increased 30–40 times. Fields of 20–30 kOe are possible with water cooling. Bitter type solenoids are capable of fields to 100 kOe, but due to the cooling and very high power requirements, are not generally used. Super-conducting solenoids are capable of 100 kOe and are in use in many industrial laboratories involved in permanent magnet development, in spite of the requirements for cryogenic containers and liquid helium. Figure 8.1

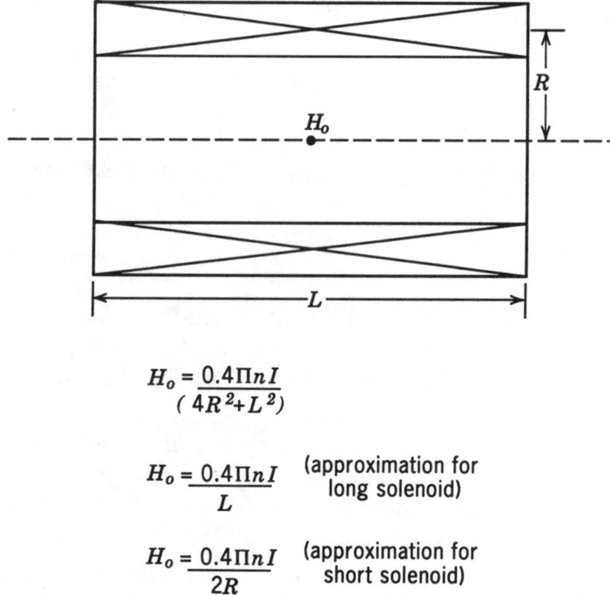

Figure 8.1 Solenoid relationships.

shows the relationship between current level and field strength for a solenoid.

The most often used apparatus for establishing a field for measurement is the electromagnet. Figure 8.2 shows a typical laboratory electromagnet with water cooling and a maximum field capability of 35 kOe in a 1 cm air gap.

8.2.2 Pulsed Fields

A very short time duration pulse can magnetize a magnet to saturation. Capacitor discharge pulse magnetizers are widely used to magnetize permanent magnets under production conditions. A pulse magnetizer is often used to saturate a high energy magnet, after which it is transfered to a d.c. electromagnet for determining its demagnetization curve.

Recently, the use of pulse techniques to completely characterize is being investigated. Very high field pulses (up to approx. 350 kOe) have been used in the research community to study Sm_2Co_{17} materials which can have $H_a > 200$ kOe. Error is small, compared to d.c. measurements, if specimen volume is small to minimize eddy currents. It is very possible that a low cost, pulsed hysteresisograph will find acceptance in the future. It may be rather imperative that measurement technology move to pulse fields as permanent magnet properties continue to improve.

A more detailed description of pulse magnetizing techniques and their advantages and disadvantages is developed in Chapter 9.

Figure 8.2 Laboratory electromagnet. From LDJ Incorporated Troy, MI, U.S.A.

8.3 MEASURING MAGNETIC FIELDS AND MAGNETIC POTENTIALS

8.3.1 Traditional Force and Torque Maeasurements

Man's first attempts to express a magnetic field in quantitative terms involved force relationships between magnetic bodies, either attractive or repulstive mode. One early system used the force resulting when the flux of the magnet under test reacted with a moving magnet or coil free to rotate and indicate relative strength. This basic magnetometer principle is shown in Figure 8.3.

Today, measurements based on force are used infrequently because of the convenience and accuracy of the induction or electromagnetic method. However, the force methods are still of interest under some measurement conditions. For example, at high temperatures or in the case of irregular specimen configuration, the freedom from having to use a coil may be very

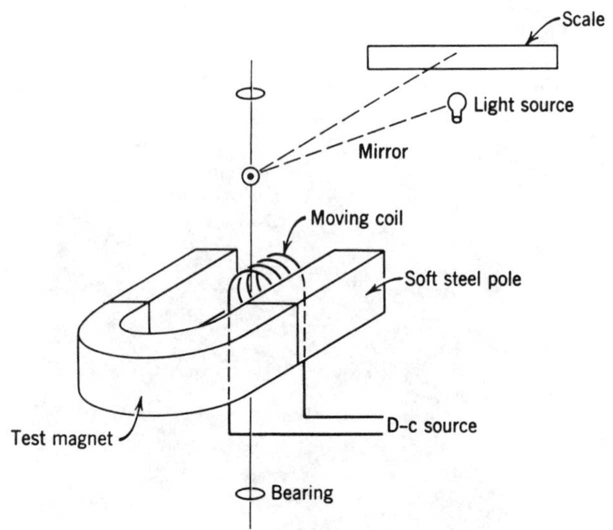

Figure 8.3 The essential elements of a moving coil magnetometer.

significant. In addition, force methods can, under some circumstances, yield a large amount of information about the properties of a magnetic material. For example with a torque magnetometer, it is possible to determine anisotropy constants, coercive force, and rotational hysteresis losses. The torque magnetometer principle is based on a magnet's rotation, if free to move, in order to align itself in a magnetic field so that its magnetic axis is parallel to the field direction. Assuming the magnet is uniform with respect to intensity of magnetization J, the polar lines from each end will be JA_m maxwells. The force exerted by the external field on each pole is JA_mH_a dynes. The resulting torque on the magnet is $T = JV_mH_a \sin\theta$ dyne-cm where θ is the angle between field direction and magnetic axis of the permanent magnet.

By rearrangement of the torque equation, the intrinsic magnetization B_i or the magnetic moment per unit volume is obtained

$$B_i = \frac{4\pi T}{V_mH_a \sin\theta} \tag{8.1}$$

8.3.2 Electromagnetic Induction Methods

The induced voltage method of measurement is based on Faraday's Law. An e.m.f. is induced in a coil when the magnetic flux within the coil changes

$$e = n\,\frac{d\phi}{dt}\,10^{-8}\ \text{V} \tag{8.2}$$

where n is the turns of the coil. Since the voltage changes with the rate of change of flux, the general practice is not to measure voltage but to measure the time integral voltage $\int e\,dt$. Equation (8.2) can be rewritten to give $d\phi = 10^8/n(e\,dt)$. By integrating

$$\int_{\phi_1}^{\phi_2} d\phi = \frac{10^8}{n} \int_0^t e\,dt \text{ maxwells}$$

Knowing the turns, one is able to calculate the total flux change from the $\int e\,dt$ measurement.

8.3.3 Measuring Magnetic Potential

In establishing the relationships between induction B and the magnetizing force H it is obviously necessary to measure H accurately. In testing arrangements, in many instances, H can be calculated from the relationship between the current in the magnetizing winding and the magnetizing force produced. For example, if a ring-shaped specimen of magnetic material is uniformly wound with a magnetizing winding, H can be determined from the expression

$$H = \frac{4\pi ni}{l} \tag{8.3}$$

Generally, in making permanent magnet measurements, the test specimens are not rings and the magnetizing current is not necessarily directly proportional to H because of air-gap effects and saturation influence encountered in the magnetizing structure. Since it is not directly possible to measure H inside the specimen, a small coil or a Hall probe adjacent to the surface of the specimen is often used to indicate H. The assumption of uniform field strength in the area of the specimen and coil is a source of error in many instances. Further consideration in obtaining reliable readings of magnetizing force are considered later in conjunction with closed circuit measurements.

In order to measure the magnetic potential between two points, an instrument known as the Chattock magnetic potential meter is used. This instrument is, in a sense, the magnetic analogy of the voltmeter. Physically, the Chattock meter consists of a flexible strip of nonmetallic material of a uniform section, uniformly wound with fine wire. The winding is connected to an indicating instrument, usually a flux meter. Consider Figure 8.4 and the potential difference between points x_1 and x_2. Fundamentally, the magnetic potential difference between two points is defined as the work done in moving a unit pole between the two points. The potential difference between x_1 and x_2 therefore is

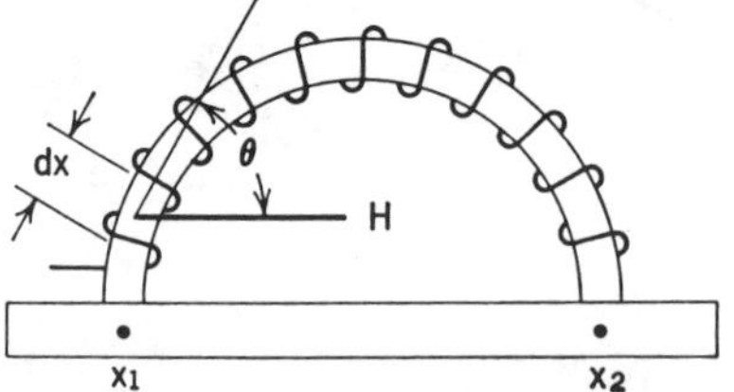

Figure 8.4 A Chattock coil to measure the magnetic potential difference between points X_1 and X_2 of a bar magnet.

$$\int_{x_1}^{x_2} H \cos\theta \, dx \tag{8.4}$$

where dx is a small increment of the path and θ is the angular displacement between the direction of H and dx. If the Chattock coil has an area A and n_u turns per unit length, the flux linkages in the length dx are equal to $An_uH \cos\theta \, dx$, and the total interlinkages from x_1 to x_2 are

$$n\phi = An_u \int_{x_1}^{x_2} H \cos\theta \, dx \tag{8.5}$$

or $\phi n = An_u$ (potential difference between two points)

$$\text{Potential difference} = \frac{\phi n}{An_u} \tag{8.6}$$

If we know the constants of the winding, the deflection of the indicating instruments can be interpreted in terms of potential difference.

8.3.4 The Hall Effect in Magnetic Measurements

The use of Hall generators in magnetic measurements has found wide acceptance. The very small size of the field sensing area and the static characteristics (no relative motion involved) enable the Hall probe to be used in areas where measurement with the induction principle would not be feasible. The basic principle of the Hall generator is shown in Figure 8.5.

When a current is flowing along one axis of a conductor or semiconductor, a voltage will be developed at right angles to the current flow when a magnetic field is applied in the plane of the current and voltage.

The relationship between current, voltage, and magnetic field is given by the formula

$$e = \frac{RHi}{d} \tag{8.7}$$

where e is the Hall voltage, R, the Hall constant, H, the Field strength, i,

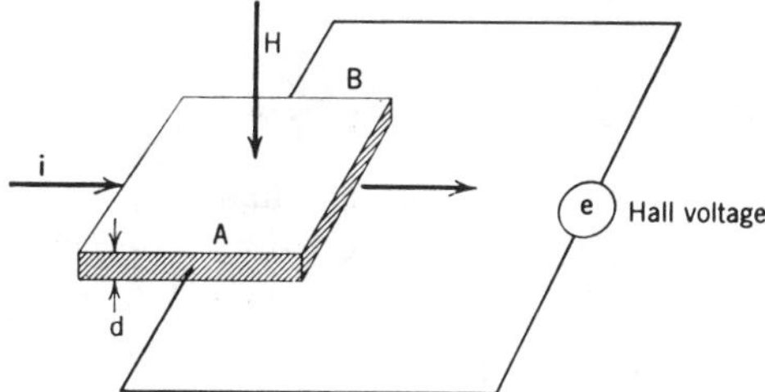

Figure 8.5 The Hall effect, with a thin plate of material such as indium arsenide, placed in a magnetic field with its plane perpendicular to the field, a coincidental longitudinal current *i*, through the material produces a proportional transverse voltage *e*, between surfaces *A* and *B*.

the drive current applied to the element and d is the thickness of the Hall element. The Hall generator can measure d.c. fields, a.c. fields or pulsed fields. The voltage developed is a function of the current and field characteristic. Because of its small size, the Hall generator is ideally suited to check field homogeneity and can be used to flux plot.

The Hall output voltage is a sine function of the angle between the lines of force and the plane of the Hall element. At any point in a field where the output is zero, it follows that the lines are parallel to the Hall element. When the Hall element is oriented for maximum output, the lines are perpendicular to the element. The Hall generator has a predominant position in high speed production measuring and comparison testing.

8.3.5 Nuclear Magnetic Resonance

Nuclear magnetic resonance (NMR) is an effect which allows the analysis of fields at the sites of nuclei. It is useful in the study of atoms and molecules and for the measurement of magnetic fields with high precision. The proton which is the nucleus of the hydrogen atom possesses a constant magnetic moment, M_p, associated with a constant spin or angular momentum. When a low level alternating field of frequency, f, is applied at right angles to the steady state field, H, it is possible to detect resonance associated with orientation of the proton moment by energy absorption from the high frequency field. The condition for resonance is

$$f = 2MpH/h \tag{8.8}$$

where h is Planck's constant. The frequency at resonance can be measured with high precision and this leads to a high precision field measurement. H can be determined accurately and in absolute terms since M_p and h are natural constants known with great accuracy.

8.4 INSTRUMENTATION FOR MAGNETIC FIELD MEASUREMENT

8.4.1 Electromechanical Integrating Devices

When a search coil is moved in a magnetic field, a voltage is generated. The magnitude of the field can be determined from the magnitude of the voltage, or more generally, from the quantity of electrical energy the voltage will circulate in a closed electrical circuit.

If a coil of n turns is linked with an instantaneous value of flux, ϕ, then the instantaneous voltage is given by the following expression

$$e = n \frac{d\phi}{dt} \tag{8.9}$$

If the coil is connected to a circuit of resistance, R, we have

$$\frac{e}{R} = \frac{n}{R} \frac{d\phi}{dt} \tag{8.10}$$

and

$$\frac{e\,dt}{R} = \frac{n\,d\phi}{R}, \qquad i\,dt = \frac{n\,d\phi}{R} \tag{8.11}$$

Integrating

$$\int_0^t i\,dt = \frac{n}{R} \int_{\phi_2}^{\phi_1} d\phi = \frac{n}{R}(\phi_1 - \phi_2) = q \tag{8.12}$$

where q is the total quantity of electricity flowing in the circuit as a result of the flux change from ϕ_1 to ϕ_2. It is apparent that if q can be measured the flux change may be determined. The ballistic galvanometer and the flux meter are instruments that will measure flux or, in other words, they are integrating electomechanical instruments.

The Ballistic Galvanometer. When magnetic flux threading a search coil to which a ballistic galvanometer is connected is changed, a momentary current flows in the circuit. The quantity of electricity is proportional to the magnitude of the flux change and is measured in terms of the deflection or throw of the galvanometer. The ballistic galvanometer differs from the ordinary galvanometer in that the moving element (usually a coil) has a large amount of inertia. Damping of the moving system is controlled by resistance in the galvanometer circuit and is an extremely important consideration in using the galvanometer. For critical damping the galvanometer coil will return to zero without oscillations in a minimium of time. If the galvanometer is underdamped, the galvanometer will, in returning to zero, exhibit an oscillator condition. The over damped galvanometer will return slowly to zero without oscillation. Generally, critical damping on overdamping of the galvanometer is used in testing. The theory of the ballistic galvanometer is based on the premise that the period is long enough so that the coil does not move appreciably until after the duration of the current impulse. Figure 8.6 shows a ballistic galvanometer and its control circuit.

The Flux Meter. The flux meter is a widely used instrument for measuring magnetic flux. The flux meter is usually a moving coil type galvanometer, and it differs from the ballistic galvanometer in that the controlling torque for the moving element is reduced to essentially zero and the deflection is independent of duration of the current impulse. As the flux changes in a search coil connected to a flux meter, a voltage is induced, current flows, and the moving coil of the flux meter is deflected. As the coil moves, a back electromotive force is set up which tends to reduce the current to zero; the flux meter coil thus follows the flux change in the search coil. Because of negligible control torque, the flux meter coil and deflection indicating system will remain at its limit of travel indicating total flux change. A zero return circuit is used with flux meters.

Figure 8.7 shows the galvanometer unit in the flux meter. This galvanometer unit is available in several sensitivity ratings (flux linkages per unit of deflection). In using the flux meter the following relationship exists

$$\phi = SD/n \tag{8.13}$$

where ϕ is the total flux change, S is the galvanometer sensitivity in linkages per unit of deflection, D is the deflection, and n, the turns in the search coil.

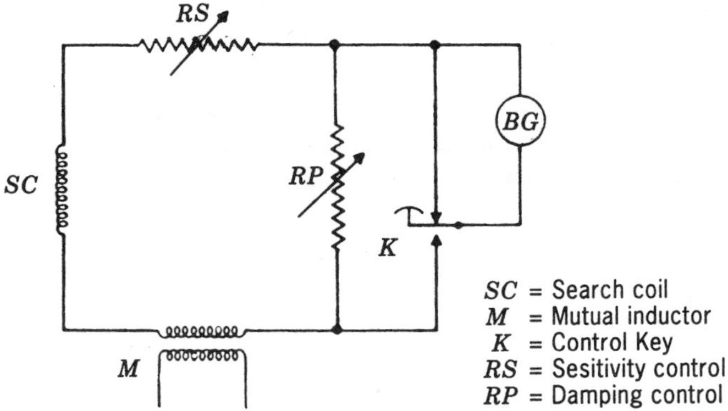

Figure 8.6 Ballistic Galvanometer and control circuit.

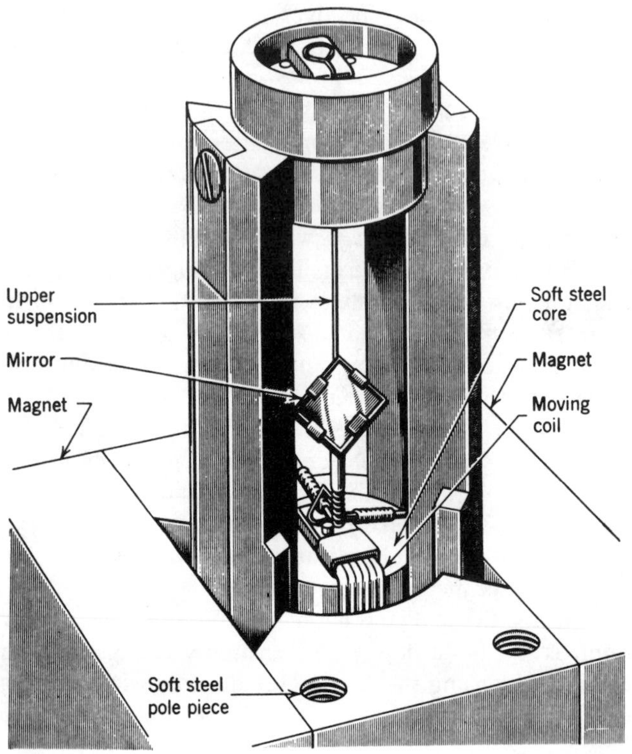

Figure 8.7 Principal parts of light beam galvanometer mechanism used in fluxmeters. From General Electric Co.

8.4.2 Electronic Integration

When a search coil is moved in a magnetic field, a voltage is generated. The e.m.f. can be expressed as

$$e = -n \frac{d\phi}{dt}$$

where e is the electromotive force in volts, ϕ is the magnetic flux in maxwells, and n is the number of turns on the coil. To obtain the change in flux ϕ we integrate

$$\Delta\phi = -\frac{1}{n}\int_0^t e\,dt \tag{8.14}$$

since $\Delta B = \Delta\phi/A$ we obtain

$$\Delta B = -\frac{1}{nA}\int_0^t e\,dt \tag{8.15}$$

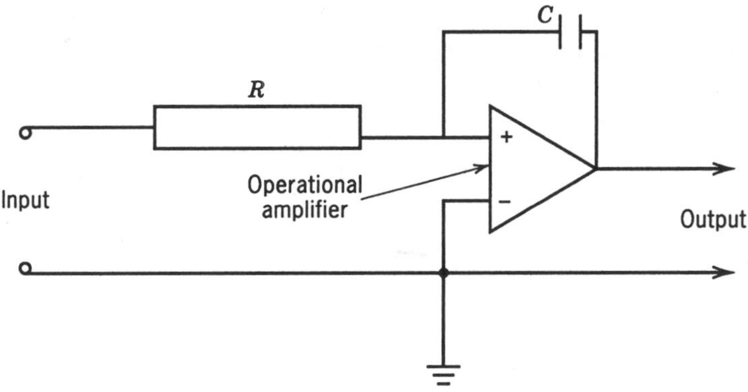

Figure 8.8 Electronic integrator.

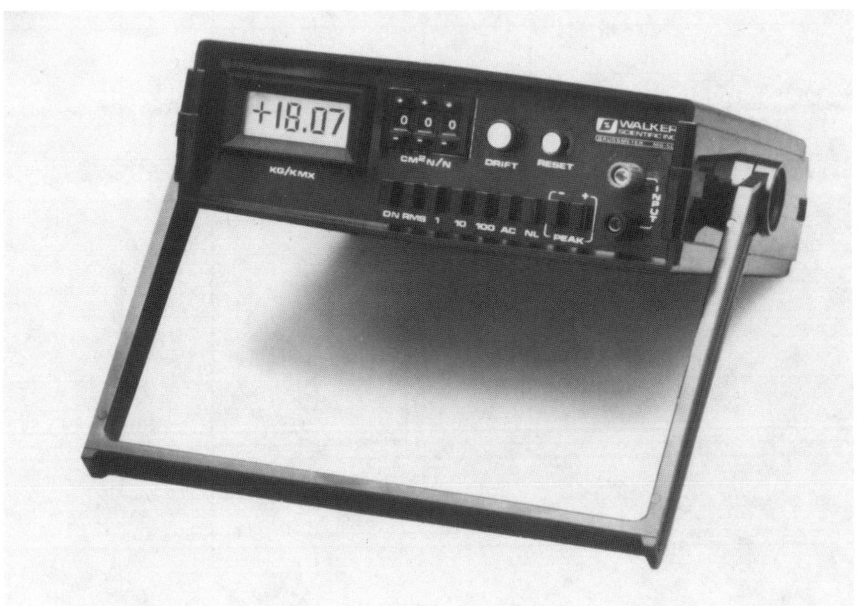

Figure 8.9 Portable electronic fluxmeter. From Walker Scientific Inc.

The flux change can be measured with an integrating instrument that gives the time integral of the voltage generated by the search coil. An integrator build with amplifier and RC feedback circuit is a very common flux measuring instrument (Figure 8.8).

If a high gain operational amplifier is used having neglectable voltage and current errors on the input, then the relationship between input e and output u is

$$u = -\frac{1}{RC}\int_0^t e\,dt \tag{8.16}$$

If a search coil is connected to such an integrating circuit, it follows that

$$u = \frac{n\Delta\phi}{RC} = \frac{nA\Delta B}{RC} \tag{8.17}$$

Figure 8.9 shows a typical electronic flux meter which is commercially available.

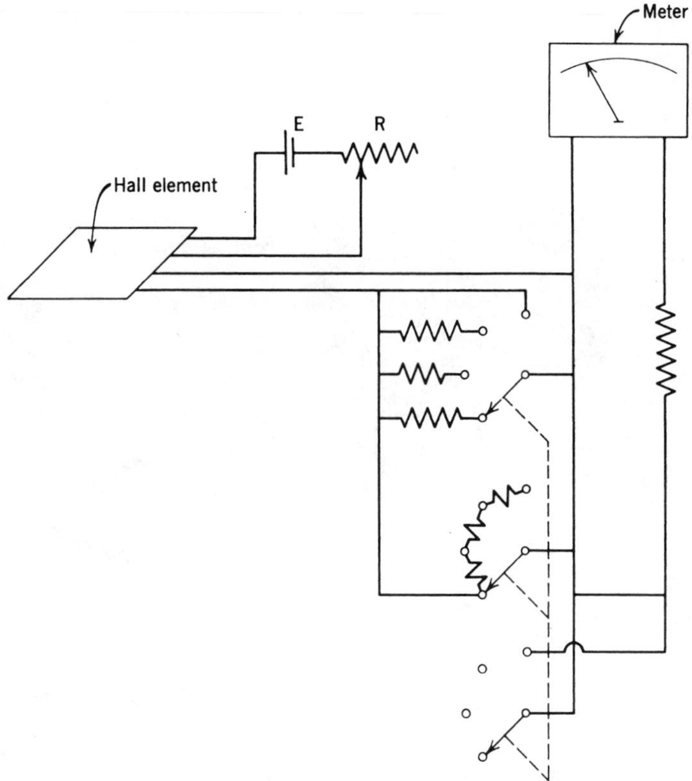

Figure 8.10 A schematic diagram of a Hall effect gaussmeter.

8.4.3 Hall Effect Gaussmeters

The Hall effect gaussmeter may be as simple as the battery operated system shown in Figure 8.10 or as sophisticated and versatile as that shown in Figure 8.11(a)(b). Modern Hall gaussmeters such as that shown in Figure 8.11 can measure both d.c. and a.c. fields. Some gaussmeters will also measure peak fields. The trend in circuit design is to use a sinusoidal control current. The output of the probe is then a sinusoidal voltage with the same carrier frequency as the current. The field intensity provides an amplitude modulation. The signal is amplified and demodulated. The carrier frequency is filtered out leaving a signal proportional to field intensity. Such an approach allows excellent accuracy and linearity over a wide range of field intensities.

8.4.4 Moving Magnet Gaussmeter

In exploring magnetic fields, the use of a search coil and the necessity of having to obtain relative motion between field and coil has obvious disadvantages. In exploring field variation in air gaps or studying leakage field patterns, use is made of a classification of instruments known as gaussmeters which read directly and continuously in terms of flux density when inserted in a magnetic field. Gaussmeters may be of the moving coil type or use a small permanent magnet as the sensitive element.

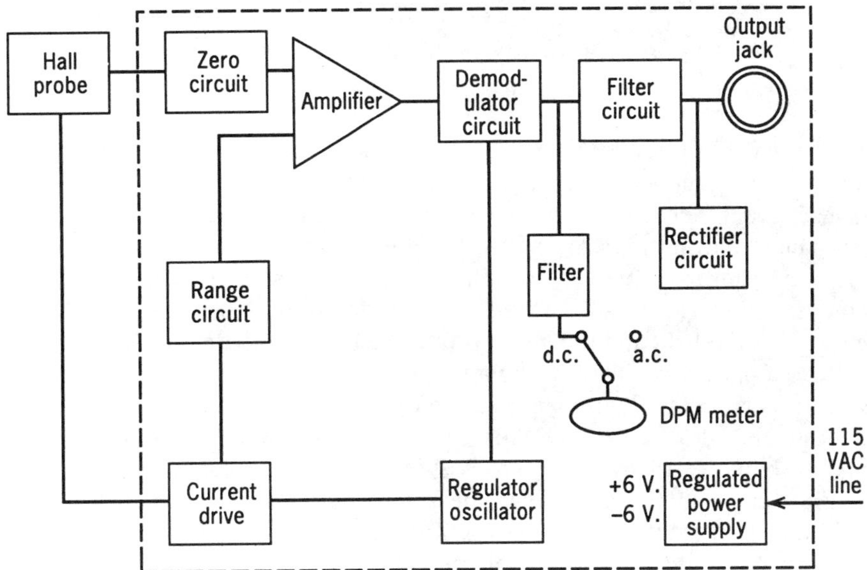

Figure 8.11 (a) Functional diagram of a gaussmeter (b) Hall Gaussmeter. From Walker Scientific Inc.

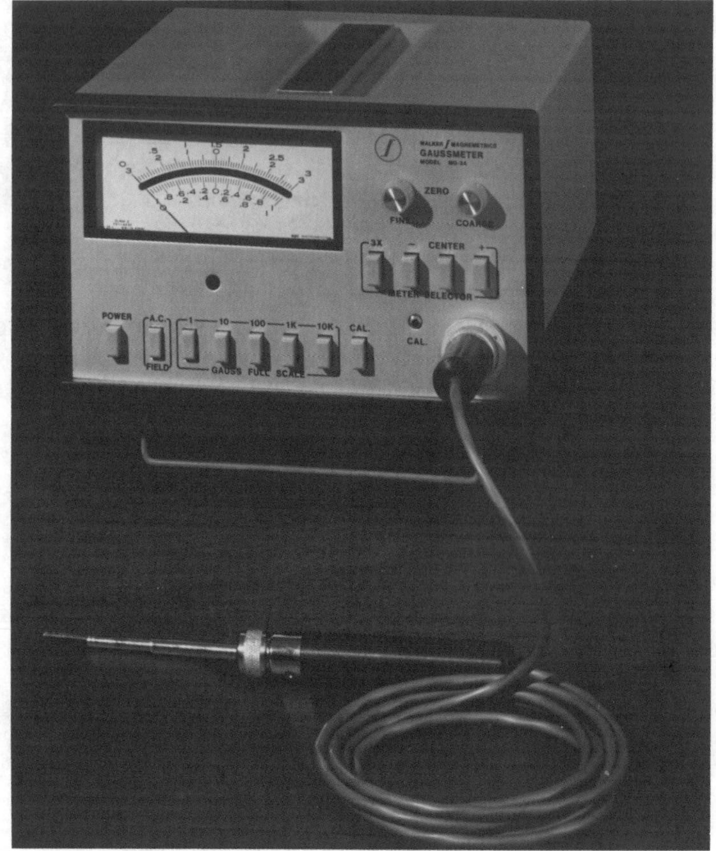

Figure 8.11 (*Continued*)

Figure 8.12 shows a cutaway section of a widely used gaussmeter employing a moving magnet. A tiny permanent magnet cylinder diametrically magnetized is located at the end of the shaft just above the pivot. The permanent magnet material has an intrinsic coercive force H_{ci} greater than the maximum field strength to be measured. Silmanal and cobalt platinum are used in this type of instrument. To read flux density, the probe is inserted in the field and the meter rotated counterclockwise until the pointer reaches a maximum. This maximum occurs when the flux of the probe is 90° with respect to the field being measured. This type of gaussmeter will also indicate the direction of the flux. In rotating the meter counterclockwise, just as the pointer leaves the zero position, the probe magnet flux coincides in direction with the field being measured; thus the pointer indicates the field direction at a particular point. This instrument is extremely useful in plotting magnitude and direction of fields, and although its accuracy is limited (5%), its ease of operation and relatively low cost have made it a widely used instrument in industry.

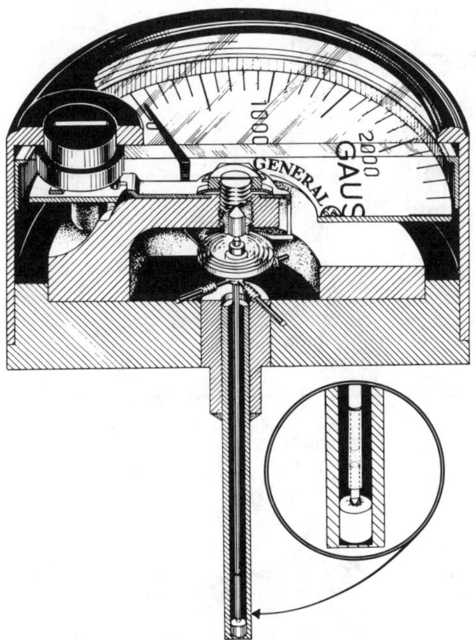

Figure 8.12 The moving magnet gaussmeter. From General Electric Co.

8.4.5 Helmholtz Pair Detection System

Helmholtz coil set can serve as the detector or pick-up coil for the induced voltage signal [3]. The magnet being measured under open circuit test conditions can be considered to be a magnetic dipole, oriented parallel to the axis of the coil set. The magnetic moment is equal to JV_m. As the magnet is moved from the center of the Helmholtz pair to a point well outside the coil, the time integrated voltage $\int e\,dt$ is measured with a flux meter or electronic integrator. Figure 8.13 shows the coil set and the voltage wave form. The value of $\int e\,dt$ is proportional to the magnetic moment of the test magnet. Since the intrinsic magnetization is the magnet moment per unit voltage, it follows that

$$B_i = \frac{c}{V_m} \int e\,dt \tag{8.18}$$

where c is a proportionality constant for the coil pair. The constant is independent of the volume and shape of magnets under test. If the test specimen is small relative to the coil size, the system can be calibrated by using a sample of known properties. The Helmholtz pair is a very convenient detector for measuring high coercive force magnets where the typical specimen is a short magnet. It allows one to be free of the tight fitting coil,

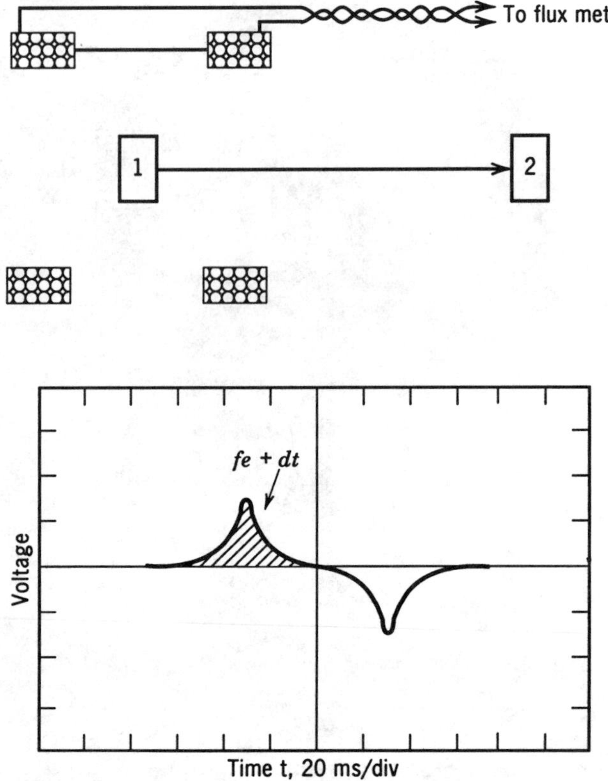

Figure 8.13 Measurement of magnetization in a Helmholtz coil. From Martin and Benz [1].

and one coil set can measure a wide range of magnets. This quality is much appreciated in a manufacturing organization where many shapes and sizes must be measured. The space between specimen and coil structure also allows measurements to be carried out at both low and high temperatures. A liquid nitrogen dewar or a small furnace can easily be located in the space between sample and coil.

The only limitation in the use of the Helmholtz pair is the need to know the load line of the magnet under test with reasonable accuracy, since it is necessary to spot the level on the intrinsic demagnetization curve and then the value of H must be subtracted from B_i to give the B output level of the magnet. This is an open circuit measurement system which can give a lot of information about a magnet with a minimum of investment in equipment and time.

8.4.6 Vibrating Sample Magnetometer

The vibrating sample magnetometer is based on vibrating a sample in a magnetic field to produce an alternating e.m.f. in pick-up coils which is proportional to the magnetic movement of the sample under test. This method is in wide use in development laboratories. It is possible to obtain properties from a wide range of sample sizes and configurations. Thin films, magnetic tapes, and dilute fine particle arrays are also examples of samples which may be measured.

The vibrating sample magnetometer is particularly well suited to allow measurements at high and low temperatures. The equipment may be used with a high temperature chamber and a low temperature dewar. In the development of rare-earth magnets, many researchers have used the vibrating sample magnetometer to measure saturation, anisotropy fields, and constants over a range of temperatures. Figure 8.14 shows, in simplified form, the basic elements of the vibrating sample magnetometer. To calibrate the instrument, usually a sample of nickel or a permanent magnet of known properties is used.

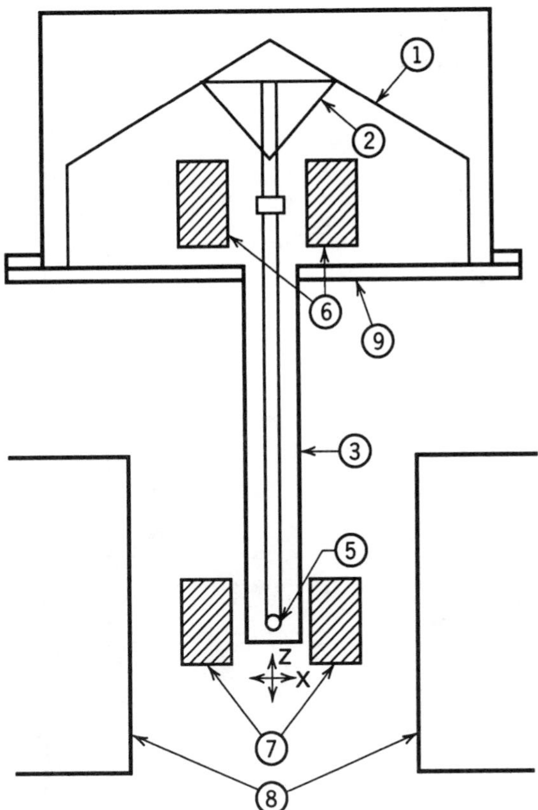

Figure 8.14 Simplified form of vibrating-sample magnetometer. (1) loud-speaker transducer, (2) conical paper cup support, (3) driking straw, (4) reference sample, (5) sample, (6) reference coils, (7) sample coils, (8) magnet poles, (9) metal container.

8.5 OPEN CIRCUIT MEASUREMENT TECHNIQUES

The measurement of the induction B of a magnet in its open circuit condition is a very useful way to determine the quality level of a magnet. A single point B_d can be obtained and, knowing the operating load line of the sample, its quality can be determined with reference to the intersection of the load line and the demagnetization curve. Such a measurement is made with the external field $H_a = 0$. If a series of B_d readings are taken in varying magnetic fields (various levels of H_a), then the complete demagnetization curve is obtained.

Depending on the size, shape, and unit properties of the magnet sample, several detection systems may be used. For example, a close fitting coil can be withdrawn from the sample. The Helmholtz coil pair can also be a useful detector. The vibrating sample magnetometer also allows for determination of the open circuit induction.

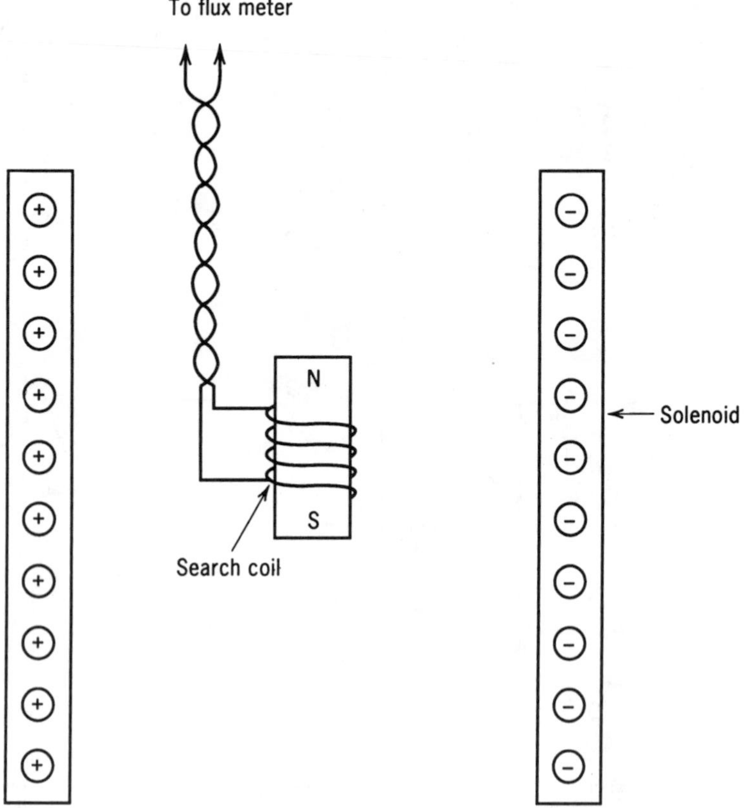

Figure 8.15 Open circuit measurements.

The sample can be magnetized in a solenoid which may be a superconducting coil, or the solenoid might be connected to steady state d.c. or a pulse power supply. The solenoid can supply the peak fields for magnetization and also allow various demagnetizing field levels to be applied so that a series of B_d values for the complete demagnetization curve are obtained. Figure 8.15 shows the solenoid, sample, and search coil arrangement. For a single B_d reading, of course, magnetization by several techniques is possible. Pulse power to a solenoid or use of a laboratory electromagnet are perhaps most often used.

For measuring B_d, the magnet sample is pulled out of a close fitting search coil. The volt-time wave form as the sample is passed through the coil is shown in Figure 8.16. The area under the curve, $\int e\,dt$, either the plus or minus pulse, can be used to determine B_d

$$B_d = (1 - N)B_i$$

$$B_d = \frac{10^8}{nA_m} \int e\,dt \tag{8.18}$$

where n is the number of coil turns and A_m is the sample area. Best results

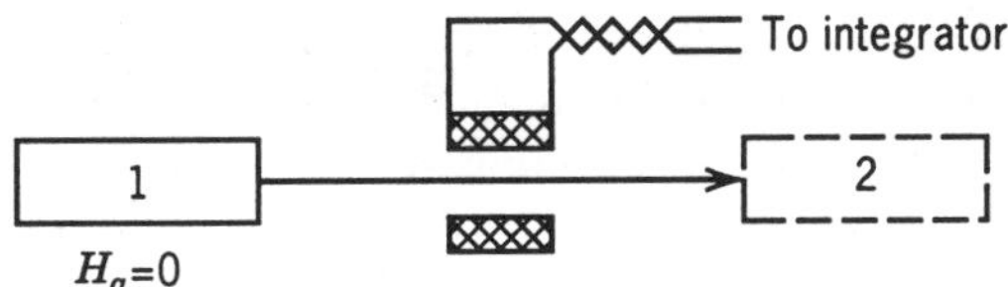

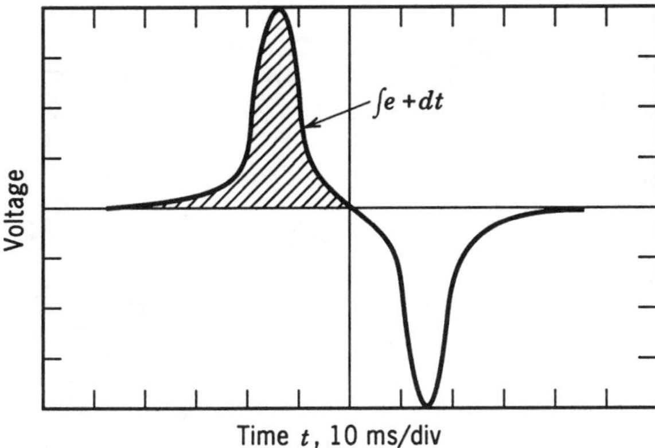

Figure 8.16 Plot of voltage generated by a pick-up coil as a specimen is moved through the coil from 1 to 2. From Martin [2].

are obtained when the sample length to diameter ratio is 2 or more. Proper consideration must also be given to the possibility of irreversible magnetic loss for short specimens where magnet properties have $H_{ci} < B_r$.

The most convenient read out instruments are the electronic flux meter or an intergrating digital voltmeter. If the electronic voltmeter is changed so as to intergrate either a plus or a minus pulse during passage of the magnet through the coil, the necessity of positioning or centering the magnet in the coil is eliminated. With the flux meter, the coil must be positioned in the center of the magnet before withdrawal.

When one is interested in measuring B_i in a constant external field, as in the solenoid of Figure 8.15, the B_i value is found from

$$B_i = \frac{1}{1-N} \frac{10^8}{NA_m} \int e\,dt \tag{8.19}$$

and the H_d value from

$$H_d = NB_i \tag{8.20}$$

By applying a series of H_a values and reading the corresponding B_i values, the intrinsic demagnetization curve can be obtained.

An interesting variation of open circuit measurements involves the use of a Hall probe to indicate intrinsic coercive force. Figure 8.17 shows the principle involved. Any magnetic material, when placed in a field, distorts the field except when the field intensity H exactly matches the intrinsic coercive force H_{ci} of the previously fully magnetized sample. H_{ci} is independent of magnet geometry. A Hall probe is located to indicate only the vertical component of the magnetic field, but not the applied field H_a.

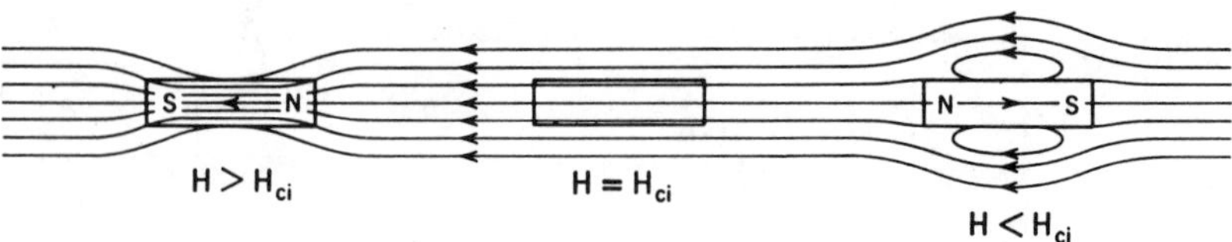

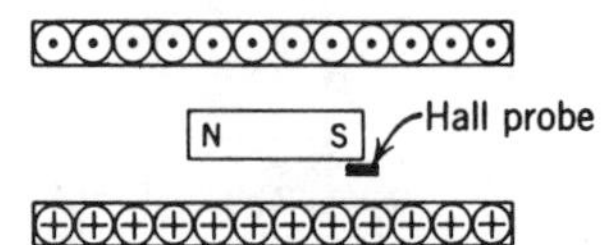

Figure 8.17 Resultant field distortion occurring except when the applied field is equal to the intrinsic coercive force (middle condition).

8.6 INSTRUMENTATION SYSTEMS FOR CLOSED CIRCUIT AND HYSTERESIS LOOP MEASUREMENTS

8.6.1 Equipment and Techniques

For efficient use of magnetizing potentials and freedom from demagnetization factors, closed circuit measurements are the most common type of measurement made on modern high coercive force permanent magnet materials. The usual practice is to place the sample between pole pieces of a laboratory electromagnet. The flux density B and the magnetization B_i of the central sample cross-section are measured. A Hall probe may be positioned adjacent to the specimen to measure H. A small search coil adjacent to the sample and connected to an intergrator may also be used to measure H. The output of the B or B_i and H measuring circuits are generally used to drive an $X - Y$ recorder so that direct recording of the demagnetization curve or complete hysteresis loop can be achieved. Figure 8.18a shows the schematic diagram of the equipment arrangement known as a hysteresisograph. Figure 8.18b is a photograph of a late model microprocessor controlled hysteresisograph.

The sample surfaces are flat and parallel, to insure low reluctance pole interfaces, so the demagnetizing factor can be neglected. With very high coercive force magnets, the laboratory electromagnet may not have sufficient field strength to saturate the sample. It is common practice, under such circumstances, to magnetize the sample in a high field intensity pulsed coil or superconducting solenoid. Then the sample is transferred to the laboratory electromagnet to obtain the demagnetizing curve of the material.

Several variations of coil arrangements for determining B, B_i and H signals are in use. In Figure 8.19, an arrangement is shown which uses pole piece coils. These coils are at the surface of a removable pole piece and pick up only a part of the total sample flux. The area of the sample need not be known. The sample can be of any size with respect to cross-section. It can even be larger than the area of the electromagnet's poles. This arrangement is very useful in exploring property variations. The sample can be reversed in the electromagnet and pole variations can also be explored. For measurement of magnetization, two pole coils must be used. One coil measures B and the other, H. The coils are connected in series opposition so as to give the value $B + (-H)$ as input to the intergrator. An alternate approach is to add the signals after electronic integration.

In measuring very high energy product magnets, the assumption that the internal field within the magnet is the same as the field adjacent to the sample, becomes a source of error. Figure 8.20 contrasts the case where the pole pieces are unsaturated and form an equipotential plane $H = H_e$, with the case where saturation leads to H values within the magnet which are less than that measured outside the magnet $H \neq H_e$. For this case, improved results can be obtained using a built-in potential coil (Figure 8.21). By

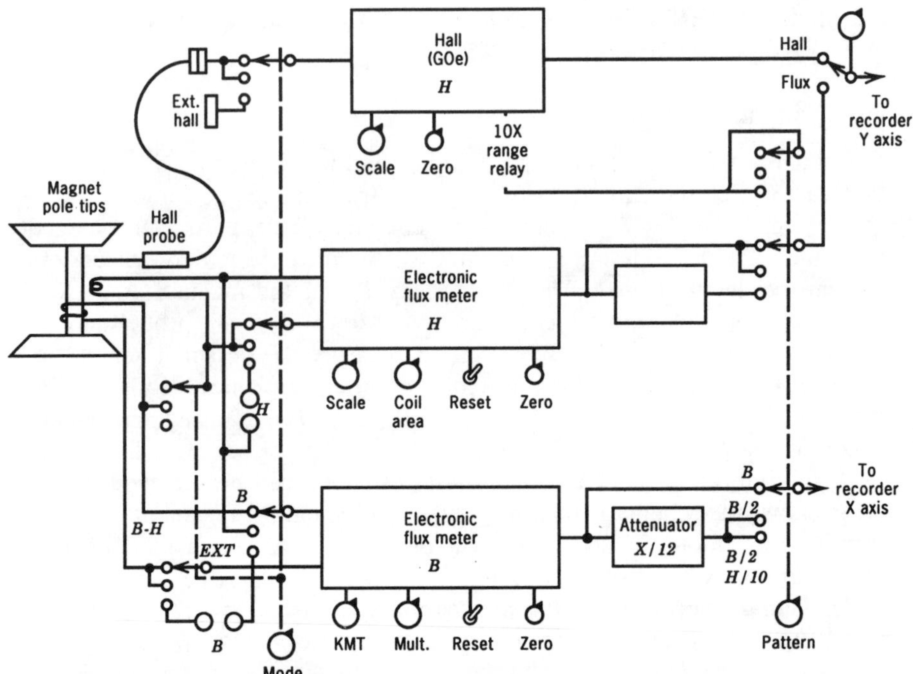

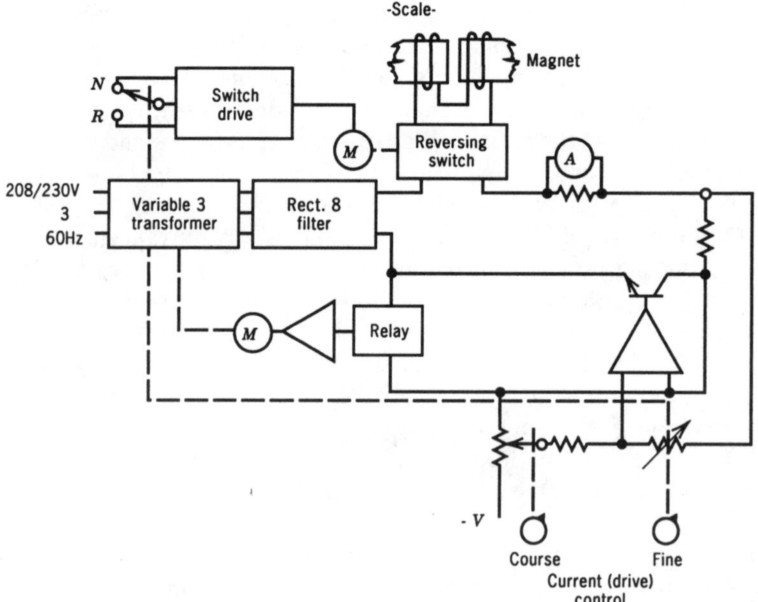

Figure 8.18 (a) Hysteresisograph block diagram; (b) Microprocessor controlled hysteresisograph. From Walker Scientific Inc.

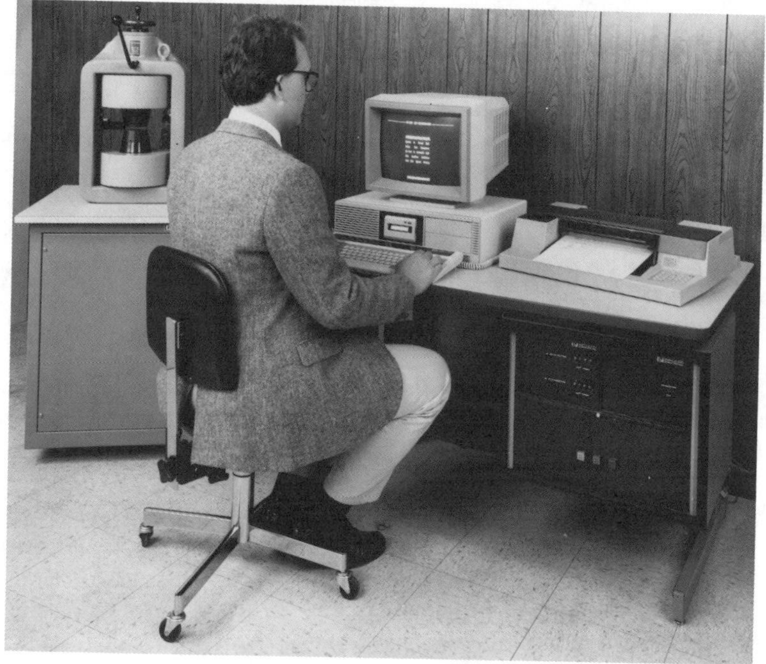

Figure 8.18 (*Continued*)

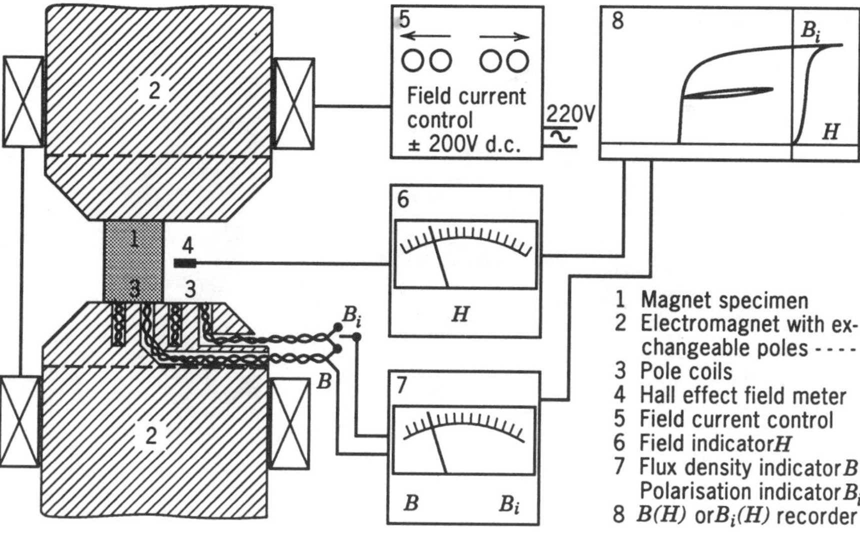

Figure 8.19 Hysteresiograph arrangement with pole coils. From Steingroever [4].

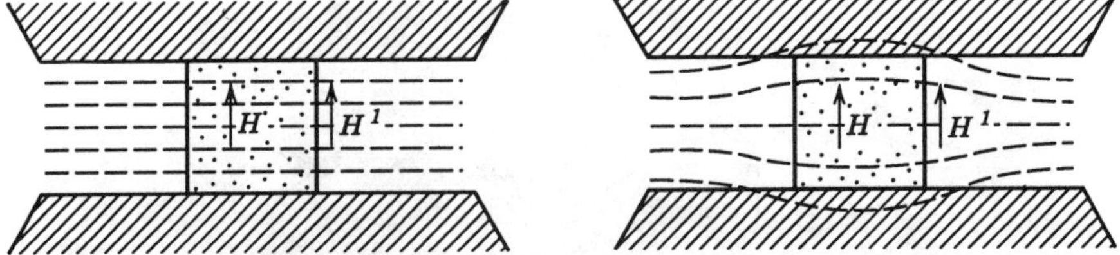

Figure 8.20 (a) Potential planes between poles of high permeability. $H = H_e$. (b) Potential surfaces between saturated poles. $H \neq H_e$. From Steingroever [4].

connecting the potential coil to a fluxmeter, $\int H\, dl$ between the two coil ends can be measured when the field changes, or when the coil is placed in the field from a field-free region.

The International Electrotechnical Commission is the international standards organization responsible for standards work in electrical and magnetic materials. Several Standard Documents have been issued relating to measuring and evaluation techniques for permanent magnets. IEC 68/12 relates to methods of measurements of the magnetic properties of permanent magnet materials. This document suggests methods to use in choosing specimen geometry in relationship to the electromagnetic structure. The uniformity of the field and level of saturation of the poles and placement of the B and H sensors are described. The intent is to offer suggestions on techniques without specifying standard equipment.

With reference to Figure 8.22 the following guidelines are listed from IEC 68/12.

1. *Geometrical Conditions*

$$D_1 \geq D_2 + L \tag{8.21}$$

$$D_1 \geq 2.0L \tag{8.22}$$

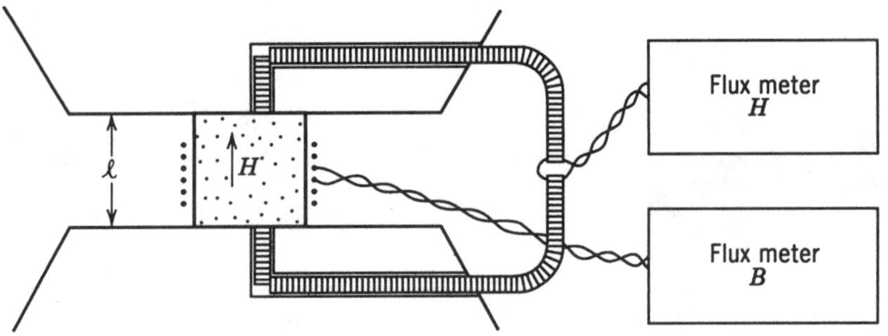

Figure 8.21 Potential pole coils measuring the potential difference $\int H \cdot dl$ between the ends of a sample. From Steingroever [4].

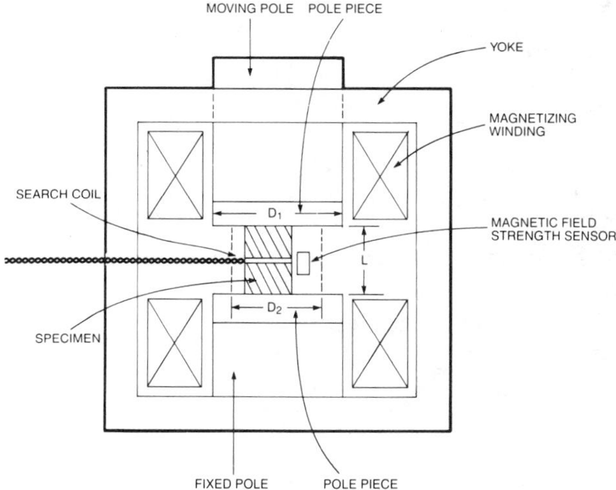

Figure 8.22 Electromagnet and specimen arrangement.

where D_1 is the diameter of a circular pole piece or the shortest dimension of a rectangular pole, L is the distance between the pole pieces and D_2 is the maximum diameter of the cylinder volume with a homogeneous field.

2. *Electromagnetic Conditions*. The flux density in pole pieces shall be maintained at a very low level compared with saturation flux density so that pole member can be near equipotential conditions. The suggested levels are 10,000 G for iron and 12,000 G for Permendur (50% Cobalt alloy). The coercivity of the yoke and pole pieces should not be more than 2 Oe. Under the above conditions, the field in the specified volume gives variations in homogenity of less than 2% both radially and axially. However, these recommended kilogauss levels are often unreasonably low at the pole tips for high coercive force materials. Consequently, field distortion must be considered when higher flux density levels are used.

3. *Specimen Geometry*. The test specimen shall have a simple shape (cylinder or parallelepiped) and its dimensions should be chosen in accordance with the above equations relating pole geometry. The length (L) of the specimen shall be greater than 5 mm to avoid self demagnetization influences.

4. *Instrumentation For Measuring B and H*. The change in flux density in the specimen is determined by integration of the induced voltage in a search coil around the specimen. The coil should be wound as closely as possible on the magnet and should be symmetrical with respect to the pole pieces. The leads must be tightly twisted to avoid voltage being induced in the lead loops. It is possible to minimize error in the flux density measurement by following the above. Acceptable error in the flux density measurement is of the order of $\pm 1\%$. The variation of flux density B in the specimen between time t_1 and t_2 is given by

$$\Delta B = B_2 - B_1 = \frac{10^{-8}}{AN} \int_{t_1}^{t_2} e\,dt \tag{8.23}$$

where B_2 is the flux density in gauss at time t_2, B_1 is the flux density in gauss at time t_1, A is the cross-section of the specimen in cm^2, N is the number of coil turns, and $\int e\,dt$ is the induced voltage in volt seconds. It is necessary to correct the change in flux density by taking into account the flux included in the search coil. The corrected change is given by

$$\Delta B_{corr} = \frac{10^{-8}}{AN} \int_{t_1}^{t_2} e\,dt - \Delta H \frac{A_t - A}{A} \tag{8.24}$$

where ΔH is the change in field strength in oersteds, A_t is the effective cross-section area of coil in cm^2 based on mean coil diameter.

The field strength at the specimen surface is equal to the field strength inside the specimen only in that part of space where the field strength vector is parallel to the side surface of the specimen. Therefore the field sensor must be placed in the homogeneous field zone as near the specimen as possible and symmetrical with respect to the poles. The field strength can be determined by using a search coil, a magnetic potentiometer, or a Hall probe. A suitable readout instrument must be used. The dimensions of the field sensor and its location must be within the area limited by diameter D_2 [refer to equations (8.21) and (8.22)]. The field sensor or transducer must be calibrated so that the total error is within $\pm 1\%$. When using Hall probes, non-linearity must be considered, especially above 10,000 G.

8.6.2 Determination of Demagnetization Curve

For the measurements described below a low-drift flux meter or a ballistic galvanometer is used to measure the voltage integral. The test specimen is assembled in the electromagnet and saturated at a high magnetic field strength H_{max}. At this field strength, the induction in the specimen is equal to B_{max} (Figure 8.23a). Then the current is switched off and the change of magnetic flux density $\Delta B_1 = B_{max} - B'_r$ can be measured. The magnetic field strength in the absence of the magnetizing current is not zero ($H'_r \neq 0$) due to the permanent magnetization of the poles and yoke. Consequently, $B'_r \neq B_r$. The values H'_r, like any other value of magnetic field strength, may be measured with the magnetic field strength sensor inserted into the space between the pole pieces. By increasing the negative magnetizing current to obtain the value $-H_{max}$, the change of flux density $\Delta B_2 = B'_r + B_{max}$ can be measured. The value of the magnetic flux density B'_r is calculated from

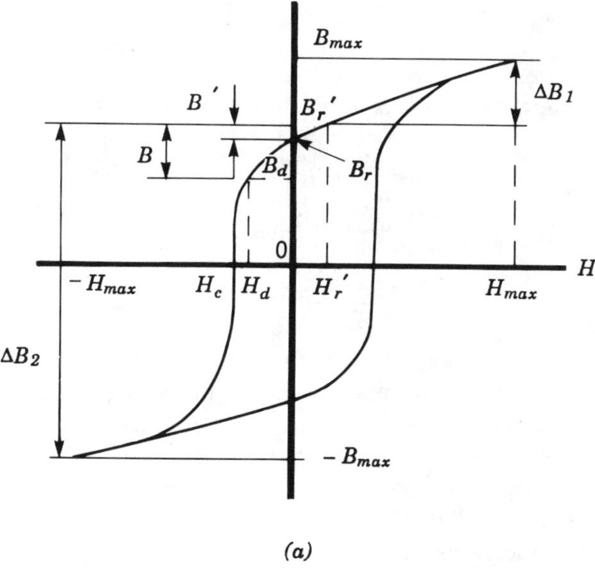

(a)

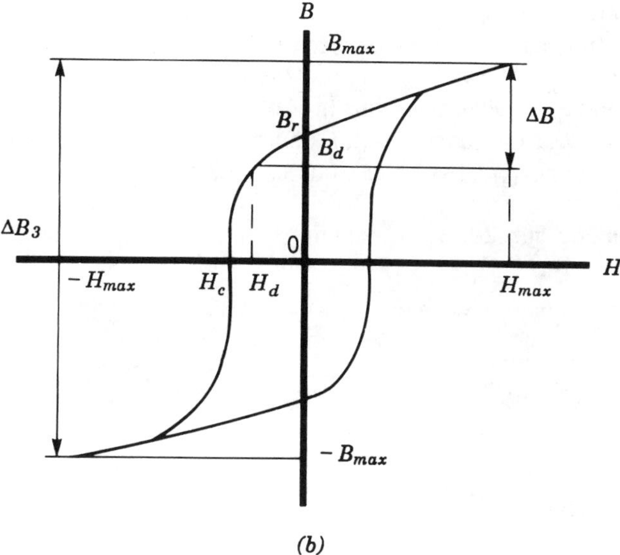

(b)

Figure 8.23 Principal points on the hysteresis loop.

$$B'_r = \frac{\Delta B_2 - \Delta B_1}{2} \tag{8.25}$$

The test specimen is again magnetized to the point B_{max}, H_{max} and after switching off the magnetizing current the magnetic flux density returns to the point B'_r, H'_r. After a negative magnetic field has been applied, the magnetic flux density at any point B_d, H_d can be calculated from the flux density change ΔB according to the equation

$$B_d = B'_r - \Delta B \tag{8.26}$$

The determination of any pair of associated values B and H on the demagnetization curve shall start from the point B'_r, H'_r and necessitates magnetizing the specimen to B_{max}, H_{max}. The actual value of the residual flux density B_r may be obtained by the linear interpolation of the nearest points. To avoid the repetition of magnetizing to B_{max}, H_{max} several points B, H can be determined by successive changes ΔB, ΔH between the points, but this method increases the measuring error.

Together with the above-mentioned procedures, the following measurement order is recommended. The value of B_{max} corresponding to the maximum magnetizing current, is determined by reversing the maximum magnetizing current without changing its value (Figure 8.23b)

$$B_{max} = \frac{\Delta B_3}{2} \tag{8.27}$$

To determine value of the magnetic flux density B at any point on the demagnetization curve, the value of the demagnetizing current should be such as to produce the field strength H corresponding to this point on the curve. Then the cyclic remagnetization of the specimen is repeated. After that the change of the magnetic flux density ΔB is determined by changing the magnetizing current from the value corresponding to the required value up to its maximum value.

The magnetic flux density is calculated from the equation

$$B_d = B_{max} - \Delta B \tag{8.28}$$

An alternative method for determining the demagnetization curve is the use of an electronic integrator. The integrator is connected to the search coil and is adjusted to zero. Then the demagnetized test specimen is put into the search coil and the assembly mounted in the electromagnet. After magnetizing to the required magnetic field strength the magnetizing current is switched off. However, the values of the magnetic flux density B'_r and the magnetic field strength H'_r may still be in the first quadrant because of the permanent magnetization of the poles and the yoke. The current is then reversed and increased until the magnetic field has passed the coercivity H_c or H_{ci}. The speed of the variation of the magnetic field strength shall be sufficiently slow to avoid producing a phase difference between H and B. With some materials there is a considerable delay between the change in the magnetic flux density and the magnetic field strength. In this case, the time constant of the flux integrator should be long enough to ensure accurate integration. Although the procedure detailed is for obtaining the $B(H)$ demagnetization curve, one can easily obtain the $B_i(H)$ by arranging to add H to the B value at every point. (In the second quadrant $B_i = B + H$).

8.7 SEARCH COIL ARRANGEMENTS AND CHARACTERISTICS

The induction principle of measurement is based on a voltage generated, proportional to magnetic flux linking the turns of a coil. Hence, the coil, commonly termed the search coil in magnetic measurements, has importance in that its constants must be known. To determine the flux within some defined area, a coil is wound having a size and shape corresponding to the defined area. With the coil located in the field and connected to an indicating instrument, a measurement of flux is obtained by moving the coil from the field or by changing the flux threading the coil. In either case, the deflection of the indicating instrument is proportional to the interlinkages between coil and field, or

$$\phi n = kD \tag{8.29}$$

where ϕ is the total flux linking coil, n is the coil turns, k is the calibration constant of the indicating instrument, and D is the deflection of the instrument. Since $\phi = BA$ where B is the flux density and
A is the area

$$B = kD/nA \tag{8.30}$$

This expression indicates that the product nA must be determined accurately if D is to represent the flux density accurately. In the case of relatively large diameter coils having a single layer winding, negligible error is encountered in calculating the product nA. However, in coils having appreciable depth of winding, the effective coil area cannot feasibly be determined, and the nA product for the coil is determined by comparison with a coil of known nA in a uniform field. A circuit suitable for the comparison is shown in Figure 8.24. If both coils are withdrawn from the uniform field simultaneously and the potentiometer is adjusted until the galvonometer does not deflect, the coils are related by the following expression

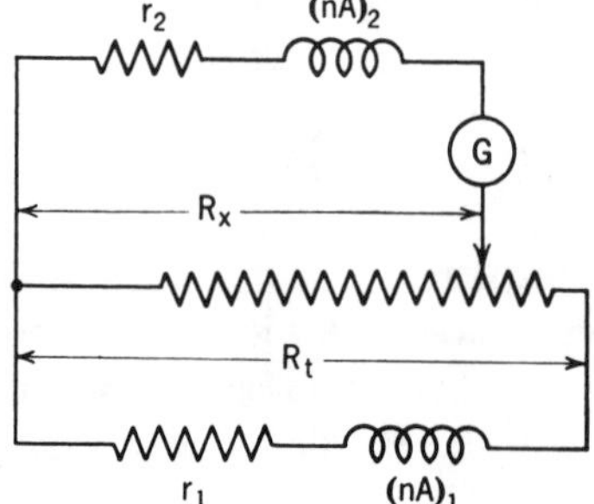

Figure 8.24 A circuit for the determination of the unknown $(nA)_1$ of a search coil, by comparing a coil of known $(nA)_2$.

$$\frac{(nA)1}{(nA)2} = \frac{R_t + r_1}{R_x} \tag{8.31}$$

The nA of the unknown coil can then be determined.

Search coils may, of course, surround ferromagnetic specimens or be located in air gaps, as the nature of the test requirement dictates. An interesting search coil arrangement is illustrated in Figure 8.25, showing the use of a differential search coil to measure the radial gap flux in a loud speaker assembly. The movement of a single search coil from the gap between pole stem and plate would measure the flux in the pole stem, which consists of useful gap flux plus leakage flux. Since it is interesting to know the gap flux only, the search coil is accurately moved from one extremity of the gap to the other. The measurement of gap flux might also be arrived at by obtaining the measurement of total flux by withdrawing the coil completely from the bottom of the gap and subtracting the measured leakage flux by removing the coil from the top of the gap. A convenient way of obtaining the gap flux in a single reading is the use of two coils arranged as those in Figure 8.25. The coils are differentially connected, and the reading obtained by the arrangement is total flux of the pole stem, less the leakage. Measurements of this nature are useful in analysis of magnetic circuits and efficiency studies.

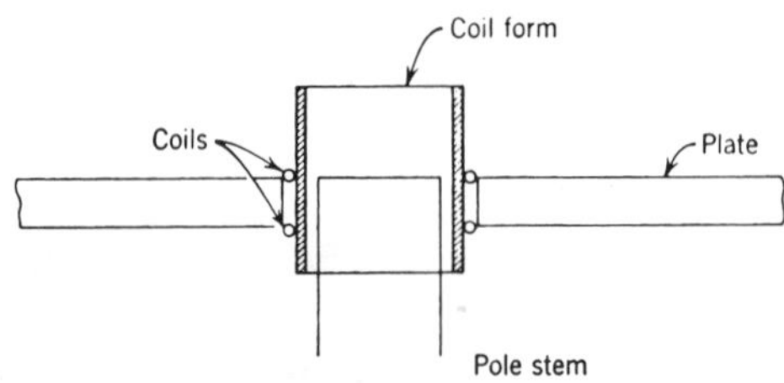

Figure 8.25 A set of two coils differentially connected to yield air-gap flux in a useful air gap.

8.8 SPECIAL MEASURING TECHNIQUES FOR PERMANENT MAGNET CIRCUIT ANALYSIS AND DESIGN OPTIMIZATION

An important category of measurements is involved in the analysis of a permanent magnet circuit. For example in the simple circuit shown in Figure 8.26 one would like to check the load line of the device. Additionally there may be questions about the potential loss in the pole pieces and the air gap flux density H_g. An indicating instrument such as a flux meter or electronic digital integrator and some simple coils are all that is needed to completely analyze the structure. As indicated in Figure 8.26a the H_g of the gap can be measured with a gaussmeter or a small coil and fluxmeter. Another test method uses a potential coil to measure total magnetomotive force F across the gap. Dividing F by l_g, the gap length, gives a value for H_g. This technique can be used when the gap is very short and a thin enough Hall probe or search coil is not available. To determine the load line, wrap a few turns of wire around the magnet and pull the magnet from the assembly and the coil (Figure 8.26b). Knowing the magnet area, B_d can be computed. H_d can be determined by moving a potential coil from point a to b (Figure

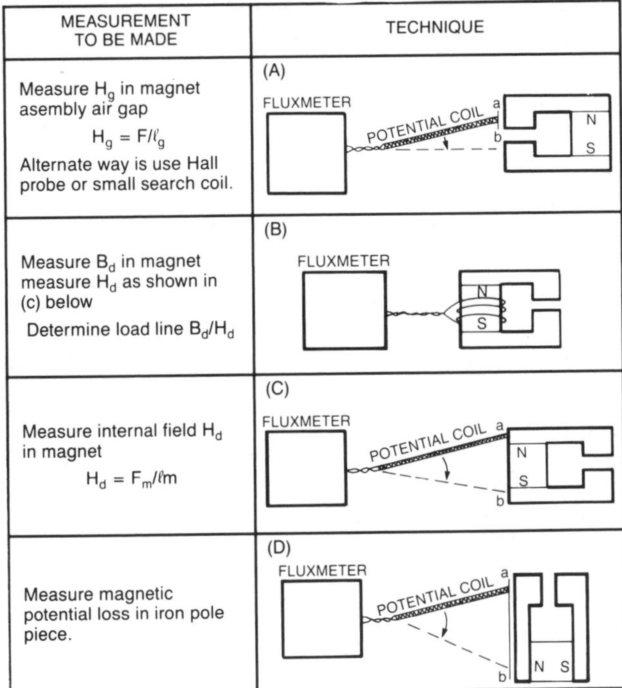

MEASUREMENT TO BE MADE	TECHNIQUE
Measure H_g in magnet asembly air gap $H_g = F/\ell_g$ Alternate way is use Hall probe or small search coil.	(A)
Measure B_d in magnet measure H_d as shown in (c) below Determine load line B_d/H_d	(B)
Measure internal field H_d in magnet $H_d = F_m/\ell m$	(C)
Measure magnetic potential loss in iron pole piece.	(D)

Figure 8.26 Measurements for design analysis.

8.26c). With known values for both H_d and B_d, the slope of the operating line, B_d/H_d, can be calculated. Perhaps there is a question about pole piece material or the magnetomotive force, F, loss in the pole pieces (has the designer selected the iron cross-section for optimum H drop in the iron?). In Figure 8.26d the use of the potential coil allows one to measure the F drop between points a and b and by dividing F by the length of iron the unit potential drop in the iron can be determined. From magnetization curves for various flux conducting materials one can then make a determination regarding how reasonable is the potential loss in the iron.

8.9 CALIBRATION, STANDARDS, PRECISION, AND ACCURACY

Flux meters are calibrated by their manufacturers in units of flux linkages. When a flux meter is used to compare magnets with respect to a reference magnet without regard to absolute values, periodic recalibration is unnecessary. However, when measurements of high accuracy with reference to absolute values are the issue, the flux meter must be calibrated periodically.

There are several methods used for calibration: [5, 6]

1. *Calibration with a Known Voltage, U_c, over a Time Interval t* (Figure 8.27). A very stable voltage source U_c and an accurate timer are involved. Commercial volt-second sources are available with accuracy of about 0.1%.

$$U = \frac{t}{RC} U_c \tag{8.32}$$

gives the relationship between intergrator output U, U_c, and t.

2. *Calibration with Mutual Inductance Standard.* When primary current I is switched on and off in a mutual inductor M, the flux in the secondary coil, MI, is equal to $N\Delta\phi$, hence

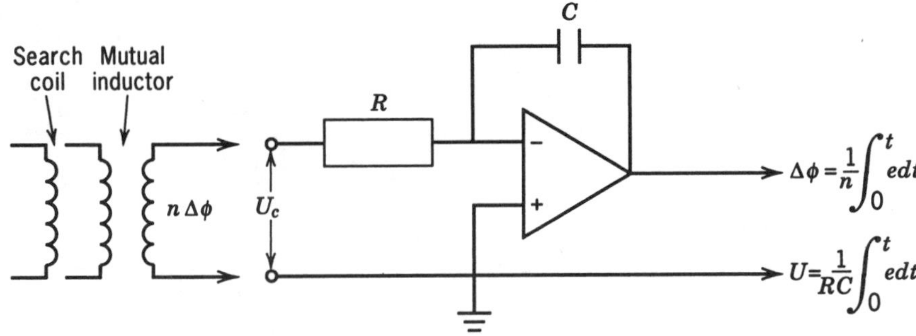

Figure 8.27 Electronic integrator calibration.

$$U = \frac{MI}{RC} \tag{8.33}$$

Using a mutual inductor is a traditional method of calibration. The mutual inductor can be certified by the Bureau of Standards to an accuracy of 0.05%. If current is read on a high accuracy digital meter, the overall calibration accuracy is of the order of 0.1%.

3. *Calibration with a Search Coil of Known Area Turns in a Known Field.* A search coil having a known nA product is placed in a known homogeneous magnetic field. A field determined by use of a nuclear resonance gaussmeter makes an excellent reference. The winding area can be calculated from coil dimensions and the number of turns. If the coil is turned 180° then $2B$ is measured and

$$U = \frac{2nAB}{RC} \tag{8.34}$$

4. *Calibration with Materials of Known Saturation Magnetization.* Iron, nickel, and cobalt are suitable references since their saturation magnetization is similar to that of permanent magnets. Nickel is, perhaps, most readily

Table 8.1 Error comparison in calibration

Mutual Inductor	
Calibration and stability inductor	0.02%
Current value and stability	0.10%
Total error	±0.12%
Nickel Reference	
Uncertainty in saturation value	0.20%
Uncertainty in physical density	0.01%
Uncertainty in cross-sectional area	0.20%
Total error	±0.41%
Using Known Field and Coil	
Flux intergrator calibration	0.12%
Search coil area turns	0.20%
Total error	±0.32%
Nuclear Magnetic Resonance	
Resonant frequency	0.010%
Gyromagnetic ratio	0.001%
Total error	±0.011%

available in pure form. The saturation magnetization of nickel is 6210 G at 20°C. The principle difficulties in using nickel are the problems of deformation and chipping.

5. *Calibration with a Reference Permanent Magnet.* A stablized permanent magnet in a shielded assembly is a very good reference for equipment calibration. Long-term stability of one part in 10^5 is possible at a fixed temperature. Corrections can also be made over a given temperature range for the low reversible loss involved. A permanent magnet reference will retain its accuracy indefinitely.

Clegg [7] has made a rather systematic analysis of the likely error accumulation for the various calibration techniques. Table 8.1 shows the error comparisons.

8.10 MEASUREMENT PRACTICE IN PRODUCTION QUALITY CONTROL

Unit properties determined by obtaining demagnetization curves are generally not used in acceptance testing. Obtaining the complete demagnetization curves is time consuming and costly. Often the closed circuit test does not identify the range of operating field conditions in a device. Samples for unit property testing must be of a critical size and shape. If the actual magnet component involved is large or nonuniform in cross-section, samples must be cut from it that are suitable for the unit property tests. This obviously leads to destructive testing. In acceptance testing, a test is required that will classify magnets into acceptable or unacceptable categories for a specific application. This kind of test is useful for quality control in the magnet producing plant and also as an incoming inspection test in the user's plant. A satisfactory acceptance test must simulate the magnetic circuit conditions and the field environment of the actual device or system in which the magnet is used. In some applications a magnet must be subjected to demagnetizing conditions set up by the end device. Often a magnet is used in a circuit where it is magnetized in place and its load line is higher than the open circuit load line. In such a situation it may be necessary to remagnetize the magnet if for any reason it is removed from its magnetic circuit. This situation is dependent on the H_{ci} of the material. Often high H_{ci} materials are specified when open circuit conditions and low load lines are involved.

Acceptance testing generally requires a fixture that can be adjusted so that magnet under test is working at the same load conditions as in the actual device.

It is suggested that a reference magnet be used in acceptance testing rather than exchanging and using absolute magnetic quantities. Reference magnets are magnets selected and approved by the producer and user to allow satisfactory function of the end device.

In addition to magnetic measurements, it is at times desirable to test for related variables; for example, force or torque between magnets or between a magnet and a soft iron armature. A generated voltage test is at times useful for testing magnets and magnet assemblies used in motors, generators, and tachometers. In multipole structures the average flux is measured. At constant speed and load conditions the generated voltage is proportional to average magnetic flux.

REFERENCES

[1] D. L. Martin and M. G. Benz, Magnetization changes for cobalt rare earth permanent magnets. IEEE Trans. Magn. 8 (1972) 35.

[2] D. L. Martin, Permanent magnet characterization measurements, General Electric Report No. 81CRD086, 1981.

[3] H. Zijlstra, Experimental Methods in Magnetism, Part 2 (John Wiley, New York, 1967).

[4] E. Steingroever, Magnetism and Magnetic Materials AIP Conference Proceedings, New York 1974.

[5] A.S.T.M. Publication A-34, Testing Magnetic Materials.

[6] A.S.T.M. Publication A-340, Standard Terminology, Symbols and Definitions Relating to Magnetic Testing.

[7] A. G. Clegg, Paper W-1, 6th International Workshop on Rare Earth Cobalt Permanent Magnets and their Application. 1982.

9

MAGNETIZATION AND DEMAGNETIZATION

9.1 Introduction
9.2 Theoretical Considerations
9.3 Requirements for Complete Magnetization
9.4 Equipment and Techniques to Magnetize
  9.4.1 d.c. Fields
  9.4.2 Pulse Fields
9.5 Equipment and Techniques to Demagnetize
9.6 Calibration and Stabilization Techniques

9.1 INTRODUCTION

Changing the state of magnetization is a very important consideration in using permanent magnets. Most permanent magnets are magnetized after assembly into a magnetic circuit. Assembly operations are simplified when working with a demagnetized magnet. In addition, the danger of collecting stray magnet particles in the useful air gap is avoided. Demagnetized magnets are less costly to handle and transport because they do not require separation and restraint.

Recent progress in property development has been largely in terms of increased coercivity. With increased resistance to demagnetization, such materials are proportionally more difficult to magnetize. To successfully use these new magnets requires magnetizing equipment capable of producing very high field levels as well as a good understanding of the magnetization process.

Present equipment and techniques for magnetization are in great contrast to early touch and shock treatments used with the earliest man-made magnets (Figure 9.1). Early permanent magnets were magnetized with rather modest field levels and when one designed a permanent magnet circuit, the tendency was to consider only how field energy was established in the useful air gap. Little consideration was necessary as to how energy input to magnetize was achieved. One way or another, it was always possible to fully magnetize.

In considering the high H_{ci} rare-earth magnets, we need to be aware of the large field requirements and to some extent, the design will be influenced by consideration of how we will magnetize. It is very easy to design structures that cannot be magnetized in place.

9.2 THEORETICAL CONSIDERATIONS

In Chapter 2 the relationships involving energy and work were developed. Various areas of the hysteresis loop were related to magnetization, demagnetization, and work in an electromagnetic system containing a perma-

Figure 9.1 Smithing an iron rod in the direction of the Earth's field, as shown in Gilbert's *De Magnete* (Septentrio, North; Auster, South).

nent magnet. We normally use electrical energy to magnetize. In the case of using one permanent magnet to magnetize another permanent magnet, we can use mechanical energy input.

When we magnetize or demagnetize a permanent magnet, we are only changing the balance of some very large force systems within the material. The familiar shaped magnetization curves and hysteresis loops are really records of the change of internal energy balances, with respect to external energy input.

From the unit property relationships, we can determine the field required to saturate a particular type of permanent magnet. Prior to the use of rare-earth magnets, one could generally say that saturation fields H_s of 3 to 5 times H_{ci} were required. With rare-earth properties came some unusual new relationships which could only be understood by considering the particular coercive force mechanisms involved. In Chapter 3 the various mechanisms for coercive force and finite particle interaction were developed. We will now look at their significance in the context of attempting to fully magnetize some specific types of permanent magnets.

In $SmCo_5$ we have a situation where three levels of field strength are necessary, depending on the sequence of thermal and prior field exposure. Figure 9.2 shows the events and the particular mechanism of magnetization involved. A "virgin" magnet that has only been exposed to elevated temperatures magnetizes along path 1 since most grains are multidomain. At saturation, the domain walls are driven out and the magnetization vectors all point in the same direction. Field level H_1 which is less than the H_{ci} of the material will fully magnetize.

To field demagnetize, path 2 is followed. At the H_{ci} point, half of the magnetization vectors are reversed. If one now attempts to remagnetize in the initial direction, it is necessary to rotate magnetization vectors, which is a higher energy process than domain wall motion. Consequently, path 3 is followed and a field level H_2 approximately equal to H_{ci} is required. If one needs to change the polarity of the magnet, applying H_2 will only leave the magnet partially magnetized. To fully magnetize requires level H_3, a very large field to reverse all of the magnetization sites. It must be remembered that when we say a magnet has an H_{ci} value of 20 kOe, that this value is a statistical average of a great number of magnetization sites and that the field to reverse some of the sites may be very much larger than the average value.

Another type rare-earth magnet is the precipitation hardening $Sm(Co, Fe, Cu, Zr)_7$. Figure 9.3 shows the magnetization curve of this material, compared to $SmCo_5$. Also shown is a higher H_{ci} version, $Sm(Co, Fe, Cu, Zr)_{7.5}$ which is extremely difficult to magnetize. Figure 9.4 shows the possible difference in terms of the mechanics of magnetization. The low H_{ci} version appears to have grains which are mostly single domain and have a rotational, relatively high energy process in which field requirements approximately equal to the H_{ci} level are required (Figure 9.4a).

In developing the high coercive version, thermal events to put domain

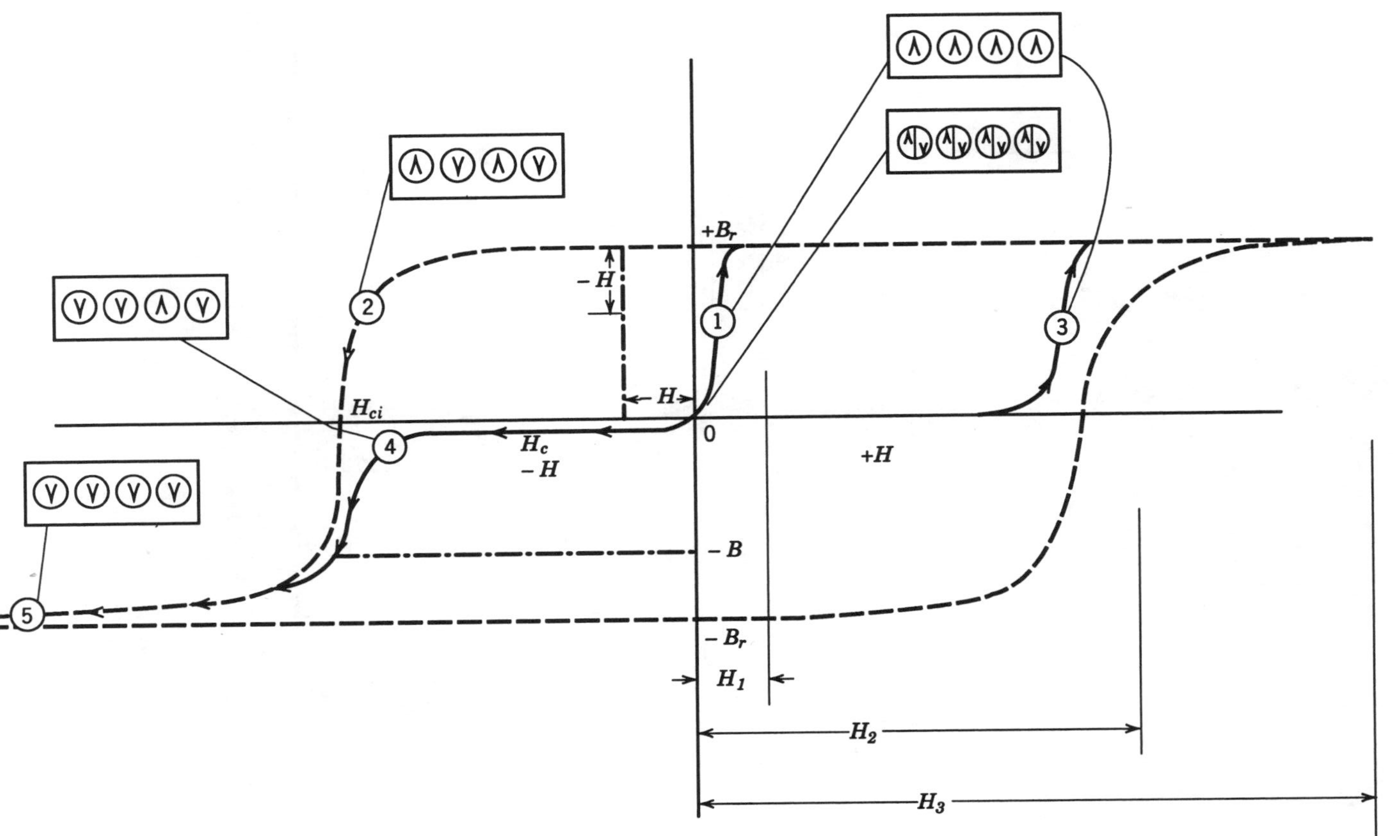

Figure 9.2 Magnetization processes in sintered RCo_5 magnets.

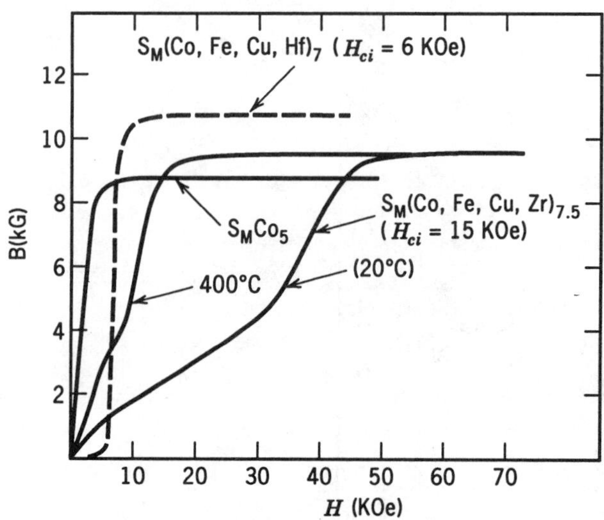

Figure 9.3 Magnetizing parameters.

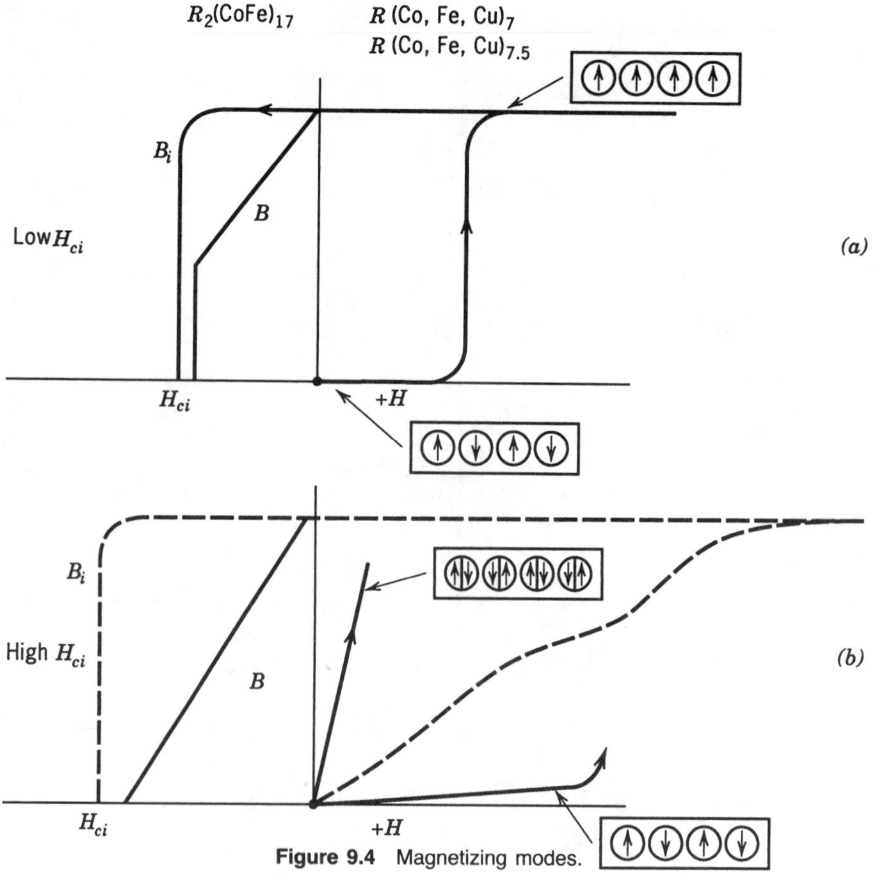

Figure 9.4 Magnetizing modes.

walls in some of the grains were used (Figure 9.4b). This material has a mixture of single and multidomain grains, and the difficulty of magnetization can be changed with heat treatments, which change the relative population of single and multidomain magnetization sites. It is interesting to note that Figure 9.3 also shows the magnetization curve for the high H_{ci} version measured at 400°C. A combination of thermal energy and field energy can often be used to advantage in magnetizing a permanent magnet. It is also of interest to note that thermal energy in itself can result in remagnetization. Kavalerova et al. [1] have reported on the reversible thermal remagnetization of a field demagnetized magnet which fully remagnetized when exposed to 450°C with a keeper in place. This thermal induced remagnetization can be understood in terms of thermal activation of domain walls, displaced during field demagnetization, which return to initial states under the influence of the internal field of unreversed domain volumes. This action has also been noted in partially demagnetized magnets that are exposed to high temperature environments. Calibrated magnets have shown a tendency to remagnetize and fall out of the calibration tolerance. This difficulty has only been noticed with high levels of field knock down. Refer to Section 5.11 for further details.

9.3 REQUIREMENTS FOR COMPLETE MAGNETIZATION

To fully magnetize, we are concerned with:

1. external field magnitudes to produce saturation;
2. the net field actually seen by the magnet after self-demagnetization and magnetic circuit influences;
3. conformance of field shape to that of the configuration being magnetized;
4. the time of the magnetization and field penetration considerations;
5. field distortion producing events after magnetization which may leave the magnet in a partially demagnetized condition.

From the magnetization curve and the hysteresis loop, we can determine the field strength to saturate a given permanent magnet material. In the magnetic property tables of Appendix 2, the field to saturate H_s is given for each material. As the field strength is increased, B_i will approach some maximum value characteristic of the material, B_{is}. The saturating field strength, H_s, will be found to be of the order of 3 to 5 times the intrinsic coercive force, H_{ci}. The values of H_s given represents the condition of a completely closed circuit magnet. The influence of self-demagnetization is developed later.

The demagnetization curves of B_i and B versus H supplied by magnet producers are measured with the material in a saturated condition. Failure to properly saturate a magnet designed on the basis of these curves will lead to disappointing performance. In order to evaluate or compare magnets, it is essential that they be fully magnetized.

Figure 9.5 shows the sensitivity of magnet properties to level of magnetizing force for $SmCo_5$. In this example, it is very clear that partial magnetization would be very wasteful and that the properties achieved are very nonlinear with field.

At this point, the question arises as to how can one ensure that saturation is achieved. It is quite easy to check field level applied with gaussmeters in the case of d.c. fields. With pulse fields, a Hall probe operated in the d.c. mode in conjunction with an oscilloscope can be used. Another technique is to check for saturation by field reversal. This check is performed by magnetizing a magnet, testing in some specified manner, and then applying the same field level in the opposite direction. The magnet is tested again and if the results are the same, the field level may be considered adequate. It is also possible to apply an initial field level and make a reference measurement. Then, a field level, larger by perhaps 25%, is used and if the magnetization of the magnet does not increase, it is safe to assume that the initial level was adequate.

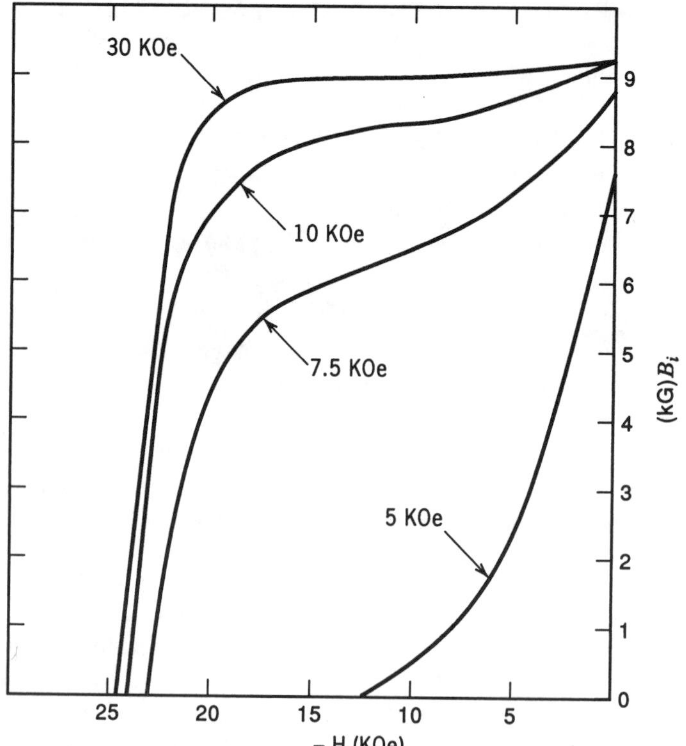

Figure 9.5 $SmCo_5$ magnetized at various levels of field.

Self-demagnetization and Magnetic Circuit Influences. The field levels suggested by the magnet producer are always the actual field level seen by the permanent magnet. In practice, the only time the applied field is approximately the same as actual field, is in the case of a high quality laboratory electromagnet. With low reluctance, high permeability iron yoke, the m.m.f. applied will be very close to the m.m.f. across the magnet. The general problem is to magnetize a magnet in its magnetic circuit. An air gap is often present and one may have to provide a total flux that can saturate a parallel return path element. Still more difficult, are parallel magnetic path structures and the need to carry high levels of saturation m.m.f. into such circuits. Consider the case of a bar magnet being magnetized in a solenoid (Figure 9.6). The permeance coefficient, B/H is determined from the bar magnet geometry. If one draws a line through the saturation point, $B_{is}H_s$ having a slope equal to the permeance coefficient $B/H+1$, then the intersection of this line with the H axis will give the total field necessary for saturation, $H_t - H_s$ being the field necessary to overcome the self-demagnetization influence of the free poles and allow a net value of H_s to be experienced by the permanent magnet.

Conformance of Field Shape to that of Configuration Being Magnetized. If the field generated does not conform to the configuration of the magnet, we will have partial magnetization. The permeability of most permanent magnets is very low and hence, the presence of the magnet does little to shape an applied field. The field should always coincide with the easy axis of the permanent magnet. When magnet configuration and field do not coincide, it is possible to have fields that are too great, which, in effect, leave regions magnetized off axis and the result appears as partial magnetization. Figure 9.7 shows the influence of fields applied at various angles to Alnico 5–7 material.

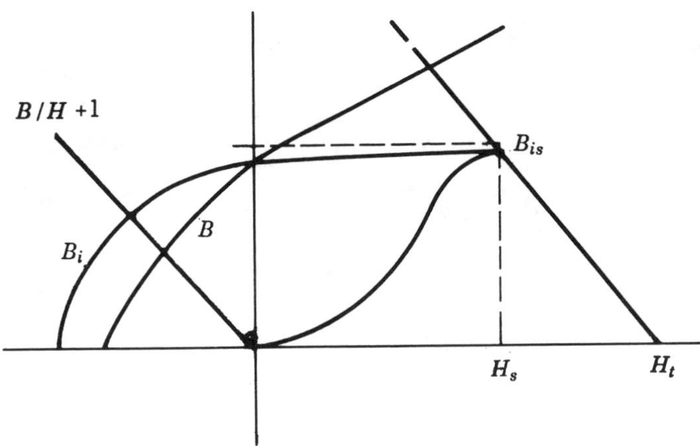

Figure 9.6 Analysis of self-demagnetization.

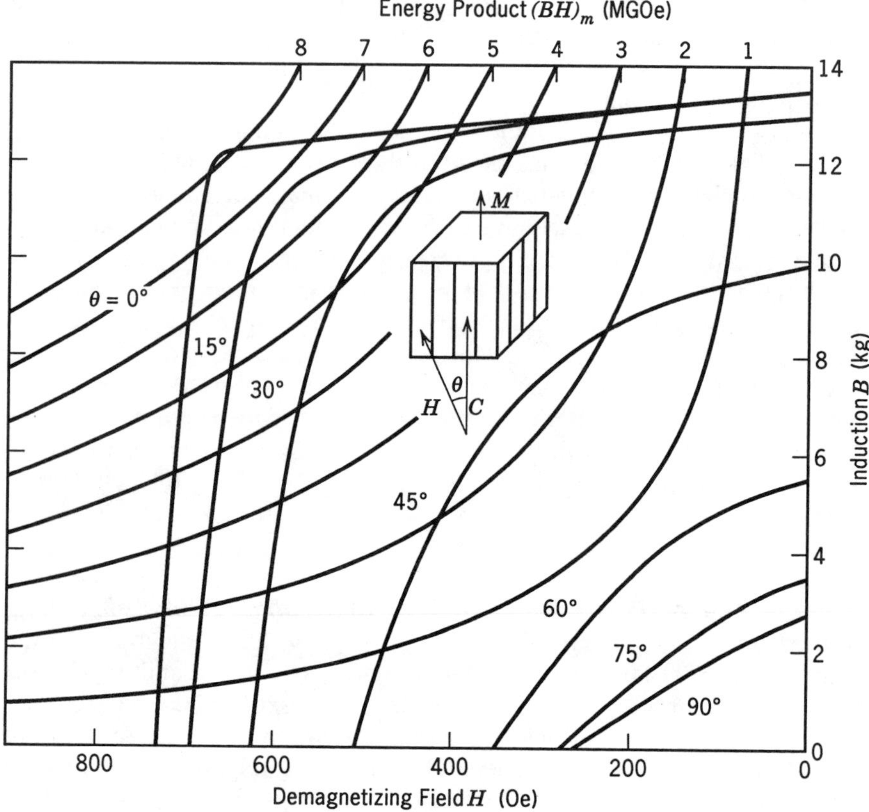

Figure 9.7 Effect of magnetizing off axis.

Time Requirements and Penetration Considerations. Although the magnetization process is essentially instantaneous, the time duration of the applied field is important because of the existence of eddy currents in metallic materials. In addition, with highly inductive electromagnets, the current rise time may be of the order of 1–2 s.

Figure 9.8 shows a relationship interrelating depth of penetration with resistivity, permeability, and frequency of wave form. In general, the frequency must be chosen so that the magnetizing pulse lasts longer than the eddy current. The eddy current path is a function of geometry and, for large metallic magnets, there are problems with penetration. The general experience with Alnico and rare-earth magnets has been to use about 10 ms minimum pulse width. This width of pulse allows a wide range of magnet configurations and sizes to be fully magnetized.

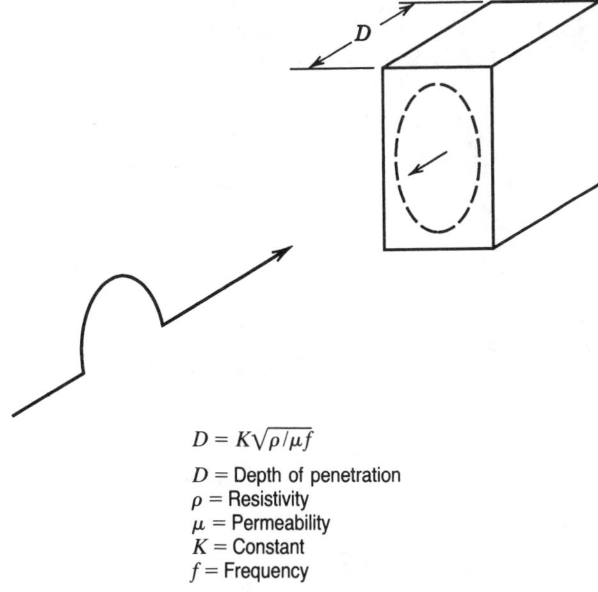

Figure 9.8 Relationship of magnetizing variables.

Field Distortion after Magnetization. After saturation or calibration it is very easy to inadvertently demagnetize a magnet with improper handling, and therefore great care must be taken to preserve the original condition of magnetization. To the inexperienced, many of the recommended precautions seem trivial and are neglected, but the proper environment after magnetization is most important for efficient utilization of the permanent magnet, as described below. In the saturated or calibrated condition, a magnet should not be closely approached or touched along its length by an extraneous ferromagnetic material. Close proximity of these materials will alter the existing flux pattern, resulting in the appearance of consequent poles on the magnet surface, tending to a varying degree, to divert useful flux from the air gap. To this point, a most revealing series of experiments have been reported by McCaig [2]. This work describes the degree of demagnetization suffered by Alnico-type magnets by various and repeated contact with ferromagnetic bodies. The methods of contact are illustrated in Figure 9.9 together with curves indicating the relative demagnetization for the contact methods shown. These data were taken on Alcomax III, an alloy similar to Alnico 5, the bars having dimensions which indicate a working point B/H of 17.8, very near the maximum energy point. It was found that a direct pull separating the magnet pole face from the steel block, Method A, or a separation by sliding the magnet pole face off the block, Method B, resulted in less than a 2% loss. Very drastic reductions in magnetization

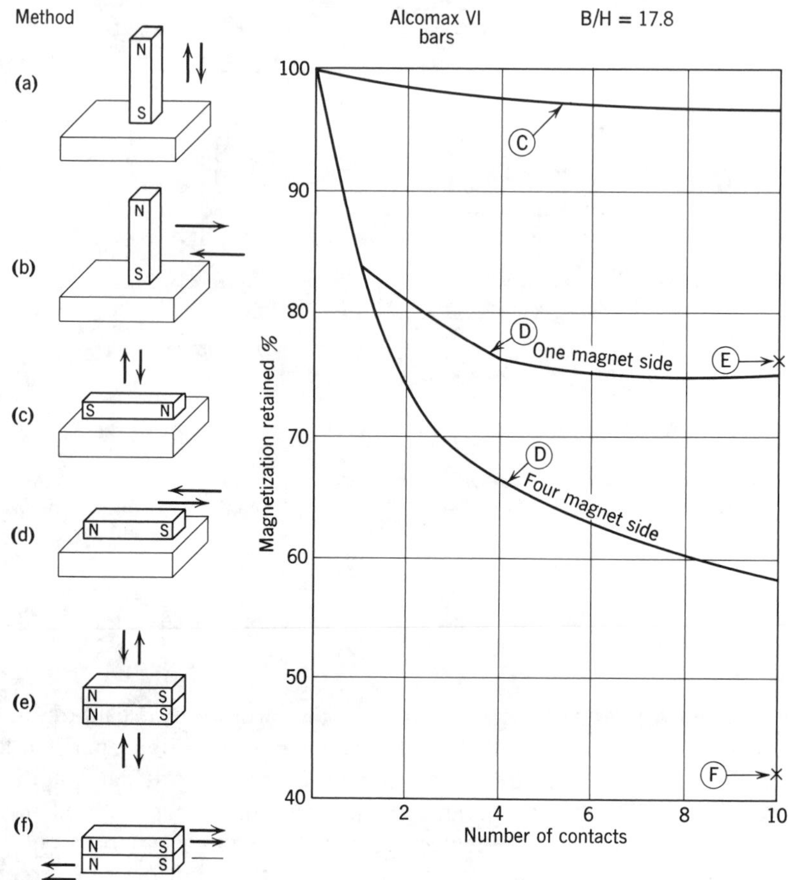

Figure 9.9 Demagnetization suffered under various types of contact. From McCaig [2].

were found to occur as a result of sliding the magnet off the block with a movement parallel to the direction of magnet magnetization. McCaig reports that most of such damage occurs as the magnet crosses the block edge with a motion parallel to the magnet axis, Method D. Sliding the magnet on the block without crossing the block edge, or crossing the edge with a motion at right angles to the magnet axis, has little effect.

For similar reasons, when two saturated magnets are in close proximity, demagnetization will also be noticed. Repeated contact with poles in the repelling position yielded a 24% loss of remanence, Method E. The most serious loss was noted when two contacting magnets, with unlike poles together, were separated by sliding in the direction of magnetization. With all four magnet sides separated in this manner, a total remanence loss of 57.5% was noted, Method F. Special attention must be given to the separation of pairs of magnetized magnets because the method requiring the

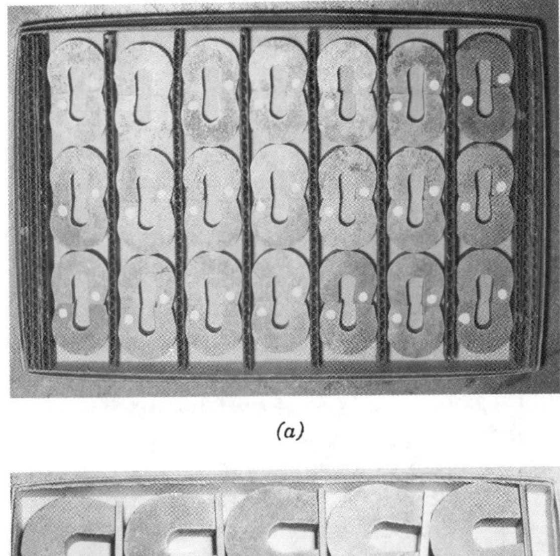

(a)

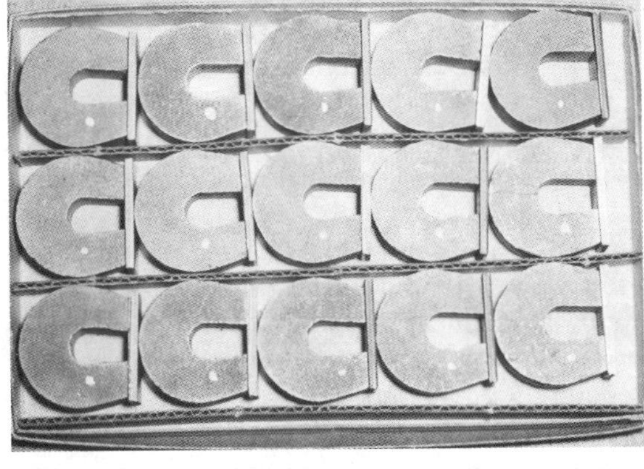

(b)

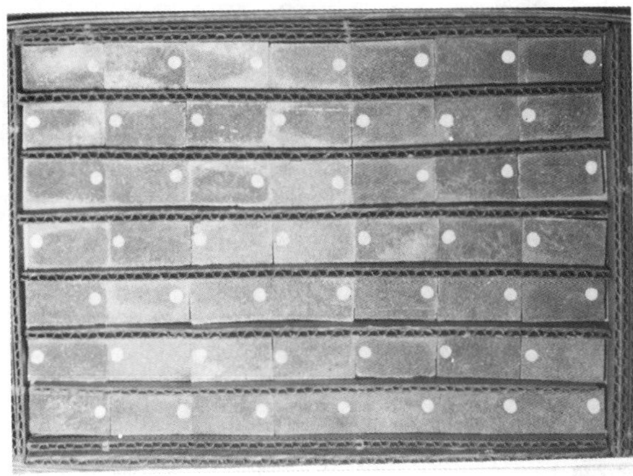

(c)

Figure 9.10 Packaging of magnetized magnets for storing or shipment. (a) Together in pairs, (b) with soft steel keepers, (c) bar magnets, spaced with unlike poles together.

Table 9.1 Protection from demagnetization by surface coating

Coating thickness (mm)	0	0.1	0.2	0.8	1.48	3.1
Magnetization remaining (%)	61	67.0	68.0	87.5	95.0	99.0

least effort, sliding the magnets apart, will cause the greatest demagnetization. Such magnets, rather, should be separated with a direct pull, with no sliding, for the least damage. Although the precise loss of remanence will vary somewhat with magnet material and the working point of the magnet, these tests serve to alert the user to the serious consequences of improper handling of saturated magnets, and the relative effect of the methods of contact.

As a protection against this type of mishandling during assembly, operation or service, many magnets are sheathed with a sufficient thickness of nonmagnetic material, such as aluminum or plastic, to maintain a relatively safe distance between the magnet and ferromagnetic material. McCaig has also investigated the required thickness of protective coating to reduce losses by a given amount with the most probable conditions (Method D in Figure 9.9). It may be seen from these data that over 3 mm of coating are required to hold magnetization losses below 1% (Table 9.1).

In the interest of conserving space or to avoid damage from fields of magnetized magnets during shipment, storage, or handling, they may be keepered with soft iron or paired in the proper polarity sense as shown in Figure 9.10a and b. These measures will be successful, provided, of course, that the proper precautions mentioned above, are observed when removing the keeper or separating the paired magnets. Only insignificant fields will be detected around adequately keepered magnets, and they may be safely located much closer together without harmful effect. As shown in Figure 9.10c, bar-type magnets may be arranged in long rows either before or after magnetization, and stored with the polarity of adjacent rows reversed and with a row spacing approximately equal to the diameter of the magnet.

The relative ease of demagnetization through improper handling noted above, has led many designers to enclose the magnet so that after assembly and magnetization, it is inaccessible. The precautions suggested above are by and large only needed for low coercive force type I magnets. With a type II magnet there is little limb area and very little limb leakage as was described in Section 2.7.

9.4 EQUIPMENT AND TECHNIQUES TO MAGNETIZE

9.4.1 d.c. Fields

Direct current equipment is important consideration in magnetizing permanent magnets. Figure 9.11 shows a general purpose d.c. magnetizer suitable

Figure 9.11 Conventional electromagnet magnetizer with variable air gap; operates from 125 V d.c. Pole pieces can be changed to shape the field to suit common magnet shapes.

for magnetizing a wide range of unit properties and magnet volumes. Such a structure is limited as to the magnet configurations it can saturate. Bars, cylinders, and curved arcs are suitable configurations. Figure 9.12 shows the field developed as a function of current and pole spacing. d.c. electromagnets can be controllable and are widely used in the hysteresisograph to characterize permanent magnets. Due to penetration problems mentioned in Section 9.3, a d.c. electromagnet can be used advantageously to magnetize very large volumes of electrically conductive permanent magnets. Some basic design considerations for electromagnetic magnetizers follow.

If F is the total m.m.f. developed in a magnetizer, and H_y, L_y are the values of H in the yoke and yoke length, the magnet to be magnetized has

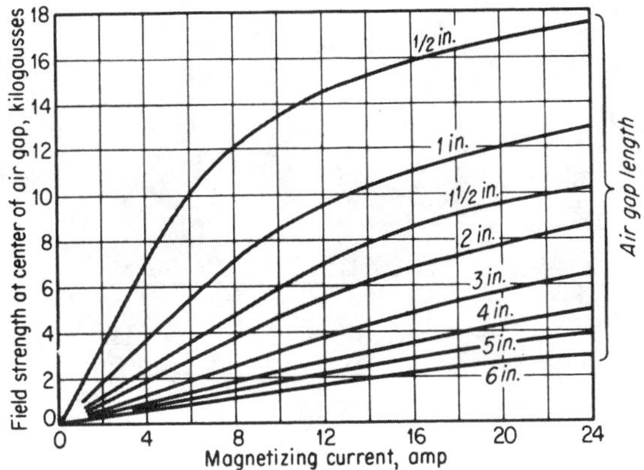

Figure 9.12 Air-gap field strength vs. magnetizing current curves for various gap lengths demonstrate the degree of control possible with an electromagnet magnetizer.

H_d and L_m for H in magnet and magnet length. The m.m.f. equation around a closed path is

$$F = H_d L_m + H_y L_y \quad \text{and} \quad H_d = \frac{F - H_y L_y}{L_m}$$

If we are to make H_d as large as possible for a given level of F, it is necessary that $H_y L_y$ be a minimum. $H_y L_y = \phi_y L_y / A_y \mu$ where A_y = yoke area, μ = the yoke permeability, and ϕ_y = yoke flux. For $H_y L_y$ to be small, A_y and μ should be large. If one neglects the m.m.f. drop in yoke, then $H_d = F/L_m$. In order to keep ϕ_y to a minimum and limit yoke potential drop, the leakage flux must be given consideration in the electromagnet design.

Figure 9.13 show a laboratory type electromagnet which has a very low yoke potential drop and low reluctance joints. Permendur pole tips are used to allow fields in excess of 30 kOe in short gaps. This electromagnet current source is designed for very low ripple current and can serve as a controllable field for a wide variety of laboratory experiments. Figure 9.14 shows field levels that can be developed with Permendur pole pieces dimensioned as shown.

Other often used d.c. magnetizing structures are the open solenoid and the iron bound solenoid shown in Figure 9.15. Open solenoids are often

Figure 9.13 Laboratory Electromagnet. From Walker Scientific Inc.

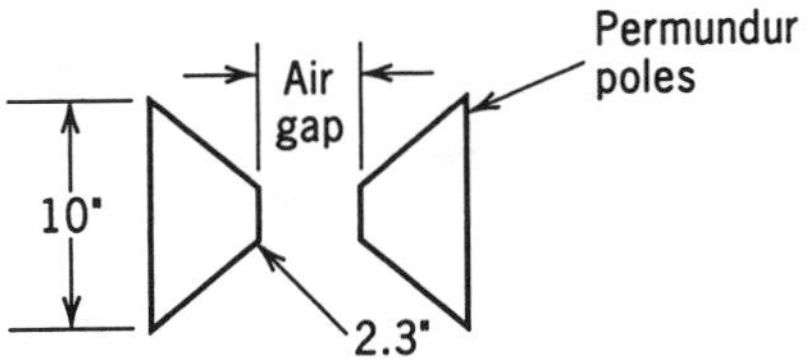

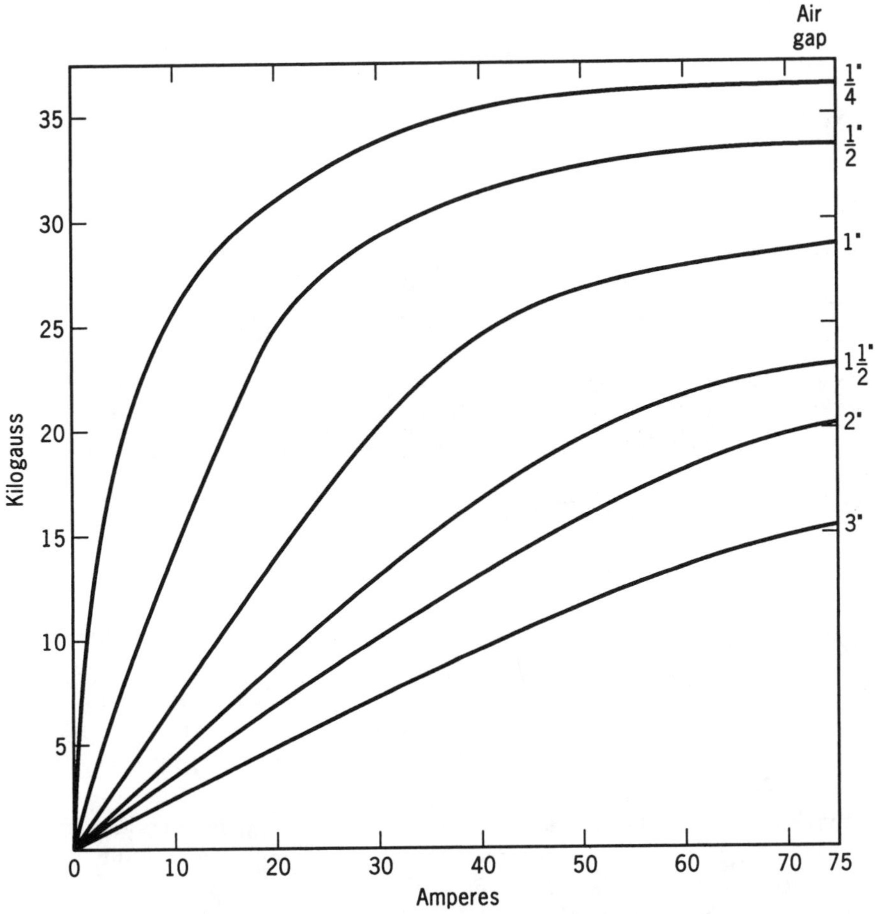

Figure 9.14 Laboratory electromagnet field strength.

used to magnetize on a production basis, since they lend themselves to flowing magnets through them rapidly. An open solenoid can easily be water cooled, and the resulting field strength can be raised by a factor of approximately 30, over an air cooled solenoid. The iron clad solenoid allows for very high fields due to the absence of leakage near the gap. Structures of 50 kOe are feasible. The geometry does limit easy access to the air gap, but

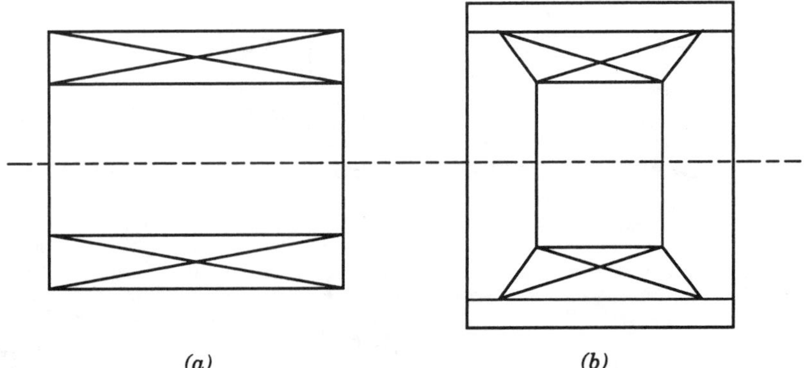

Figure 9.15 Solenoids. (a) Simple solenoid; (b) ironbound solenoid.

still it is possible to arrange for a sample to be positioned in the gap by entering through a hole along the axis of the gap.

In some instances, steady state fields from one permanent magnet are used to magnetize another permanent magnet. There are problems in avoiding distortion as the magnet to be magnetized is removed from the permanent magnet supplying the field. Here the energy to magnetize is supplied by mechanical energy input from the person removing the magnetized magnet.

9.4.2 Pulse Fields

Since the time interval required is extremely short, magnetizing can be achieved by a current impulse, provided its magnitude is sufficient to deliver the peak H required. The developent of materials with high coercive force and high available energy has led to relatively intricate magnet configurations. The length of magnet limbs has decreased and circular shapes and parallel circuits are common. Many of the newer shaped permanent magnet arrangements cannot be magnetized by placing them in contact with conventional electromagnets. Instead, they are magnetized by the flux field around a conductor threading through the window of the magnetic circuit. Others must be wound with several turns of heavy wire. In fact, many magnet designs are materially influenced by how the magnetizing conductor or conductors can be arranged. Impulse magnetization has become popular not only for these newer shapes, but also for nearly all types of permanent magnets, because equipment to produce extremely high instantaneous currents requires relatively low investment. The basic components of a capacitor discharge impulse magnetizer are shown in Figure 9.16. The capacitor is charged to voltage v_{dc} at a rate determined by R_0. The capacitor is switched to discharge into the magnetizing coil having inductance L and resistance R. As long as R is greater than $2\sqrt{(L/C)}$, the current pulse will

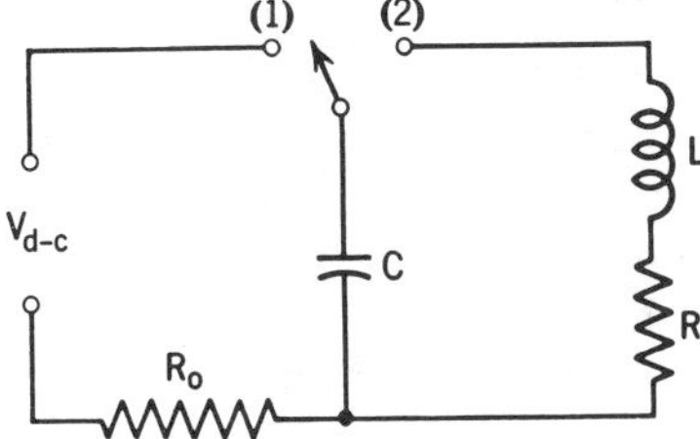

Figure 9.16 Elements of capacitor discharge magnetizer.

Figure 9.17 Capacitor discharge magnetizer. From Walker Scientific Inc.

be unidirectional without oscillation. A high energy impulse magnetizer is shown in Figure 9.17. This unit uses 600 V electrolytic capacitors and 16 kJ of energy is available for magnetizing.

Energy transfer relationships involving the capacitor discharge magnetizer are shown in Figure 9.18. These relationships allow the estimation of the capacitor energy that must be stored to magnetize a given volume of magnet, taking into account the magnet materials unit properties.

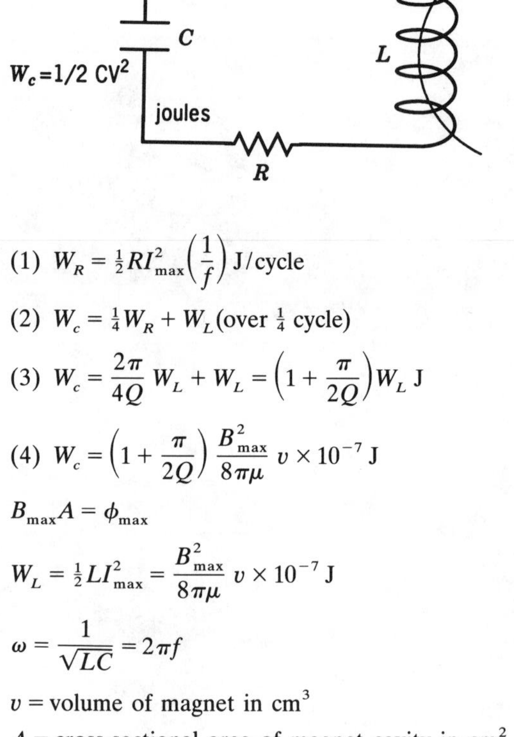

$$(1)\ W_R = \tfrac{1}{2} R I^2_{max} \left(\frac{1}{f}\right) \text{ J/cycle}$$

$$(2)\ W_c = \tfrac{1}{4} W_R + W_L (\text{over } \tfrac{1}{4} \text{ cycle})$$

$$(3)\ W_c = \frac{2\pi}{4Q} W_L + W_L = \left(1 + \frac{\pi}{2Q}\right) W_L \text{ J}$$

$$(4)\ W_c = \left(1 + \frac{\pi}{2Q}\right) \frac{B^2_{max}}{8\pi\mu} v \times 10^{-7} \text{ J}$$

$$B_{max} A = \phi_{max}$$

$$W_L = \tfrac{1}{2} L I^2_{max} = \frac{B^2_{max}}{8\pi\mu} v \times 10^{-7} \text{ J}$$

$$\omega = \frac{1}{\sqrt{LC}} = 2\pi f$$

v = volume of magnet in cm^3

A = cross-sectional area of magnet cavity in cm^2

B_{max} = lines/cm^2

μ = permeability

I = amperes

V = volts

f = frequency

$$Q = \frac{\omega L}{R} = 2\pi \frac{W_L}{W_R} \text{ (efficiency factor)}$$

Figure 9.18 Energy relationships in a capacitor discharge type of magnet charger.

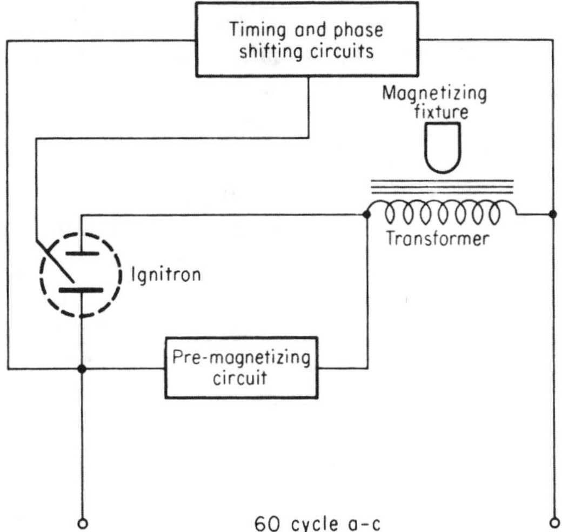

Figure 9.19 Half-cycle magnetizer circuit passes half-cycle current impulses whose amplitude and duration is controlled by the firing of a thyratron in the ignitor circuit. For currents above 10,000 A a transformer is used. To prevent flux reversal, a premagnetizing circuit saturates the transformer with d.c. pulses of opposite polarity to the main pulse.

Another type of impulse magnetizer is the half cycle magnetizer shown in Figure 9.19. This unit has the disadvantage of drawing large current surges directly from the power line. It does, however, offer a very fast repetition rate. The timing and phase shifting control feature allows the current magnitude to be changed. Figure 9.20 shows voltage, current, and flux changes for various firing points of a half cycle magnetizer.

9.5 EQUIPMENT AND TECHNIQUES TO DEMAGNETIZE

Achieving the demagnetized state is very important to both users and producers of permanent magnets. In the producing organization, magnets must be magnetized for measurement. Nearly all magnets are shipped

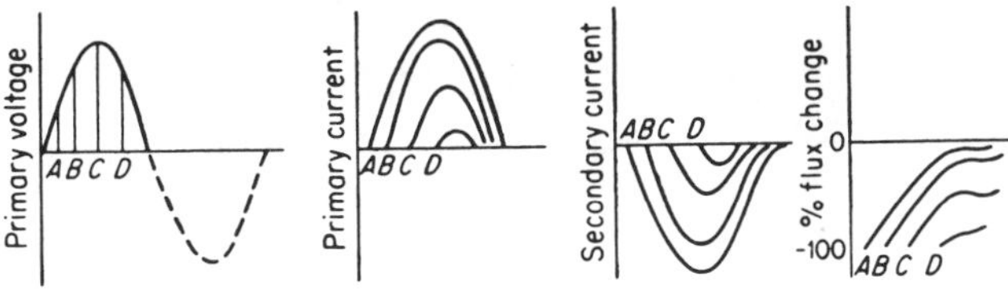

Figure 9.20 Voltage, current and flux change for various firing points of a half-cycle magnetizer with transformer.

demagnetized to facilitate shipping and meet the need of assembling demagnetized components in the user's plant. There is the serious problem of magnetic chips and dirt being attracted into the air gaps of magnetic assemblies if the permanent magnet is not well demagnetized. Demagnetizing energy may be in the form of external field energy or thermal energy or a combination of these two energy forms. A d.c. field can be applied, which will demagnetize; however, it must be determined rather precisely for a particular magnet. As shown in Figure 9.21, the induction will be reduced as the demagnetizing field is applied to a saturated magnet. When the field is removed the magnet induction will recoil along a minor hysteresis loop. It is possible to arrange field conditions so that upon recoil the magnet will be left of zero induction. The exact field strength H_{cr} required is termed relaxation coercive force and is just slightly greater than H_{ci}. Demagnetizing by recoil is slow and difficult to achieve since H_{cr} for each magnet must be found. For demagnetizing a large number of magnets, an a.c. field is usually used. With this reversal method, the applied a.c. field is slowly and continuously reduced so that the magnet is cycled through continuously smaller hysteresis loops and eventually reaches a state close to $B = 0$. It is necessary to reduce the field slowly so that the interior hysteresis loops remain symmetrical about the origin or else an appreciable remanence will remain when H is finally reduced to zero. The magnet can be slowly withdrawn from the field or the field can be reduced with respect to a fixed

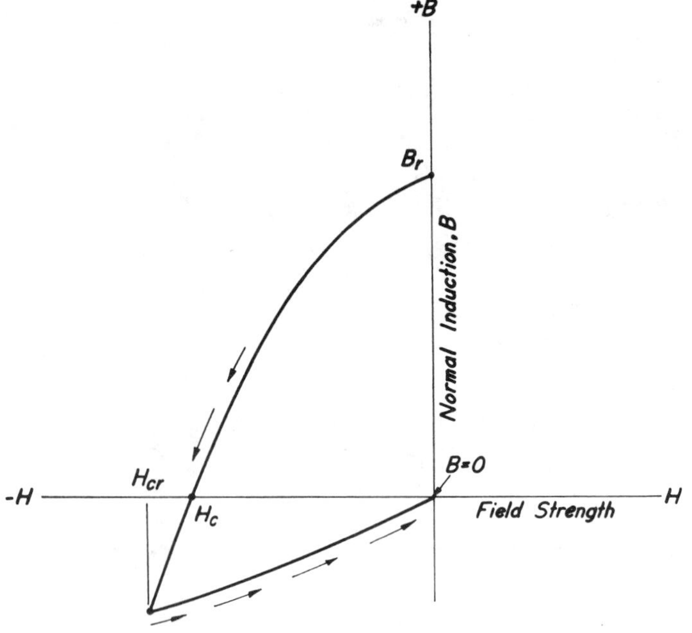

Figure 9.21 Demagnetization by recoil.

magnet. The starting field needs to be approximately equal to the H_{ci} of the material involved. In large metallic permanent magnets, eddy currents effectively shield the magnet. A reversing frequency of 5–10 Hz has proven effective on very large cast magnets.

Thermal demagnetization is used extensively with ferrite magnets since with the low Curie temperature the ferrite magnet is not damaged structurally. With rare-earth magnets, at times thermal demagnetization is used. However, the original heat treatment must be used after the thermal cycle if full properties are to be achieved.

Initial experiences with rare-earth magnets indicate that by heating magnets to 300–400°C, a substantial reduction in a.c. field magnitude can be achieved.

9.6 CALIBRATION AND STABILIZATION TECHNIQUES

In several applications the flux of the magnet must be adjusted within close limits. For example a microwave tube might need to have the field supply adjusted to ±1% from unit to unit. In a typical production run of magnets, the saturated fully magnetized flux values might deviate by ±5%. The task is to treat each magnet and bring it within the desired limit. In Figure 9.22,

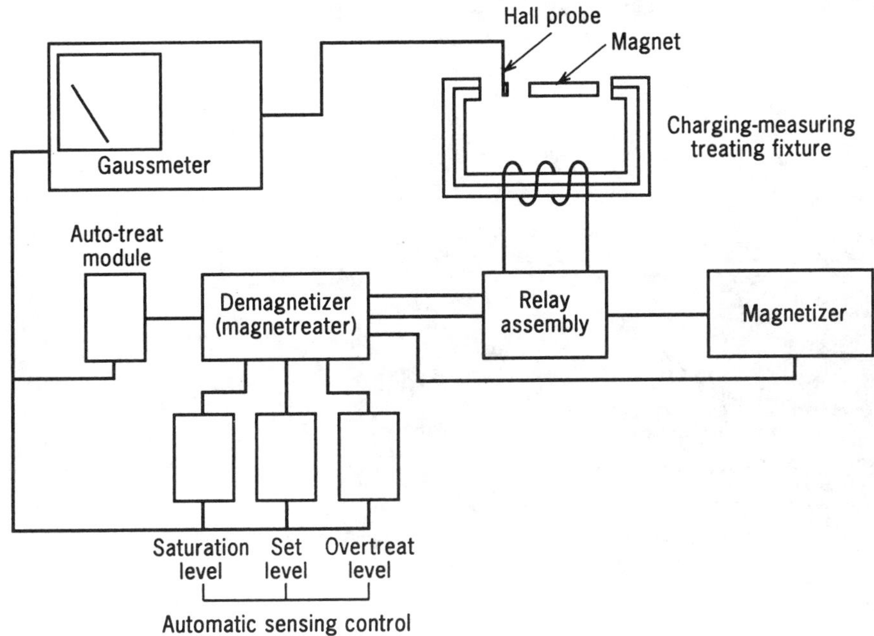

Figure 9.22 Calibration system. From RFL Industries.

the elements of a calibration system are shown. The demagnetizing energy is stored in a capacitor bank over a period of a few seconds, then discharged through a coil as an oscillatory damped wave with a time duration of a few milliseconds. The level of flux density may be monitored during the calibration process by using a gaussmeter. The operation is stopped when the desired level is achieved. Resolution of the order of 0.1% is possible with such a system. The system can be semi-automatic or fully programable by using automatic switching modules.

REFERENCES

[1] L. A. Kavalerova, B. G. Livshitz, A. S. Lileev, and V. P. Menushenkov, "The Reversibility of Magnetic Properties of Sintered SmCo, Permanent Magnets," IEEE Trans. on Magnetics, *MAG-11*, 1673 (1975).

[2] M. McCaig, "Demagnetization of Magnets Due to Contact with Ferromagnetic Bodies," *J. Sci. Instrums.*, **33**, 311 (1956).

APPENDIX 1

GLOSSARY OF TERMS, DEFINITIONS, SYMBOLS, SPECIFICATIONS, AND STANDARDS

TERMS AND DEFINITIONS

Air Gap: A nonmagnetic discontinuity in a ferromagnetic circuit. For example, the space between the poles of a magnet, although filled with brass or wood or any other nonmagnetic material, is nevertheless called an air gap.

Ampere-turn: A unit of magnetomotive force.

Anisotropy: Relates to the importance of direction in determining properties of a magnetic material.

Area: A, the cross-sectional area of a particular volume of space.

Circuit, Closed Magnetic: A circuit where the magnetic flux is conducted continually around a closed path through ferromagnetic materials; for example, a steel ring.

Circuit, Open Magnetic: When a magnet does not have a closed external ferromagnetic circuit and does not form a complete conducting circuit itself, the magnet is said to be open circuited; for example, a permanent magnet ring interrupted by an air gap.

Coercive Force: H_c, the magnetizing force required to bring the induction to zero in a magnetic material which is in a symmetrically cyclically magnetized condition.

Coercive Force, Intrinsic: H_{ci}, the magnetizing force required to bring to zero the intrinsic induction of a magnetic material which is in a symmetrically cyclically magnetized condition.

Coercivity: That property of a material measured by the maximum value of the coercive force.

Coercimeter: An instrument for measuring the normal and intrinsic coercive properties of a magnetic material or of a sample taken from a basic magnet.

Curie temperature: T_c, the temperature at which a material changes from a ferromagnetic to a paramagnetic state.

Cyclically Magnetized Condition: A magnetic material is in a cyclically magnetized condition when, under the influence of a magnetizing force which varies cyclically between two specific limits, its successive hysteresis loops are identical.

Demagnetization Curve: That portion of the hysteresis loop which lies between the residual induction point, B_r and the coercive force point, H_c. Points on this curve are designated by the coordinates B_d and H_d.

Demagnetizing Force: H_d, a magnetizing force applied in such a direction as to reduce the remanent induction in a magnetized body.

Diamagnetic Material: A material having a permeability less than that of a vacuum.

Dyne: The force producing an acceleration of one centimeter per second per second when applied to a one gram mass.

Eddy Current Loss: That portion of the core loss due to currents circulating in the magnetic material as a result of electromotive forces induced by varying induction.

Electromagnet: A magnet, consisting of a solenoid with an iron core, which has a magnetic field existing only during the time of current flow through the coil.

Energy-Product Curve, Magnetic: The curve obtained by plotting the product of the coordinates of the demagnetization curve, B_dH_d, as abscissas against the induction B_d.

Erg: The work done by a force of one dyne whose point of application is moved through one centimeter in the direction of the force.

Ferromagnetic Material: A paramagnetic material which exhibits a high degree of magnetizability.

Gauss: The CGS unit of magnetic induction, (see Induction, Magnetic).

Gilbert: The CGS unit of magnetomotive force (see Magnetomotive Force).

Hysteresis, Magnetic: The property of a magnetic material by virtue of which the magnetic induction for a given magnetizing force depends upon the previous conditions of magnetization.

Hysteresis Loop: A curve (usually with rectangular coordinates) which shows, for a magnetic material in a cyclically magnetized condition, for each value of the magnetizing force, two values of the magnetic induction, one when the magnetizing force is increasing, the other when it is decreasing.

Hysteresis Loss: The power expended in a magnetic material, as a result of magnetic hysteresis, when the magnetic induction is cyclic.

Induction Curve, Normal: A curve depicting the relation between normal induction and magnetizing force.

Induction, Intrinsic (or Ferric Induction): B_i, the excess of the induction in a magnetic material over the induction in vacuum, for a given value of the magnetizing force. The equation for intrinsic induction or magnetization is $B_i = B \pm H$.

Induction, Magnetic (or Magnetic Flux Density): B, flux per unit area through an element of area at right angles to the direction of the flux. The CGS unit of induction is called the gauss and is defined by the equation $B = d\phi/dA$.

Induction, Normal: B, the limiting induction, either positive or negative, in a magnetic material which is in a symmetrically cyclically magnetized condition.

Induction, Remanent: B_d, see Remanence.

Induction, Residual: B_r, the magnetic induction corresponding to zero magnetizing force in a magnetic material which is in a symmetrically cyclically magnetized condition.

Induction, Saturation: B_s, the maximum intrinsic induction possible in a material.

Keeper: A magnetic conductor used to complete the magnetic circuit of a permanent magnet to protect it against demagnetizing influences.

Kilogauss: One kilogauss is equal to 1000 G.

Leakage, Flux: F, that portion of the flux which does not pass through the air gap, or useful part of the magnetic circuit.

Linkage, Flux: ϕN, the product of the number of turns in an electric circuit by the average value of the flux linked with the circuit.

Load Line: Graphic representation of permeance.

Magnetic Field Strength: H, see Magnetizing Force.

Magnetic Flux: A condition in a medium produced by a magnetomotive force, such that when altered in magnitude a voltage is induced in an electric circuit linked with the flux. The CGS unit of magnetic flux is called the maxwell and is defined by the equation $e = -N(d\phi/dt) \times 10^{-8}$ where e = induced e.m.f. in volts, and $d\phi/dt$ = time rate of change of flux in maxwell per second.

Magnetic Line of Force: An imaginary line in a magnetic field which at every point has the direction of the magnetic flux at that point.

Magnetizing Force: H, magnetomotive force per unit length. The CGS unit is called the oersted and is defined by the equation $H = dF/dL$ where F is in gilberts and L in centimeters. For a toroid, or at the center of a long solenoid, the magnetizing force in oersteds may be calculated as follows, $H = 0.4\pi NI/L$ where I is in amperes and L is in centimeters.

Magnetomotive Force: That which tends to produce a magnetic field. In magnetic testing it is most commonly produced by a current flowing through a coil of wire, and its magniture is proportional to the current, and to the number of turns. The CGS unit of magnetomotive force is called the gilbert and is defined by the equation $F = 0.4\pi NI$ where I is in amperes. Magnetomotive force may also result from a magnetized body.

Maxwell: The CGS unit of magnetic flux.

Oersted: H, the CGS unit of magnetizing force.

Paramagnetic Material: A material having a permeability which is slightly greater than that of a vacuum, and which is approximately independent of the magnetizing force.

Permanent Magnet Material: Shaped piece of ferromagnetic material which once having been magnetized, shows definite resistance to external demagnetizing forces, i.e., requires a high coercive force to remove the resultant magnetism.

Permeability, Differential: μ_d, the slope of the normal induction curve.

Permeability, Incremental: The ratio of the cyclic change in magnetic induction to the corresponding cyclic change in magnetizing force when the mean induction differs from zero.

Permeability, Initial: μ_i, the slope of the normal induction curve at zero magnetizing force.

Permeability, Recoil: μ_r, the average slope of the minor hysteresis loop.

Permeance: P, the ratio of the flux through any cross-section of a tubular portion of a magnetic circuit bounded by lines of force and by two equipotential surfaces to the magnetic potential difference between the surfaces, taken within the portion under consideration.

Poles, Consequent: Additional magnetic poles which are present at other than the ends of a magnetic material.

Poles, North and South Magnetic: The north pole of a magnet, or compass, is attracted toward the north magnetic pole of the earth, and the south pole of a magnet is attracted toward the south magnetic pole of the earth. This is based upon tradition and not physics, as, actually, two unlike poles will attract each other while like poles repel. However, the north seeking pole of a magnet is designated by the letter N, and the other pole by S. The N pole of one magnet will attract the S pole of another magnet.

Reluctance: R, the reciprocal of permeance $R = /F/\phi$. For uniform L and A, $R = L/A$ where A is area in square centimeters and L is length in centimeters.

Remanence (or Remanent Induction): B_d, the magnetic induction which remains in a magnetic circuit after the removal of an applied magnetomotive force.

Retentivity: The property of a magnetic material measured by the maximum value of the residual induction.

Saturation: A condition where all of the available elementary magnetic moments in a ferromagnetic material are aligned in substantially the same direction.

Soft Magnetic Material: Shaped piece of ferromagnetic material which once having been magnetized is very easily demagnetized, i.e., requires a slight coercive force to remove the resultant magnetism.

Stabilization: A treatment of a magnetic material designed to increase the permanency of its magnetic properties or condition.

Symmetrically Cyclically Magnetized Condition: A magnetized material is in a symmetrically cyclically magnetized condition when it is cyclically magnetized and the limits of the applied magnetizing forces are equal and of opposite sign, so that the limits of induction are equal and of opposite sign.

SYMBOLS

A	Area
a	Area, number of paths
B	Magnetic flux density; magnetic induction
B_d	Remanence
B_g	Flux density in air gap
B_i	Intrinsic magnetization
B_m	Flux density in magnet at any operating point on normal magnetizing curve
B_r	Residual induction
B_s	Saturation induction
B_{is}	Intrinsic saturation magnetization
$(BH)_{max}$	Maximum energy product (also written as $(B_dH_d)_{max}$
B/H	Load line, unit permeance
C	Capacitance
D	Diameter
d	Distance; diameter
E	Energy
e	Electron charge
F	Force; magnetomotive force
f	Frequency
G	Air gap; gauss
H	Magnetic field stength; magnetizing force
H_c	Normal coercive force
H_{ci}	Intrinsic coercive force
H_d	Demagnetizing force
H_g	Magnetizing force in air gap

H_s	Magnetizing force to saturate magnet
h	Planck's constant
I	Current
J	Intensity of magnetization; current density
K	Anisotropy constant
k	Boltzmann's constant; leakage and reluctance factors
L	Length dimension; inductance
L_g	Length of air gap
L_m	Length of magnet
l	Length
M	Mass
MGOe	Million gauss-oersteds
m	Unit magnetic pole
m.m.f.	Magnetomotive force
N	North pole; demagnetization factor
NI	Ampere turns (also written as AT)
n	Number of turns
Oe	Oersteds
P	Power; permeance
P_g	Gap permeance
P_t	Total permeance
p	Packing factor; number of poles
Q	Electric quantity
R	Resistance; radius; reluctance
r	Radius; resistance
S	South pole
S	Speed
T	Time; temperature; torque
t	Thickness; time
T_c	Curie temperature
V	Volume; voltage; velocity
W	Weight; work
ϕ	Magnetic flux
ϕ_g	Gap flux
ϕ_m	Magnet flux
μ	Permeability
μ_0	Magnetic constant
μ_d	Differential permeability
μ_i	Initial permeability
μ_r	Recoil permeability
μm	Micrometers (micron)
ω	Angular frequency
δ	Density; thickness
ρ	Resistivity

SPECIFICATIONS AND STANDARDS

The International Electrotechnical Commission (IEC) is the international standards organization responsible for magnetic and electrical materials, standards and specifications. Table A.1.1 shows the IEC classification of magnetic materials. In IEC, Working Group Five (WG5) is responsible for standards work on permanent magnets or hard magnetic materials. The working group is made up of representatives of the various countries. Members may come from a national standards group or may represent a producers trade association. In the United States, for example, most magnet producers below to The Magnetic Materials Producers Association (MMPA), 800 Custer Avenue, Evanston, Illinois 60202. The United States member of WG5 represents MMPA. This association publishes its own standards document. MMPA Standard No. 0100-87 lists the various material grades and their properties, both magnetic and physical as well as a cross-reference to IEC Standards. In Tables 4.2 and 4.3 as well as in Appendix 2 and Appendix 3, the MMPA brief designation is also used. This designation contains information regarding energy product and intrinsic coercive force. For example, $RECo_5$ which has properties of 16 MGOe and 18 kCe is designated as 16/18. The IEC code reference for rare-earth cobalt alloys is R5 from Table A.1.1. In terms of IEC documents, $RECo_5$ would be described as R5 16/18. MMPA also publishes *A Guide to Understanding, Specifying and Using Permanent Magnet Materials*.

In Europe to date there is no equivalent to MMPA, although the Magnet Centre at Sunderland Polytechnic, Chester Road, Sunderland, SR135D, U.K. acts as a central clearing house for magnet standards activity. Other countries in Europe also have active national standards programs.

In Japan two trade groups are concerned with permanent magnets. (i) The Japan Electronic Materials Manufacturers Association (EMA) Kotohirakaikan Building, Toranomon 1 Chome, Minato-Ku, Tokyo, Japan 105 and (ii) Japan Society of Plastic Technology, Hakau Building, 4-1, 2 Chome, Akasaka, Minato-Ku, Tokyo, Japan, which represents the producers of bonded magnets.

Table A.1.1 IEC classifications of magnetic materials

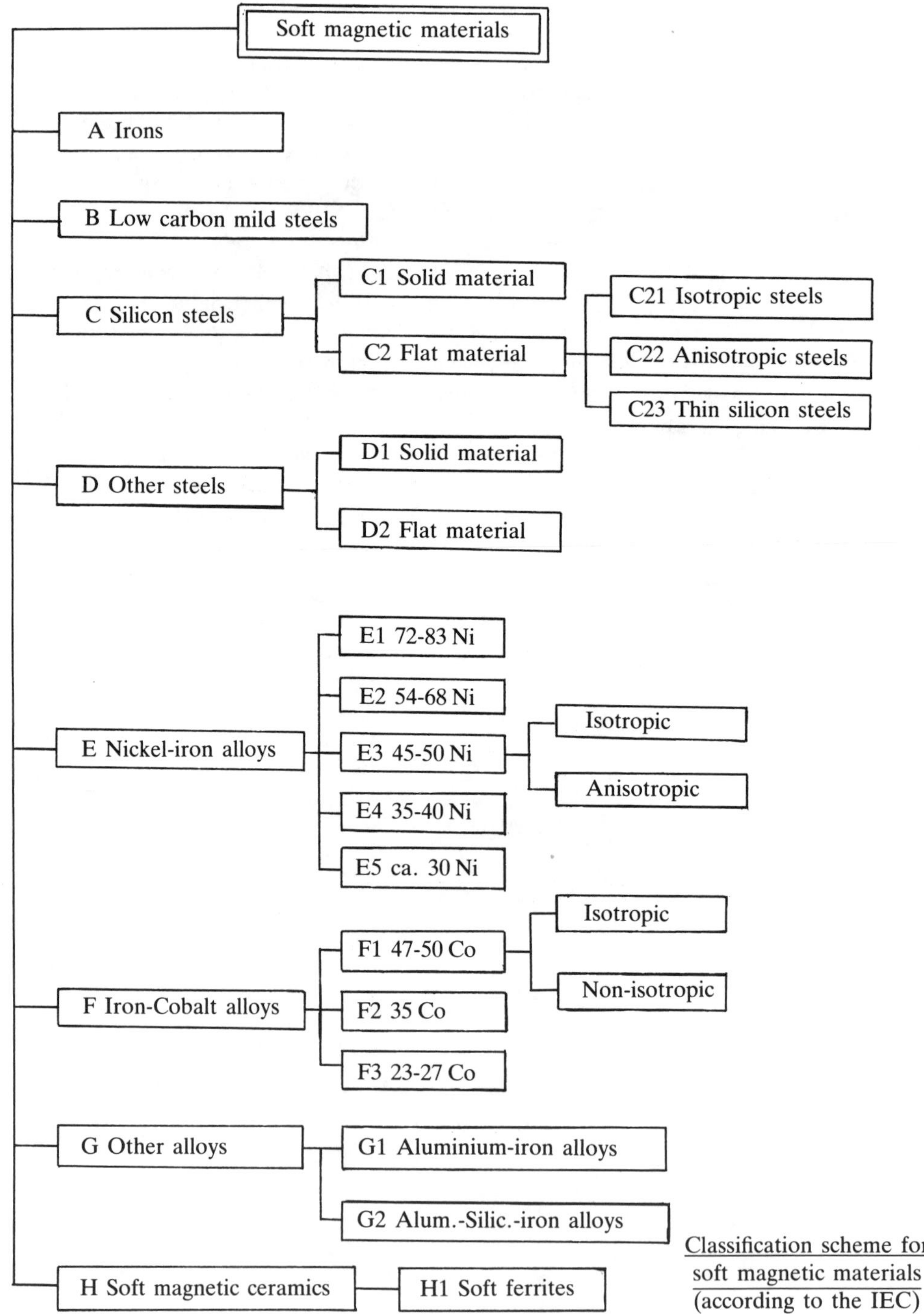

Classification scheme for soft magnetic materials (according to the IEC)

Table A.1.1 (*Continued*)

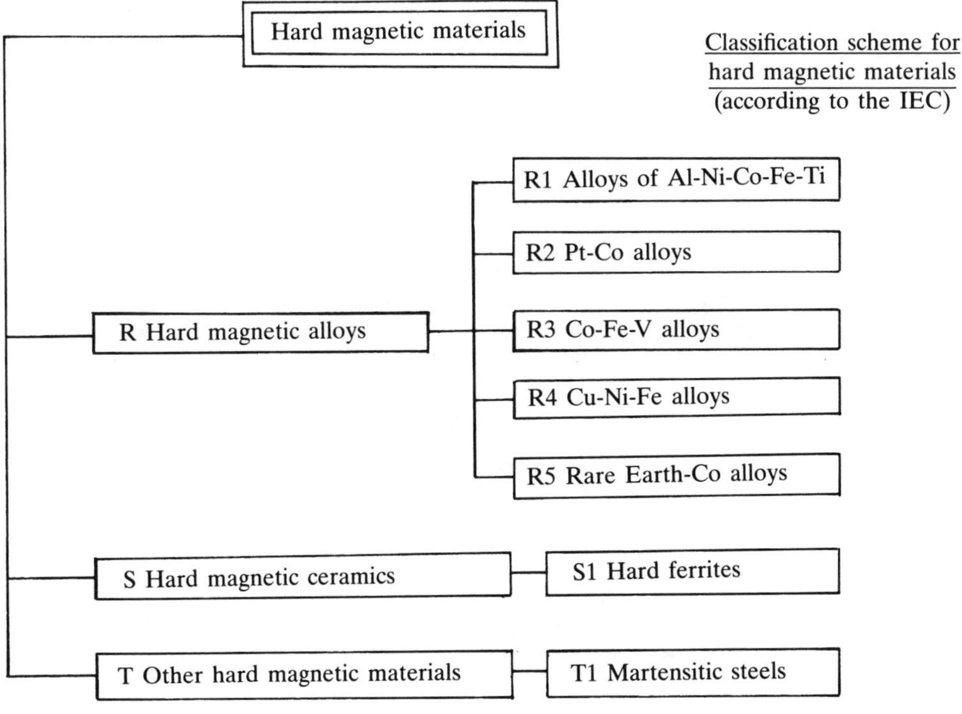

APPENDIX 2

MAGNETIC AND PHYSICAL PROPERTY TABLES

Table A.2.1 Magnetic propeties of Alnico permanent magnets

Alloy or Grade	Alnico 2 cast	Alnico 2 sintered	Alnico 5 cast	Alnico 5 sintered	Alnico 5 DG	Alnico 5-7	Alnico 6 cast	Alnico 6 sintered	Alnico 8 cast	Alnico 8 cast HC	Alnico 8 sintered	Alnico 8 sintered HC	Alnico 9 cast
B_r(G)	7500	7100	12.4000	10,500	12,800	13,500	10,500	9400	8200	7400	7600	6500	10,600
H_c(Oe)	560	550	640	600	670	740	780	790	1650	1900	1500	1800	1500
H_{ci}(Oe)	580	570	680	620	685	750	800	820	1860	2150	1690	2020	1550
$(BH)_{max}$(MGOe)	1.7	1.5	5.5	3	6.25	7.5	3.9	2.9	5.3	5.0	4.5	4.5	9.2
B/H at $(BH)_{max}$	12	12	18	18.4	17	17	13	12	5.0	3.9	5	3.7	7.0
B_d at $(BH)_{max}$(G)	4600	4300	10,000	8000	10,300	11,100	7000	5900	5200	4500	4700	4100	8000
H_d at $(BH)_{max}$(Oe)	370	350	550	435	600	650	540	500	1000	1150	960	1100	1150
μ_r	6.4	6.4	4.3	4.0	4.0	3.8	5.3	4.5	2.0	1.9	2.1	1.9	–
H_S(Oe)	2500	2500	3000	3000	3000	3500	4000	4000	8000	8000	8000	8000	8000
Max. service Temp. (°C)	520	520	520	520	520	520	520	520	520	520	520	520	520
Temp. coeff. of B_r (%/°C)	0.02	0.02	0.02	0.02	0.02	0.02	0.02	0.02	0.02	0.02	0.02	0.02	0.02

Table A2.2 Physical properties of Alnico permenent magnets

Alloy or Grade	Alnico 2 cast	Alnico 2 sintered	Alnico 5 cast	Alnico 5 sintered	Alnico 5 DG	Alnico 5-7	Alnico 6 cast	Alnico 6 sintered	Alnico 8 cast	Alnico 8 cast HC	Alnico 8 sintered	Alnico 8 sintered HC	Alnico 9 cast
Density lb/in^3	0.256	0.247	0.264	0.253	0.264	0.264	0.268	0.249	0.262	0.262	0.252	0.252	0.252
Tensile (lb/in^2)	3000	65,000	5400	50,000	5200	5000	23,000	55,000	39,000	39,000	50,000	50,000	–
Transverse modulus of rupture (lb/in^2)	7200	70,000	10,500	56,000	9000	8000	45,000	100,000	30,000	30,000	55,000	55,000	–
Coefficient of thermal expansion (°C × 10^{-6})	12.4	12.4	11.6	11.3	11.4	11.4	11.4	11.3	11.0	11.0	11.6	11.6	11.2
Resistivity at 25°C μ-ohms/cm per cm^2)	65	68	47	50	47	47	50	44	50	50	53	53	53
Rockwell hardness	45C	43C	50C	44C	50C	50C	50C	53C	56C	56C	43C	43C	43C
Curie temperture (°C)	810	810	900	900	900	900	860	860	860	860	860	860	860
Composition possible elements or Wt%	Al Ni Co Cu Ti Fe												

Table A.2.3 Magnetic properties of ferrite permanent magnets

Alloy or Grade	Ceramic 1	Ceramic 5	Ceramic 7	Ceramic 8	Ceramic 9	Ceramic 10	Bonded Flex. (i)	Bonded Flex. (a)	Bonded Rigid (i)	Bonded Rigid (a)
B_r(G)	2300	3800	3400	3850	3800	4100	1700	2500	1400	3000
H_c(Oe)	1850	2400	3250	2950	3400	2800	1600	2200	1050	2400
H_{ci}(Oe)	3250	2550	4000	3050	3900	2900	3000	3000	–	2800
$(BH)_{max}$(MGOe)	1.05	3.40	2.75	3.50	3.40	4.00	0.7	1.50	0.40	2.00
B/H at $(BH)_{max}$	1.2	1.1	1.1	1.1	1.1	1.1	1.2	1.2	1.4	1.2
B_d at $(BH)_{max}$(G)	1100	1900	1700	1900	1900	2000	850	1250	750	1550
H_d at $(BH)_{max}$(Oe)	900	1800	1700	1900	1900	2000	850	1250	550	1250
μ_r	1.1	1.1	1.1	1.1	1.1	1.1	1.05	1.05	1.05	1.05
H_S(Oe)	10,000	10,000	10,000	10,000	10,000	10,000	10,000	10,000	10,000	10,000
Max. service temp. (°C)	400	400	400	400	400	400	–[a]	–[a]	–[a]	–[a]
Temp. coeff. of B_r (%/°C)	0.20	0.20	0.20	0.20	0.20	0.20	0.20	0.20	0.20	0.20
Temp. coeff. of H_{ci} (%/°C)	0.27	0.27	0.27	0.27	0.27	0.27	0.27	0.27	0.27	0.27

[a] Limited by binding material.

Table A.2.4 Physical properties of ferrite permanent magnets

Alloy or Grade	Ceramic 1	Ceramic 5	Ceramic 7	Ceramic 8	Ceramic 9	Ceramic 10	Bonded Flex. (*i*)	Bonded Flex. (*a*)	Bonded Rigid (*i*)	Bonded Rigid (*a*)
Density lb/in^3	0.18	0.17	0.17	0.17	0.17	0.17				
Tensile (lb/in^2)	–	–	–	–	–	–				
Transverse modulus of rupture (lb/in^2)	–	–	–	–	–	–				
Coefficient of thermal expansion (°C × 10^{-6})	10	10	10	10	10	10				
Resistivity at 25°C (μohms/cm per cm^2)	10^{10}	10^{10}	10^{10}	10^{10}	10^{10}	10^{10}				
Rockwell hardness	–	–	–	–	–	–				
Curie temp. (°C)	450	450	450	450	450	450				
Composition possible elements or wt%	–[b]	–[b]	–[b]	–[b]	–[b]	–[b]				

[a] Physical properties vary with bonding material.

[b] Composition is $MO6Fe_2O_3$ where M represents barium, strontium or a combination of the two.

Table A.2.5 Magnetic properties of rare-earth permanent magnets

Alloy or Garde	$RECo_5$ 14/14	$RECo_5$ 16/18	$RECo_5$ 20/15	$RECo_5$ 22/15	RE_2TN_{17} 24/18	RE_2TN_{17} 26/11	RE_2TN_{17} 28/7	$RE_2TN_{14}B$ 27/17	$RE_2TN_{14}B$ 30/18	$RE_2TN_{14}B$ 35/12	MQ I	MQ II	MQ III
B_r(kG)	7.5	8.3	9.0	9.5	10.2	10.5	10.9	10.8	11.5	12.0	6.1	7.9	11.7
H_c(kOe)	7.0	7.5	8.5	9.0	9.2	9.0	6.5	9.0	11.0	10.5	5.3	6.5	10.5
H_{ci}(kOe)	14	18	15	15	>18	11	7	>17	>18	>12	>15	>16	>13
$(BH)_{max}$(MGOe)	14	16	20	22	24	26	28	27	30	35	8.0	13.0	32
B/H at $(BH)_{max}$	1.1	1.1	1.1	1.1	1.1	1.1	1.1	1.1	1.1	1.1	1.1	1.1	1.1
B_d at $(BH)_{max}$(kG)	3.7	4.0	4.4	4.7	4.9	5.1	5.3	5.2	5.5	5.9	2.8	3.7	5.7
H_d at $(BH)_{max}$(kO_e)	3.7	4.0	4.4	4.7	4.9	5.1	5.3	5.2	5.5	5.9	2.8	3.7	5.7
μ_r	1.05	1.05	1.05	1.05	1.05	1.05	1.05	1.05	1.05	1.05	1.05	1.05	1.05
H_S(kOe)	20	20	20	30	30	30	30	25	25	25	30	30	30
Max. service temp. (°C)	250	250	250	250	300	300	300	150	150	150	–	150	150
Temp. coeff. of B_r (%/°C)	0.045	0.045	0.045	0.045	0.030	0.030	0.030	0.10	0.10	0.10	0.10	0.10	0.10
Temp. coeff. of H_{ci} (%/°C)	0.40	0.40	0.40	0.40	0.20	0.20	0.20	0.58	0.58	0.60	–	–	–

Table A.2.6 Physical properties of rare-earth permanent magnets

Alloy or Grade	$RECo_5$ 14/14	$RECo_5$ 16/18	$RECo_5$ 20/15	$RECo_5$ 22/15	RE_2TN_{17} 24/18	RE_2TN_{17} 26/11	RE_2TN_{17} 28/7	$RE_2TN_{14}B$ 27/17	$RE_2TN_{14}B$ 30/18	$RE_2TN_{14}B$ 35/12	MQ 1	MQ 2	MQ 3
Density lb/in^3	0.300	0.298	0.300	0.300	0.300	0.300	0.300	0.300	0.268	0.268	0.217	0.271	0.271
Tensile (lb/in^2)	6000	6000	6000	6000	5000	5000	5000	11,000	11,000	11,000	–	11,000	11,000
Transverse modulus of rupture (lb/in^2)	–	–	–	–	–	–	–	–	–	–	–	–	–
Coefficient of thermal expansion (°C × 10^{-6})	6 (‖) 13 (⊥)	6 (‖) 13 (⊥)	6 (‖) 13 (⊥)	6 (‖) 13 (⊥)	9 (‖) 11 (⊥)	9 (‖) 11 (⊥)	9 (‖) 11 (⊥)	3 (‖) −5 (⊥)	3 (‖) −5 (⊥)	3 (‖) −5 (⊥)	–	–	–
Resistivity at 25°C (μ-ohms/cm per cm^2)	55	55	55	55	86	86	86	150	150	150	18,000	160	160
Rockwell hardness	53	53	53	53	55	55	55	58	58	58	36C	60C	60C
Curie temp. (°C)	700	700	700	700	800	800	800	310	310	310	310	310	310
Composition possible elements or wt%	Sm Nd MM Co	Sm Nd Co	Sm Pr Nd Co	Sm Pr Nd Co	Sm Ce Fe Cu Co Zr Hf	Sm Ce Fe Cu Co Zr Hf	Sm Ce Fe Cu Co Zr Hf	Sm Ce Fe Cu Co Zr Hf	Nd Pr Dy Tb Fe Co	Nd Pr Dy Tb Fe Co	Nd Pr Dy Tb Fe Co	Nd Pr Dy Tb Fe Co	Nd Pr Dy Tb Fe Co

Table A.2.7 Magnetic properties of miscellaneous permanent magnets

Alloy or Grade	1% C Steel	6% w Steel	3% Cr Steel	35% Co Steel	CoPt	Remalby	Vicalloy	Cunife	Lodex (A)	Lodex (I)	CrFeCo	MnCAl
B_r(G)	9000	10,500	9800	9000	6000	8500	7500	5500	6800	4800	13,000	5700
H_c(Oe)	50	65	50	250	4600	250	230	530	1000	950	550	2700
H_{ci}(Oe)	–	–	–	–	–	–	–	–	–	–	–	3000
$(BH)_{max}$(MGOe)	0.20	0.30	0.22	0.95	9.10	1.00	0.80	1.50	2.42	1.25	5.00	7.00
B/H at $(BH)_{max}$	17	–	17	45	1.2	30	30	9.4	7.5	3.9	20	1.7
B_d at $(BH)_{max}$(G)	6000	–	6500	6300	3200	5400	5000	3800	4700	2400	10,000	3500
H_d at $(BH)_{max}$(Oe)	35	–	38	140	2850	220	160	405	650	625	500	2000
μ_r	–	39	30	12	1.1	–	–	1.7	3.1	1.6	4.2	–
H_s(Oe)	300	300	300	1000	10,000	1000	1000	2500	5000	5000	3000	10,000
Max. service temp. (°C)	120	120	120	150	350	500	500	350	100	100	500	100

Table A.2.8 Physical properties of miscellaneous permanent magnets

Alloy or Grade	1% C Steel	6% w Steel	3% Cr Steel	35% Co Steel	CoPt	Remalloy	Vicalloy	Cunife	Lodex (A)	Lodex (I)	CrFeCo	MuCAl
Density lb/in^3	0.281	0.293	0.278	0.291	0.565	0.295	0.296	0.311	0.335	0.355	0.274	0.184
Tensile (lb/in^2)	300,000	300,000	300,000	300,000	200,000	125,000	150,000	100,000	1000	1000	28,300	–
Transverse modulus of rupture (lb/in^2)	–	–	–	–	200,000	50,000		–	4000	4000		–
Coefficient of thermal expansion (°C × 10^{-6})	13.3	14.5	13	17.2	11.4	9.3	11.2	14	18	18	11	18
Resisitivity at 25°C (μ-ohms/cm per cm^2)	18	30	28	27	28	45	60	18	120	120	68	80
Rockwell hardness	C65	C65	C62	C62	C26	C60	C60	B200	–	–	C36	C50
Curie temp. (°C)	770	760	750	890	480	900	860	410	990	990	630	300
Composition Possible elements or wt%	Fe C	Fe w	Fe Cr	Fe Co	Co Pt	Fe Cu Co Mo	Fe Co V	Fe Cu Ni	Fe Co Pb	Fe Co Pb	Fe Cr Co Ti	Mn Al C

APPENDIX 3

DEMAGNETIZATION CURVES FOR DESIGN ANALYSIS

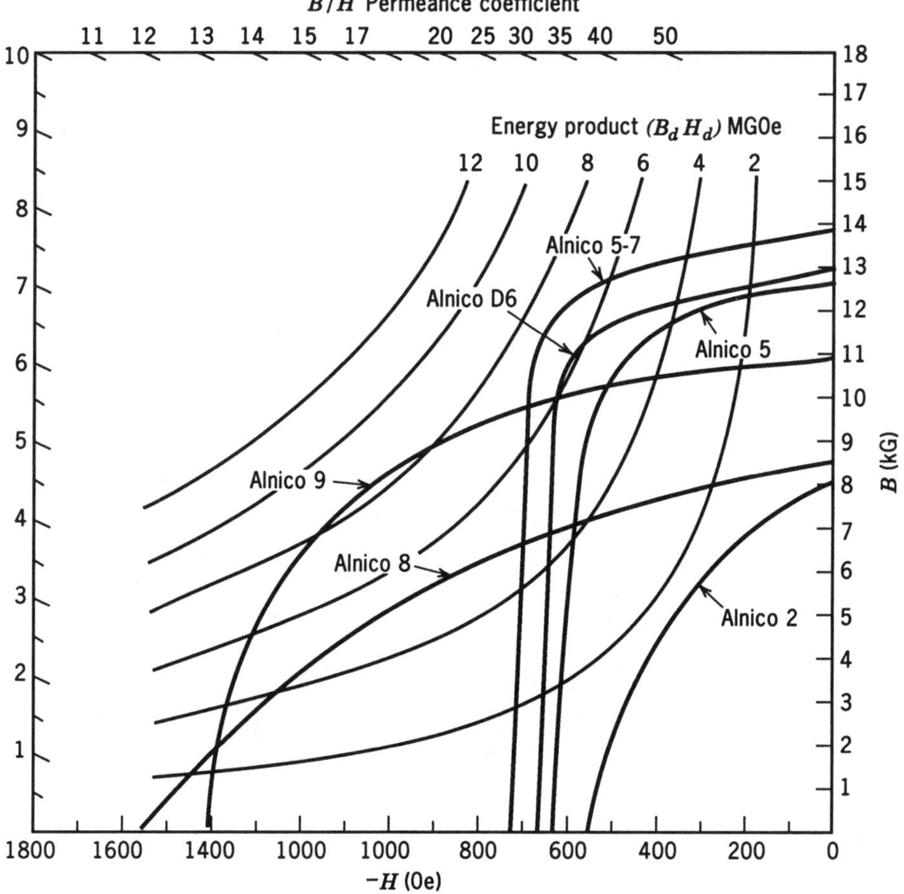

Figure A.3.1 Properties of cast Alnico magnets.

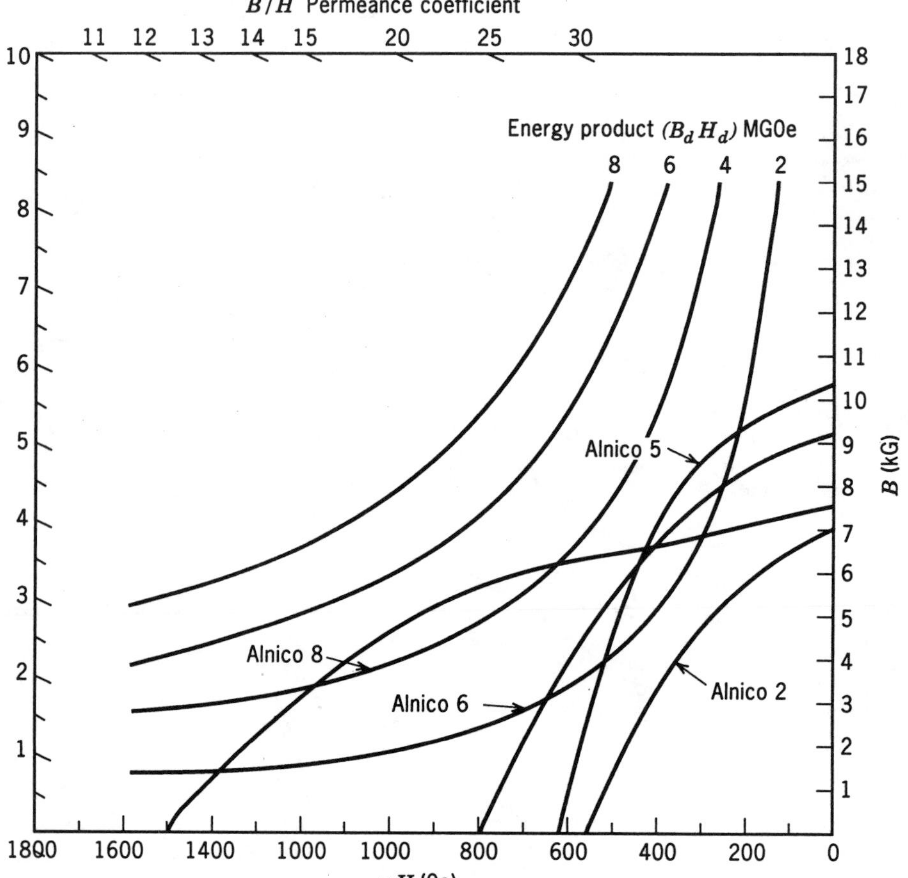

Figure A.3.2 Properties of sintered Alnico magnets.

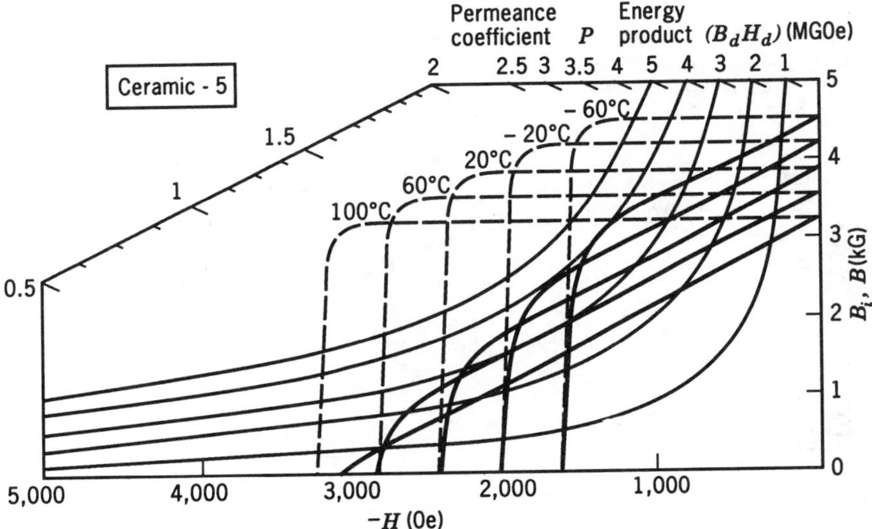

Figure A.3.3 Properties of Ceramic 5.

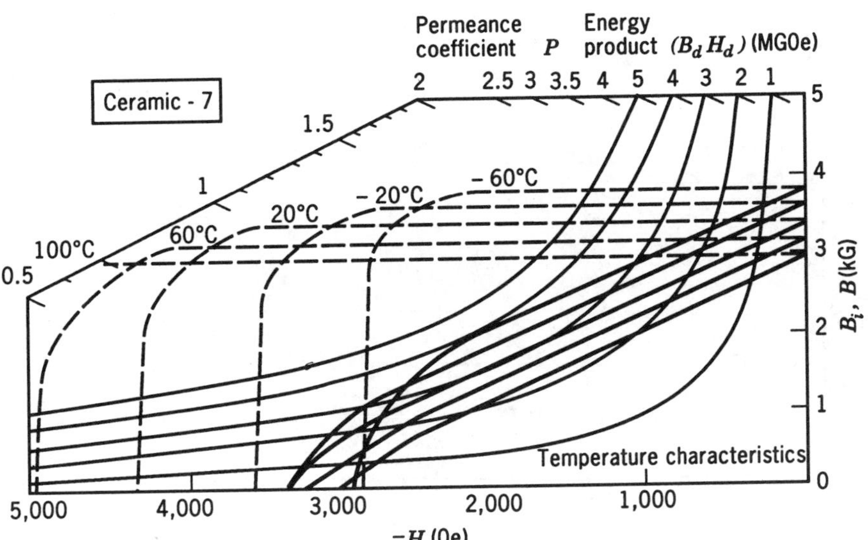

Figure A.3.4 Properties of Ceramic 7.

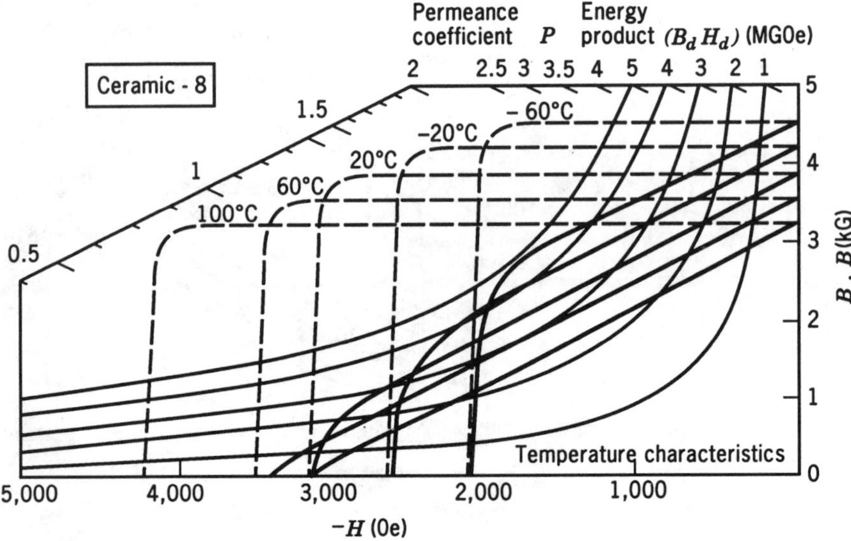

Figure A.3.5 Properties of Ceramic 8.

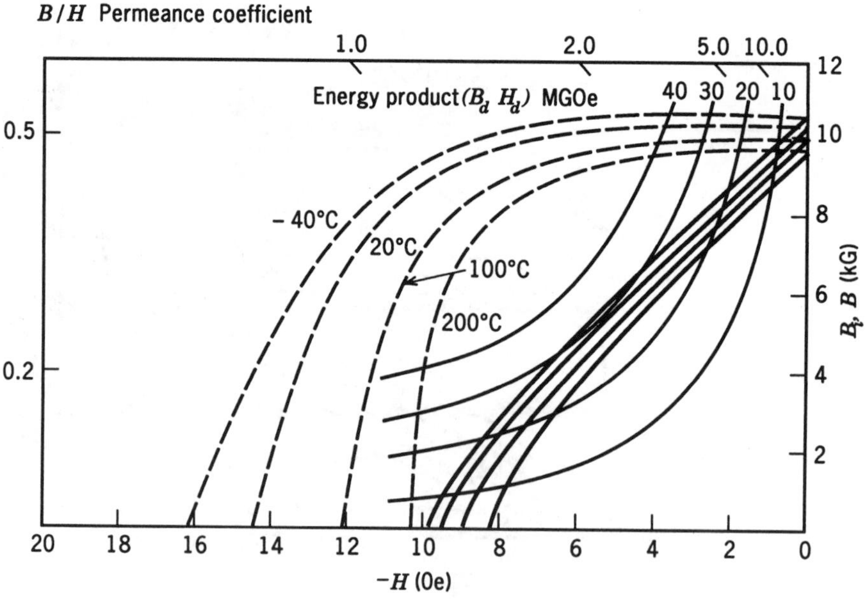

Figure A.3.6 Properties of RE_2TN_{17}–26/11.

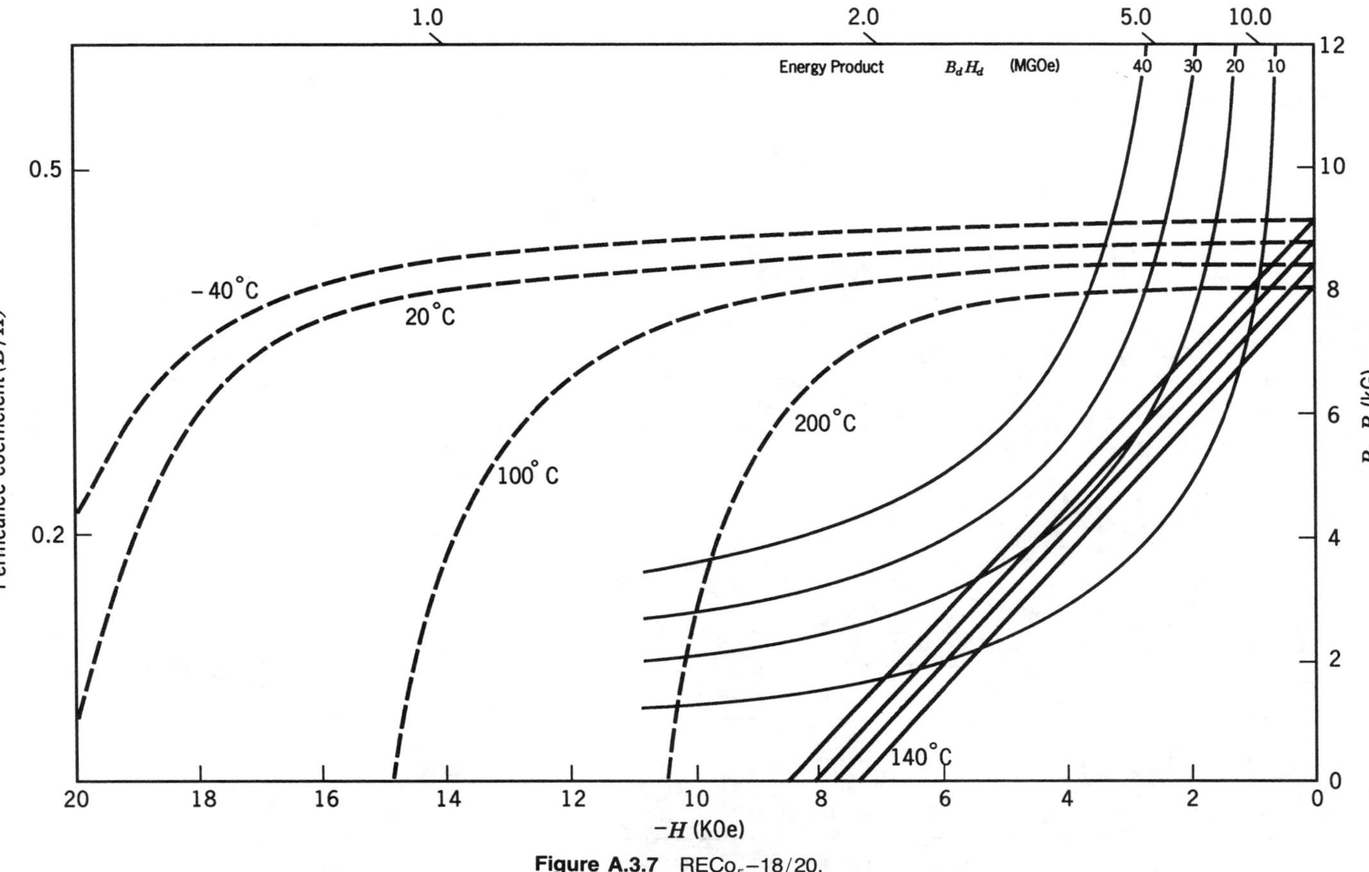

Figure A.3.7 $RECo_5$–18/20.

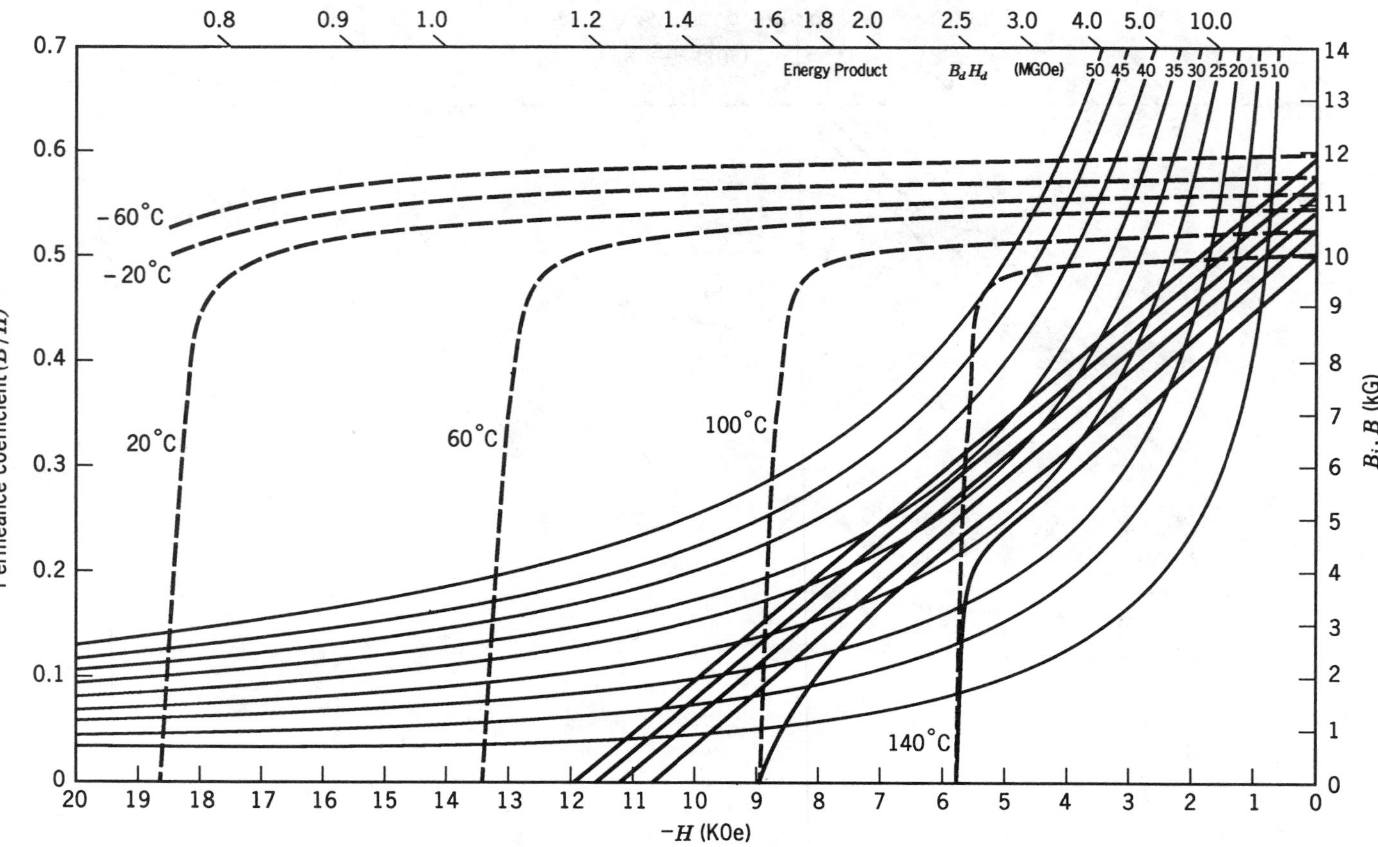

Figure A.3.8 Properties of $RE_2TN_{17}B$–30/18.

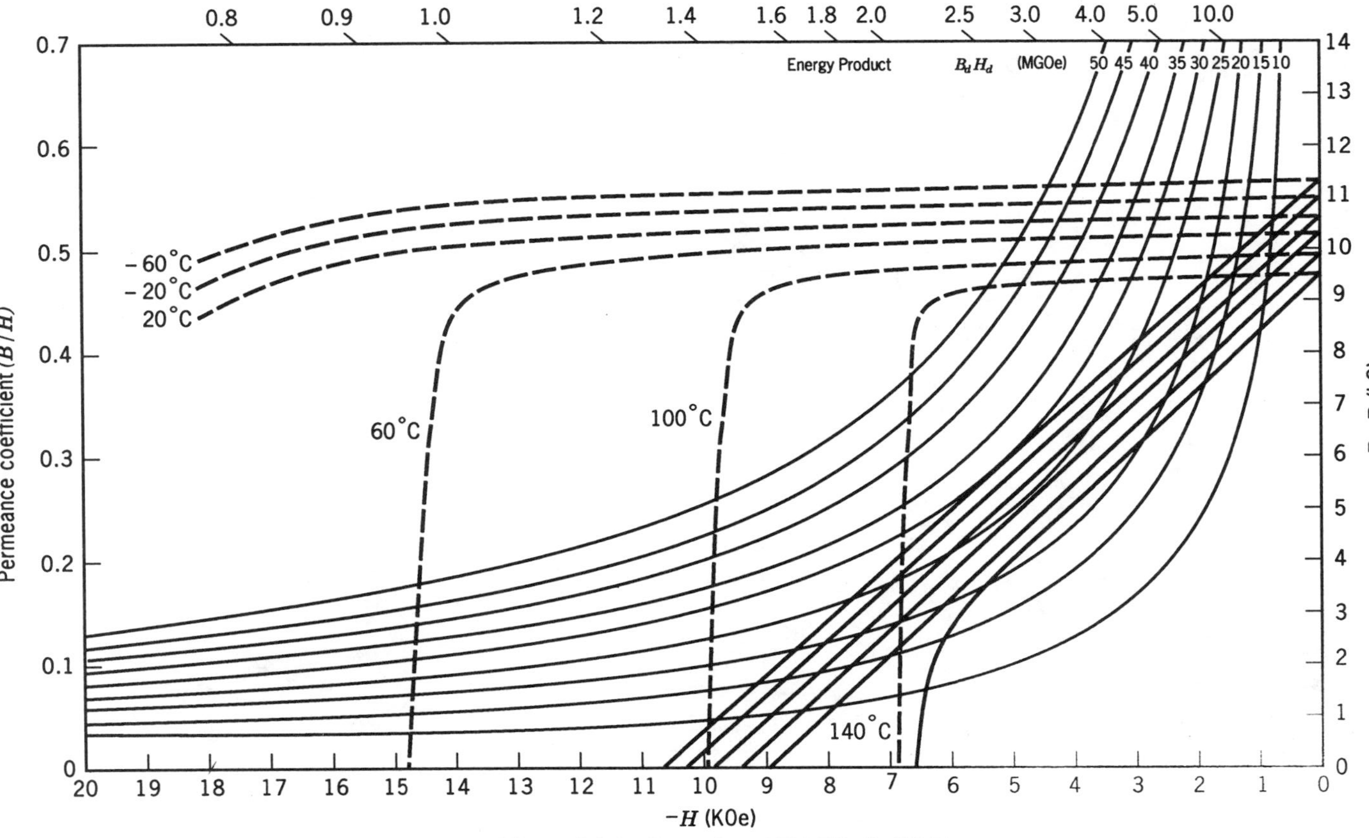

Figure A.3.9 Properties of $RE_2TN_{14}B$–27/17.

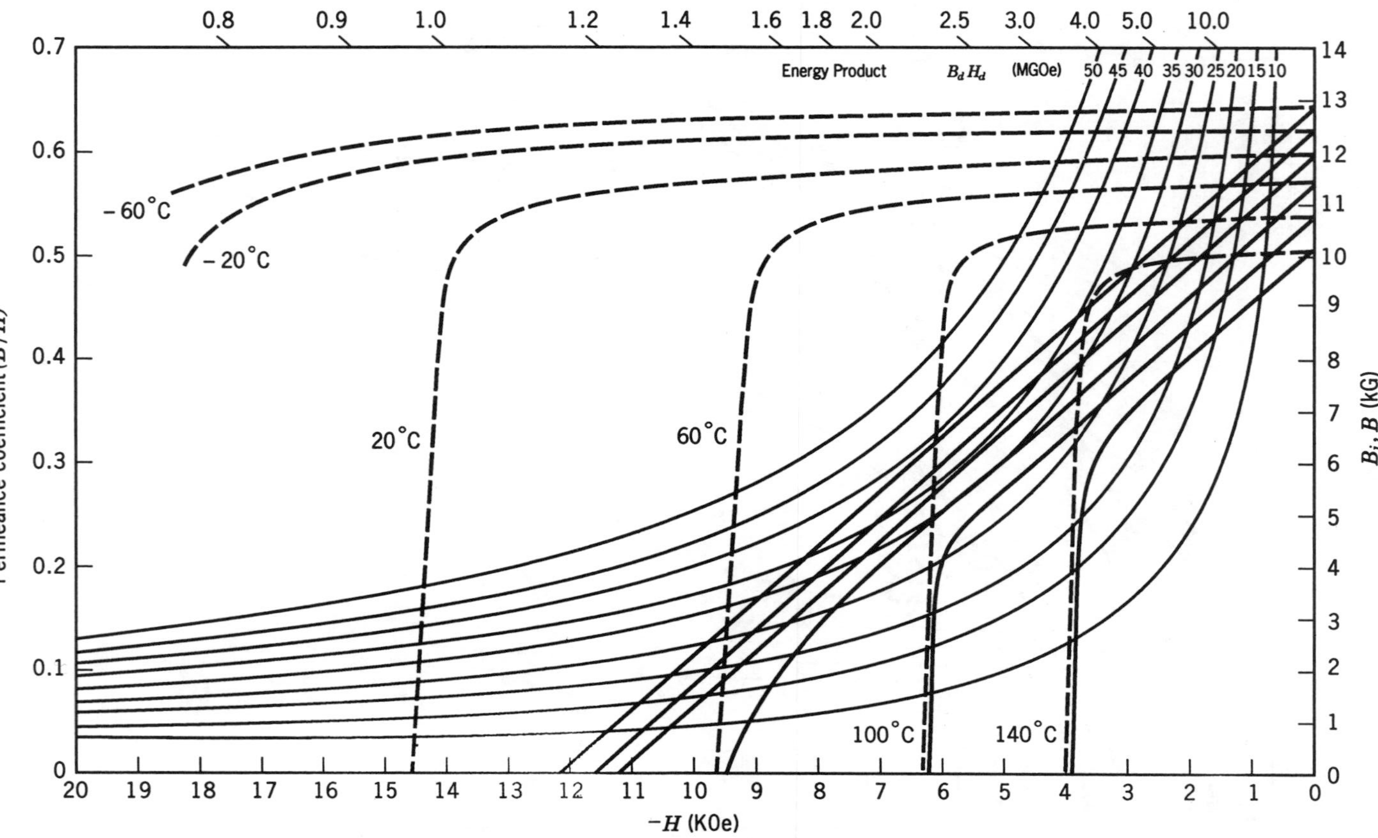

Figure A.3.10 Properties of $RE_2RN_{14}B$–35/12.

APPENDIX 4

CHRONOLOGY OF DISCOVERY OF PERMANENT MAGNETS

Table A.4.1 Chronology of Discovery of industrial permanent magnets

Present Designation and Original Constituents	Year First Reported	Reference	*Typical Properties* $(BH)_m$ (MGOe)	H_{ci} (Oe)	B_r (G)
Lodestone (Fe_3O_4)	ca. 600 B.C.	Thales of Miletus [1]	–	–	–
Iron	ca. 1200 A.D.	Guyot de Provins [1]	–	–	–
Carbon steel (C, Fe)	ca. 1600	Gilbert [2]	0.2	50	9000
Fe_3O_4 powder (Fe_3O_4, binder)	ca. 1750	Knight [3]	–	–	–
Tungsten steel (W, C, Mn, Fe)	ca.1855	Hannack [4]	0.3	70	10,500
Chrome steels (Cr, C, Mn, Fe)	ca. 1870	Hadfield [5]	0.3	65	9500
Cobalt steel (Co, W, Cr, C, Fe) (Honda or KS steel)	1916, 1917	Honda et al. [6]	0.9	230	9600
Remalloy (Mo, Co, Fe) (Comol)	1931	Seljesater et al. [7] Köster [8]	1.1	230	10,000
Silmanal (Al, Mn, Ag)	1931	Potter [9]	0.08	6300	590
Alnico 3 (Al, Ni, Fe) (Mishima alloy of MK)	1931	Mishima [10]	1.4	490	6800
Alnico 4 (Al, Ni, Co, Fe)	1931	Mishima [11] Ruder [12]	1.4	700	5500
Vectolite (Co, Fe oxides)	1933	Kato and Takei [13]	0.5	900	1600
Alnico 2 (Al, Ni, Co, Fe, Cu)	1934	Horsburgh and Tetley [14]	1.7	560	7300
New KS (Al, Ni, Co, Fe, Ti) (Alnico 12)	1934	Honda et al. [15] Ruder [16]	2.0	790	7200
Cobalt platinum (Co, Pt)	1936	Jellinghaus [17]	7.5	4300	6000
Iron powder (Fe, binder)	1937	Honda [18]	1.0	600	5000

Table A.4.1 (*Continued*)

Present Designation and Original Constituents	Year First Reported	Reference	Typical Properties		
			$(BH)_m$ (MGOe)	H_{ci} (Oe)	B_r (G)
Cunife (Cu, Ni, Fe)	1937	Neumann et al. [19]	1.8	590	5700
Cunico (Cu, Ni, Co)	1938	Dannöhl and Neumann [20]	1.0	450	5300
Alnico 5 (Al, Ni, Cu, Co, Fe) magnetic field treatment	1938	Oliver and Shedden [21] Jonas [22]	5.5	640	12,500
Vicalloy (Fe, Co, V)	1940	Nesbitt and Kelsall [23]	3.0	450	10,000
Iron-cobalt powder (Fe, Co, binder)	1942	U'gine Co. [24]	1.5	625	6000
MT magnet (Fe, Al, C)	1947	Mishima [25]	0.45	200	5000
Alcomax 3 (Al, Ni, Co, Fe, Nb) or 4	1947	Hadfield [26]	5.0	670	12,500
Alnico 5 (Al, Ni, Co, Fe, Cu) crystal orientation	1948	McCaig [27] Bemius [28] Ebeling [29] Swift Levick et al. [30]	6.5	680	13,000
P-6 (Fe, Co, Ni, V)	1952	Martin and Geisler [31]	0.5	58	4000
Hexagonal ferrites: isotropic	1952	Went et al. [32]	1.0	1800	2200
(BaFe oxides), anisotropic	1954	Stuijts et al. [38]	3.4	2200	3800
(SrFe oxides)			3.6	2200	4000
Lodex® (elongated Fe, Co, particles with binder)	1955	Luborsky et al. [34–36]	1.4	845	5300
			3.5	940	7350
Alnico 8 (Al, Ni, Co, Fe, Cu, Ti) and 9	1956	Koch et al. [37]	4.5	1450	8500
			9.2	1500	10,600
$RECO_5$	1966	Strnat et al. [38]	16.0	7000	8000
	1968	Buschow et al. [39]	20.0	–	–
	1969	Das [40]	20.0	–	–
RE $(Co, Fe, Cu)_z$	1968	Nesbitt et al. [41]	8.0	–	–
		Tawara et al. [42]	8.0	–	–
$RECo_5$	1970	Benz and Martin [43]	19.0	8000	8800
RE_2TM_{17}	1981	Mishra et al. [44]	30.0	10,000	11,000
$RE_2TM_{14}B$	1984	Sqawa et al. [45]	35.0	11,000	12,000
$RE_2TM_{14}B$	1985	Croat et al. [46]	14.0	6500	7900
$RE_2TM_{14}B$	1985	Lee [47]	32	10,500	11,600
$RE_2TM_{14}B$	1986	Sqawa et al. [48]	50.5	10,000	14,200

Source: A part of this table is from General Electric Information Report No. 66C252, October 1966, by F. Luborsky and R.J. Parker.

Table A.4.1: References

[1] E. Andrade, Endeavour **17** (1958) 22.

[2] W. Gilbert, De Magnete, translated by P. Mottelay, (Dover Publications Inc., New York, 1958).

[3] B. Wilson, Phil. Trans. Roy. Soc. London, **69** (1779) 51.

[4] G. Hannack, Stahl Eisen, **44** (1924) 1237.

[5] R. Hadfield, J. Iron Steel Inst. (London) **42** (1892) 49.
[6] K. Honda and S. Saito, Sci. Rep. Tôhoku Imp. Univ. **9** (1920) 417.
[7] K. Seljesater and B. Rogers, Trans. Am. Soc. Steel Treating **19** (1932) 553.
[8] W. Köster, Arch. Eisenhüttenw **6** (1932) 17.
[9] H. Potter, Philos. Mag. [7] **12** (1931) 255.
[10] T. Mishima, Ohm **19** (1932) 353.
[11] T. Mishima, U.S. Patent 2, 027, 996, 1935 (Appl. 8-27-31); Brit. Patent 392, 661 (Appl. 1931).
[12] W. Ruder, U.S. Patent 1, 968, 569, 1934 (Appl. 6/3/33).
[13] Y. Kato and T. Takei, J. Inst. Elect. Engrs. (Japan) **53** (1933) 408.
[14] G. Horsburgh and F. Tetley, Brit. Patent 431, 660, 1953 (Appl. 5/23/34).
[15] K. Honda, H. Masumoto and Y. Shirakawa, Sci. Rep. Tôhoku Univ. **23** (1934) 365.
[16] W. Ruder, U.S. Patents 1, 968, 569 and 1, 947, 274 (1934).
[17] W. Jellinghaus, Z. Tech. Phys. **17** (1936) 33.
[18] K. Honda, Nippon Kinzoku Gakkaishi **1** (1937) 3, 19.
[19] H. Neumann, H. Büchner and A. Reinboth, Z. Metalk., **29** (1937) 173.
[20] W. Dannöhl and H. Neumann, Z. Metallk. **30** (1938) 217.
[21] D. Oliver and J. Shedden, Nature **142** (1938) 209.
[22] B. Jonas, U.S. Patent 2, 156, 019, April 25, 1939.
[23] E. Nebsitt and G. Kelsall, Phys. Rev. **58** (1940) 203; U.S. Patent 2, 190, 667
[24] Société . . . d'Ugine, Brit. Patent 590, 392, July 16, 1947 (Appl. France 4/7/42); U.S. Patent 2, 651, 105 Sept. 8, 1953 (Appl. France 4/7/42).
[25] T. Mishima and N. Makino, J. Iron Shell Inst., Japan **43** (1957) 556, 647, 726.
[26] D. Hadfield, Brit. Patents 634, 686 and 634, 700 (Appl. 1947).
[27] M. McCaig, Proc. Phys. Soc. (London) **B62** (1949) 652.
[28] R. Bemius, German Patent 811, 976, Jan. 21, 1949.
[29] D. Ebeling, U.S. Patent 2, 578, 407, Dec. 11, 1951 (Appl. 1/10/48).
[30] Swift Levick and Sons, Ltd. and D. Horsburgh, Brit. Patent 652, 022, April 11, 1951 (Appl. 11/18/48).
[31] D. Martin and A. Geisler, Trans. Am. Soc. Metals **44** (1952) 461.
[32] J. Went, G. Rathenau, E. Gorter and G. Van Oosterhaut, Philips Tech. Rev. **13** (1952) 194.
[33] A. Stuijts, G. Rathenau and G. Weber, Philips Tech. Rev. **16** (1954) 141.
[34] U.S. Patent 2, 974, 104, March 7, 1961 (Appl. 4/8/55).
[35] L. Mendelsohn, F. Luborsky and T. Paine, J. Appl. Phys. 26 (1955) 1274.
[36] F. Luborsky, L. Mendelsohn and T. Paine, J. Appl. Phys. **28** (1957) 344.
[37] A. Koch, M. van der Steeg and K. deVos, Proc. Conf. on Magnetism and Magnetic Materials, AIEE Publ. T-91 (1956) 173; A. Luteijn and K. deVos, Philips Res. Rep. **11** (1956) 489; Brit. Patent 821, 624 (Oct. 14, 1959); U.S. Patent 2, 837, 452 (1958).
[38] K. J. Strnat, and G. Hoffer, USAF Materials Lab. Report AFML TR-65-446, 1966.

[39] K. H. J. Buschow, R. A. Naastepad and F. F. Westendorp, J. Appl. Phys. 40 1969 4029.

[40] D. K. Das, IEEE Trans. Magn. 5 (1969) 214.

[41] E. A. Nesbitt, R. H. Willens, R. C. Sherwood, E. Buehler and J. H. Wernick, Appl. Phys. Lett. **12** (1968) 361.

[42] Y. Tawara and H. Senno, Jpn. J. Appl. Phys. 7 (1968) 966.

[43] D.L. Martin and M.G. Benz, Cobalt samarium permanent magnets prepared by liquid phase sintering, Appl. Phys. Lett. 15 (1970) 176.

[44] R. K. Mishra, G. Thomas, T. Yoneyama, A. Fukumo and T. Ojima, J. Appl. Phys. 52 (1981) 2517.

[45] M. Sagawa, S. Fujimura, N. Togawa, H. Yamamoto and Y. Matsuura, J. Appl. Phys. 55 (1984) 2083.

[46] J. Croat, J. F. Herbst, R. W. Lee and F. E. Pinkerton, J. Appl. Phys. 55 (1984) 2078.

[47] Lee, R. W., Appl. Phys., Letter 46(8) 790 (1985).

[48] Sagawa, M., Hirosawa, S., Yamamoto, H., Matsuura, Y., Fugimura, S., Tokuhara, H. and Hiraga, K., (1986) IEEE Trans. Mag. 22, 910.

INDEX

Air-gap, 307
Alnico magnets, 5
  demagnetization curves, 323
  magnetic properties, 317
  physical properties, 317
  processing, 64
Alternator, 196
Ampere-turn, 307
Anisotropy, 47
  constants, 52
  crystalline, 54
  shape, 52
Antiferromagnetism, 45

Ballistic galvanometer, 256
Bearings, 193
Block wall, 48
Bonded magnets, 97

Calibration, 280, 305
Capacitance, 156
Chattock potential coil, 254
Chucks, 191
Coercimeter, 268
Coercive force, 19
Coercivity, 308
Compass, 2
Corrosion, 128
Coulomb force law, 185
Couplings, 194
Cunife, 321
Curie temperature, 46

Decibel, 209
Demagnetization, 284
Demagnetization coefficients, 25
  for cylinders, 151
  for rectangles, 153
  for tubular magnets, 152
Demagnetization curves, 20
  for design use, 323
  determination of, 274
Demagnetization factor, 22
Demagnetization field strength, 308
Design equations, 137
Design problems, 162
Diamagnetism, 44
Dipole, magnetic, 28
Domains, 46
  alignment of, 55
  buckling, 53
  coherent rotation, 53
  critical diameter, 50
  curling, 53
  fanning, 53

Economic design considerations, 175
Eddy currents, 201
Electrical circuit analogy, 165
Electrical indicating instruments, 234

Electromagnet, 249
Electron beam focusing, 237
Electronic integrators, 258
Elongated single domain magnet, 69
Energy products, 37
  contour curves, 38
  intrinsic, 40
  maximum available, 38
  useful, 40
Energy relationships, 26

Faraday's law, 194
Ferrimagnetism, 45
Ferrite magnets, 74
  magnetic properties, 318
  physical properties, 319
  process, 74
Ferromagnetism, 46
Field plotting, 146
Figure of merit, 37, 40
Flux, 309
  density, 309
  leakage, 139
  useful, 139
Fluxmeter, 257
Force on free electron charges, 236

Gauss, 308
Gaussmeter, 261
Generator, 196
Gilbert, 308
Grain oriented magnet, 67

Half cycle magnetizer, 303
Hall effect, 254
Hard magnetic material, 22
Helmholtz pair system, 263
Heusler alloys, 4
High permeability materials, 173
Holding and attracting devices, 186
Hysteresis loops, 19
  measurement of, 269

Ignition system, magneto, 196
Impulse magnetizer, 300
Inclusion hardened magnets, 62
Induced voltage, 25
Induction, 309
Intensity of magnetization, 16
Intrinsic magnetization, 16
Iron chrome cobalt magnet, 73
  magnetic properties, 321
  physical properties, 321
  process, 73
Irreversible changes, 49

Japanese production, 5

Kirkhoff's Law, 137

Leakage factors, 140
Levitated transport, 245
Linear force devices, 210
Load line, 23
Lodestone, 2
Lodex, 6
Lorentz force law, 185
Loudspeakers, 204

Magnequench, 87
Magnetic constant, 17
Magnetic field, 16
Magnetic field strength, 309
Magnetic flux, 309
Magnetic recording, 246
Magnetic resonance imaging, 243
Magnetic viscosity, 113
Magnetization, 284
  equipment and techniques in use, 296
  partial, 290
  requirements for, 289
Magnetron, 237
Manganese aluminum carbon, 79
Measurements, 248
  of magnetic fields, 251
  of magnetic potential, 253
  production quantities, 286
Mechanical force applications, 185
Meters, 234
Microphone, 200
Motors, 204
  alternating current, 222
  direct current, 214
  electronically commutated, 227
  family tree display, 206
  hysteresis, 231
  linear, 210
  servo, 222
  stepping, 233

Neodymium iron boron, 86
  discovery and development, 86
  magnetic properties, 86
  physical properties, 86
  process, 88
Nuclear radiation, 129
Nucleation, 57

Oersted, 310
Orientation, 66
Oxidation, 128

Paramagnetism, 44
Particle, 55
  alignment, 55
  packing density, 55
Permeability, 21, 310
  differential, 21
  initial, 21, 310
  maximum, 21
  reversible, 21
  of space, 17
Permeance determination, 143
  for air gaps, 154
  by flux plotting, 144
  by formula, 147
  by polar radiation model, 148
Pinning, 56
Platinum cobalt alloys, 82
  magnetic properties, 321
  physical properties, 321
Polar radiation model, 149
Pole coils, 271

Quadrapole structure, 242

Rare-earth cobalt magnets, 82
  magnetic properties, 320
  physical properties, 320
Rare-earth iron magnets, 86
  magnetic properties, 320
  mechanical properties, 320
Recoil energy contours, 142
Reed switch, 191
Remalloy, 321
Remenance, 311
Repulsion mode devices, 191
Residual induction, 311
Return path arrangements, 173

Saturation magnetization, 20, 54
Scaling law, 179
Search coils, 271
Semi-hard materials, 97
Separators, 188
Single domain particle, 50
Solenoids, 249
Stability, 101
Stabilization techniques, 130
Steel alloy magnets, 62
Synchronous torque drive, 194

Tachometer, 196
Temperature changes, 102
  irreversible, 102
  reversible, 102
  structural, 103
Temperature compensation, 123
  by composition changes, 124
  in device circuit, 123
Termperature effects, 106
  at constant temperature, 113
  due to cycling, 115
  on magnetization and coercive force, 106
Traveling wave tube, 241

Useful energy, 143

Vibrating sample magnetometer, 264
Vicalloy, 321

Watt hour meter, 193
Weber, 17